W9-AEZ-941

The Geography of Human Conflict

Approaches to Survival

"Neville Brown has used his vast knowledge of history and geography in a fascinating way to give us deeper insights into how these key influences on human life have combined to shape our course. At this time we should be thinking widely and challenging familiar boundaries of thought. Brown leads the way with a very important book. I commend it warmly." *Robert O'Neill, former Chichele Professor of the History of War, Oxford University and former Director of the International Institute for Strategic Studies*

"'Geopolitics' is a much misunderstood term that is over-used in modern politics. Neville Brown puts geopolitics back where it belongs – as a scientific and particular way of interpreting world politics that offers both explanation and meta-prediction. And he does so in a way that is both delightful and impressive. On the basis of a lifetime of scholarship and an eye for the fascinating and amusing he offers a sweep of history, culture and science that is as breathtaking as it is riveting. If students of global politics are frightened of being changed simply by reading one book, they should stay away from this one. It will stretch and convert them in a single reading." *Professor Michael Clarke, Director, Royal United Services Institute for Defence and Security Studies*

"Professor Brown's scope is extremely wide. Its historical span extends from proto-human to modern times, its geographical throughout our troubled planet and the Inner Space around it. Many threads are brought together to consider present and future circumstances: the strategic balance shifting eastwards; the displacement of Cold War rivalries with new antagonisms; accelerating change in technology, ecology and demography; failed or failing states. . . . Those involved in or otherwise concerned about the difficult decisions we face, strategic and economic, will be far better informed for having read this impressive book." *General Sir Mike Jackson, formerly Chief of General Staff*

"Neville Brown was already a leading figure in Strategic Studies when today's decision-makers were undergraduate students. A lifetime in the field has qualified him, almost uniquely, to paint 'the bigger picture' by integrating history, geography and strategic analysis into a continuum which broadens horizons as it deepens understanding." *Dr Julian Lewis MP, Shadow Defence Minister*

Montante Family Library
D'Youville College

The Geography of Human Conflict

Approaches to Survival

NEVILLE BROWN

sussex
ACADEMIC
PRESS

BRIGHTON • PORTLAND

Copyright © Neville Brown, 2009

The right of Neville Brown to be identified as Author of this work has been asserted in accordance with the Copyright, Designs and Patents Act 1988.

2 4 6 8 10 9 7 5 3 1

First published in 2009 in Great Britain by
SUSSEX ACADEMIC PRESS
P.O. Box 139
Eastbourne BN24 9BP

and in the United States of America by
SUSSEX ACADEMIC PRESS
920 NE 58th Ave.   Suite 300
Portland, Oregon 97213-3786

All rights reserved. Except for the quotation of short passages for the purposes of criticism and review, no part of this publication may be reproduced, stored in a retrieval system or transmitted in any form or by any means, electronic, mechanical, photocopying, recording or otherwise, without the prior permission of the publisher.

British Library Cataloguing in Publication Data
A CIP catalogue record for this book is available from the British Library.

Library of Congress Cataloging-in-Publication Data
Brown, Neville.
The geography of human conflict : approaches to survival / Neville Brown.
  p. cm.
Includes bibliographical references and index.
ISBN 978-1-84519-169-6 (h/c : alk. paper) —
ISBN 978-1-84519-170-2 (p/b : alk. paper)
1. Geography—Philosophy. 2. Geography—Social aspects. 3. Human geography. I. Title.
G70.B82 2009
355.4′7—dc22

                                                                    2009000073

The paper used in this book is certified by The Forest Stewardship Council (FSC), a non-profit international organization established to promote the responsible management of the world's forests. Products carrying the FSC label are independently certified to assure consumers that they come from forests that are managed to meet the social, economic and ecological needs of present and future generations.

Typeset and designed by SAP, Brighton & Eastbourne.
Printed by TJ International, Padstow, Cornwall.
This book is printed on acid-free paper.

570
382
2009

Contents

Preface

Recourse to a geographical framework in a study of how one might address a manifold global crisis derives from a consuming conviction that the salient truth about the contemporary world is that it is shrinking perceptually. The three dimensions of linear Space thereby assume added significance. So, too, does that of Time. Sound anticipation and energetic response have never mattered more. The credit crunch of 2008 has been notable for a fair measure of the latter but a near absence of the former.

Granted, a question the geographers of yesterday saw as fundamental, "possibilism" versus "determinism" in respect of the impact of their subject matter on human affairs, no longer looks that cogent. But right across the humanities explanatory links can seldom be established with the assurance once claimed. Even in Natural Science, indeterminacy is the ultimate norm. A further justification for coming in from Geography is that at least it starts from a broad base of hard fact — physical, biological, historical . . . Moreover, the subject is especially suitable as a setting for the discussion of lethal violence, the form of human conflict this work pays particular attention to.

Historical Geography on time scales ranging from a few decades to several million years can be a singularly valuable approach to illuminating how our psychological and cultural evolutions have been influenced by our interaction with an ever-changing environment, an interaction involving many trials and much error. Having said that, the end game for this text is problem analysis and policy prescription in relation to our present discontents.

Our world in December 2008 presents a mix of novel challenges. There has never been more need for lateral thinking; for joining up subject areas; for meaningful debates between professionals and laity, tutors and students, old and young . . . The paradigm change herein called for from Strategic Studies to more inclusive Survival Studies needs be effected with more resolve than comparable translations have been in the past.

What must be admitted, however, is that three big influences on world survival near term are tied to geographic specifics only in limited or derivative ways. One has especially in mind at present the international financial order. One also has in mind the global information explosion and the looming menace of biowar. On the other hand, all three are governed by that almost overpowering feeling that, for good or ill, everything and everybody is closing in. The Earth we inhabit has long been of constant mass, save for a very slow accumulation still of meteoritic dust. The World we live in gets smaller by the day. And the first and last of the three influences just cited therefore stand urgently in need of tighter international management.

✦ ✦ ✦

This particular formulation proceeds to a juxtaposing of the Eastern and the Western Question. The Eastern has long been recognized as being, for the British at least, the underlying foreign policy issue from 1814 to 1907 and again from 1943 to 1990. How does one make room for progress in the newly emergent world without affording Russia/the USSR scope for expansion? The Western Question is of my coining. How does the West avail itself of the opportunities created by successfully concluding the Cold War without generating an unmanageable backlash across the world at large?

As regards the substantial agendae, a requirement remains to integrate concern about climate change directly into multinational regional studies. There is hardly less of a need to look at ecological threats to peace which are not climate driven. So is there to consider as part of the global syndrome, instabilities evident or latent within the advanced nations. Another imperative is the integration of biowarfare concerns into mainstream survival thinking, this in good enough time for once. Procrastination followed by impulsive overreaction has brought on many a war.

As we go to press, the import of this truism can be seen in the tragedy unfolding in and about Gaza. The West's procrastination over the Israeli–Palestinian peace prospect contributed materially to the ascent of Hamas as a political and military force. Israel's heavy counter-offensive during this crunch time has owed too much to the illusion that armed might can break Hamas and everything it represents, not just militarily but politically as well. With its Islamic authoritarianism Hamas does not speak for the great majority of Gazans. With its burning sense of historical injustice it does. May one further insist that this episode vindicates absolutely the approach to this whole question encapsulated in Chapter 16. Promote a sovereign Palestinian democracy resolutely but over a decadal time scale. Do not seek the quick fix, modish of late.

✦ ✦ ✦

The account of our present discontents here rendered contains manifold criticisms of errors by those who manage and also those who commentate. So in fairness one does well to recognize some of the misjudgements and oversights one has perpetrated oneself. Not that acting the cavalier is in order. Samuel Johnson warned posterity that self-censure is often self-celebration in camouflage. Someone will flaunt certain shortcomings "in order to show how much he can spare".

Yes, but one has been prone to lapses downright detrimental to what one really was seeking to achieve. Two of mine may suffice. Throughout all the work on missile defence (1983–2002), one underestimated the progress in train on the compression of pulse duration in lasers in order to maximize pulse intensity. This shortfall did not, I would claim, materially effect conclusions drawn but could well have weakened their presentation. Then again, one was obviously right to treat climate change as a strategic issue in *The Future Global Challenge* (1977). None the less, one should have firmly endorsed the *a priori* grounds for believing the change would be dominated by a secular warming trend.

This ought not to have been difficult for an ex-aviation meteorologist blessed with academic freedom. After all, authoritative admonitions on this score go way back. In 1827, the French mathematician, Jean-Baptiste Fourier, concluded that, in those terms, air acted like a glass cover. He warned this tendency would increase as more and more carbon dioxide was released into the air by coal combustion. One could and should have tracked the on-going consequences. This was done more or less sufficiently in *The Strategic Revolution* (London: Brassey's, 1992, Chapter 6). But fifteen good years had been lost.

My settled opinion is that something akin to Survival Studies as here delineated will be a prerequisite for sustaining a tenuous peace and fragile freedom through mid-century and beyond. Whether this paradigm shift transpires or not, however, War Studies *per se* will remain a pertinent part of the whole, provided its remit broadens in various ways.

Not least of these must be seeing armed conflict in its totality. Well over a thousand books are said to have been written on Waterloo. So how many tell how the respective armies were put together? How many give the slightest inkling what happened to the 110,000 who fell dead or wounded on that fateful day or shortly before or after? And do not these rhetorical questions expose a persistent tendency not to record or predict warfare's impact in the round? How many studies of recent wars have properly examined their effects on civil society? And so on.

❖ ❖ ❖

For purposes of enumeration, Anglo-American units are employed interchangeably with metric ones (*Système Internationale*) because that is how the literature pans out. The prefixes indicating orders of magnitude (i.e. factors of ten) are used in the normal way. Thus "mega-" is a million (10^6); "giga-", a billion (10^9); and "tera-", a trillion (10^{12}). Likewise, "milli-" means a thousandth; "micro-", usually a millionth; and "nano-", a billionth. May one add that Chinese personal and place names are in *pin yin* except if some alternative is much better known.

The previous studies one has essayed this decade all bear on this analysis. Lateral referencing is therefore employed not infrequently in this text. For "EC" read *Engaging the Cosmos* (Sussex Academic Press, 2006). For "GISC" read *Global Instability and Strategic Crisis* (Routledge, 2004). For "HCC" read *History and Climate Change* (Routledge, 2001). Cross references within this text are made by chapter numbering. Thus the cross reference "14(1)" refers to the first numbered section of Chapter 14.

As regards specific word usage, my preference has been once again for "erraticism" as opposed to "variability" in allusions to acute fluctuations in climate at the interannual level. Also the word "secular" is used the way economists do — as a synonym for "long term", typically meaning a few decades. An "order of magnitude" means, as usual, a ten-fold contrast. With ordnance, the term "warhead" is reserved for one particular destructive charge and the shell that encases it. If a missile, say, delivers multiple warheads, the aggregate thereof is spoken of as the "warload".

A principle that hopefully informs this and other studies is outreach. During my wartime childhood, my ultra-liberal father had insisted on our tiny family listening once or twice a week to the evening news broadcast in English from the Third Reich. Subsequent applications of the reaching out precept were to include participating in a seminar on the world outlook held in Muammar Al Gadafi's Libya in the Spring of

1986, a time of warlike crisis between Tripoli and Washington-cum-London. Later that year, my wife Yu-Ying and I did a comprehensive two-person tour of South Africa sponsored by Dr Denis Worrall, the liberal-minded ambassador in London. We gained exposure to a wide spectrum of opinion, not least an evening's private conversation with a very radical ANC activist. After it all, we two had some sense of having contributed a little to the debate then warming up about South Africa's ending Apartheid.[1] At the Benghazi seminar just alluded to, one's first duty had been to listen. But perhaps one also did a bit to feed in modes of argument more rooted in reason and moderation than the vapourings Muammar was then giving vent to. Chairing a morning's plenary session afforded a particular opportunity.

❖ ❖ ❖

Mansfield College as a whole is a genial and supportive ambience in which to have one's being. But I would wish to extend special thanks to two of my colleagues — Professor Steven Blundell and, of course, Dr Tony Lemon, our senior geography tutor. Two Oxford University libraries have been especially supportive — the History Faculty's and Radcliffe Science. The staff at the London Library have given me solid backing as ever. Very helpful, too, have been the librarians at the Royal United Services Institute, the Royal Institute of International Affairs, the International Institute for Strategic Studies, and the Royal Astronomical Society.

As always, Mrs Jill Wells, my Literary Secretary, has been a tower of strength. Admiral Richard Hill has proffered, as the designated reader, a wealth of most constructive criticism. A goodly part of the text has again been read to my considerable advantage by my cousin, Dr Michael Brown. Yu-Ying has also been positively involved, her own most important project at the British Library notwithstanding. Nigel James of the Bodleian Map Room is warmly thanked for his apposite cartography.

The formal dedication marks the passing of a particularly close friend. But since the concept informing this study first germinated in my mind hard upon what we customarily term our formative years, I would like to recognize several early mentors too. Our small Tudor grammar school was hit hard, not least in the Natural Sciences, by the dearth of quality teaching and other deprivations during those austere post-war years. Among the masters who unselfconsciously upheld educational excellence in spite of everything were Basil Bevan, Fred Seal and Rex Thomas.

May I also express appreciation of Sergeant Bartle of the Middlesex Regiment, my platoon NCO during a month or two of pre-induction training at Depot the Buffs, under a University Training Corps programme. He was a veteran infanteer, great instructor and perfect gentleman.

❖ ❖ ❖

This last tribute, in particular, comes from an erstwhile youngster then deeply exercised about the whole question of the role of the military in a nuclear age. No doubt this concern had various sources. But among the more obvious was his being deeply shocked and more than a little scared by the film version of Erich Remarque's great 1929 classic, *All Quiet on the Western Front*. Another was disgust at what proved to be the mercifully transient upsurge in the USA of Senator Joseph McCarthy's corruptive anti-Communism. Also playing in, however, was how, shortly before entering

University College London, I managed to attach myself to a group dispatched by Britain's National Union of Students to observe a jumbo "World Peace Congress" in East Berlin.

The journey there via Austria was, for reasons not unconnected with McCarthyism, what I have previously called a medley of *mittelEuropean* spy clichés and Ealing Studios/B movie farces. This phase of the experience is on the record.[2] So, to an extent, is the encounter with Stalinism in Berlin and elsewhere.[3]

Soviet Communism had become under Stalin far more Orwellian than my students or younger colleagues today would easily credit. One instance outside of Berlin may now suffice. Circumventing the endeavours of the US Army of Occupation in Austria to block, sometimes rather physically, our lawful progress eastwards involved several dozen of us spending the best part of one day hosted by a small Soviet army station located just beyond the Anglo-Soviet zonal border across from Linz. The foregathering was thoroughly pleasant, thanks not least to the jovially avuncular company commander. But the barrack rooms these riflemen occupied were comprehensively festooned with the most vitriolic cartoons, viciously anti-Western and more than a little anti-Semitic. Your top-hatted Wall Street Shylock, cadaverously clutching an atomic bomb in each of his bloodied hands.

Not until many months after, however, did the full import sink into this youthful mind, habituated as it had become even in infancy to celebrating with relief how the USSR built back from a whole succession of military catastrophes to break the strength of Hitler's Wehrmacht. And that achievement still looks priceless and amazing. Hardly since in 480 BC Leonides of Sparta gave Classical Greece invaluable extra time and inspiration by fighting Persia to the death in the narrow pass at Thermopylae have comparatively open societies owed so much to a thoroughly closed one.

NEVILLE BROWN

Notes

1 Neville and Yu-Ying Brown, *South Africa: Sanctions or Targeted Aid?*, Muirhead Paper No. 1 (Birmingham: University of Birmingham, 1987), 50 pp.
2 *The Innsbruck Story* (London: National Council for Civil Liberties, 1951).
3 *GISC*, pp. xiii–xiv.

About the Author

Since 1994 Professor Neville Brown has been a senior member of Mansfield College, Oxford. Thanks to several strokes of extreme good fortune, his career has come to be based on an unusual interaction between the humanities and physics, especially the sky sciences. After majoring in the physical sciences, he read economics with geography at University College London (UCL) followed by modern history at New College, Oxford. For about half of the time he then spent as a forecasting officer in the meteorological branch of the Fleet Air Arm (1957–60), he specialized in regional upper air analysis. But other assignments included extensive experience on a front-line coastal air station plus some with the Mediterranean Fleet in a gunnery trials cruiser. He was a field meteorologist on two British Schools expeditions to sub-polar regions.

In 1980, he was elected to a chair in International Security Affairs at the University of Birmingham. He has held Visiting Fellowships, or the equivalent, at the UK National Defence College, then at Latimer; the School of Physics and Astrophysics at the University of Leicester; the International Institute for Strategic Studies in London; the Stockholm International Peace Research Institute; the Australian National University, Canberra; and (2001–4) the Defence Engineering Group, UCL. From 1965 to 1972, he worked part-time though quite proactively as a defence correspondent in the Middle East and South-East Asia, successively accredited to several leading journals.

From 1981 to 1986, Professor Brown was the first Chairman of the Council for Arms Control, a British all-party body drawn from parliament, the churches and other professions and committed to a serious but multilateral approach to arms control. He thus became involved in the multinational debate about Ballistic Missile Defence (BMD). In 1985 and again in 1987, he paid extended visits to the Strategic Defense Initiative Organization (SDIO) in the Pentagon. The first was at the invitation of Allan Mense, then Acting Chief Scientist; and the second as the guest of O'Dean Judd as Chief Scientist. From April 1994 to the summer of 1997, he was attached half-time to the Directorate of Sensors and Electronic Systems (within the Procurement Executive, UK Ministry of Defense) as the Academic Consultant to the official Pre-Feasibility Study on BMD policy. A declassified version of an 87,000-word *Fundamental Issues Study* he wrote in this connection was published by Mansfield College in 1998. Throughout, the vexed question of Space-based missile defence has been on the agenda.

He has authored twelve books or major reports. Among them has been *The Future of Air Power* (Beckenham: Croom Helm, 1986). But with the award-winning *Future Global Challenge* (New York: Crane Russak, 1977), he had begun to give economic, cultural and ecological factors some salience in the quest for a peaceable world. This thrust has continued with *New Strategy Through Space* (Leicester: Leicester University Press, 1990) through to *Global Instability and Strategic Crisis* (London: Taylor and Francis, 2004). His chief contribution to date to the scholarly study of the atmosphere has been *History and Climate Change, a Eurocentric Perspective* (London: Taylor and Francis, 2001). It reviews the last two millennia.

A fairly radical departure was effected with the publication of *Engaging the Cosmos: Astronomy, Philosophy and Faith* (Brighton, Portland, Vancouver: Sussex Academic Press, 2006). It is likely that his future work will mainly have a philosophical bent.

In 1990, Professor Brown was elected a Fellow of the Royal Astronomical Society, partly in recognition of his work at Leicester and on *New Strategy Through Space*. In 1995, the University of Birmingham conferred on him an official Doctorate of Science in Applied Geophysics.

List of Maps

The Roots of Modern Strife

To the memory of Milton Cummings (1933–2007),
Professor Emeritus at Johns Hopkins.
As scrupulous a scholar and steadfast a friend as
one could wish to know.

1 | Setting the Scene

1 Contingent Aggression

Perhaps the most archetypal image of generalship would be several of the fraternity poring over an outspread map to "situate their appreciation" of the landscape in view. They will further be pondering what, as the Duke of Wellington put it, lies "the other side of the hill". He was himself adept at positioning main forces on reverse slopes ready for surprise deployment.

Moreover, his narrow yet so decisive victory at Waterloo exemplified the importance of geographic ambience in determining battle outcome. Chosen in the light of technologies and doctrines then current, the field contested (bounded as it was by buildings and woodland) covered eight square kilometres. There 140,000 troops were amassed, averagely one every 57 square metres. Such a density could easily have precluded an outcome the first day.

Singularly influential, however, was the weather factor. The night through it had rained "cats and dogs". Therefore Napoleon felt constrained against starting his attack before 1300 hours, his hope being that this would allow the ground to dry enough to enable his artillery to move in close support. This delay meant Marshal von Blucher and his Prussians aligned with Wellington's left wing just in time to tip the balance.[1]

Global Projection

Napoleon had misjudged the relationship between time and space on the given day, something he was more often a consummate master of. Other instances of such error are familiar, not least that made by King Harold of England marching too forcefully from Stamford Bridge to Hastings in 1066. What we tend to overlook, however, is how wrong our not so distant forebears could be about the wider world that grand strategies must evolve within. Nor do we easily appreciate how ignorance levels can vary as between what may be quite similar cultures.

The post-Pythagorean Greek, Parmenides (born *c.* 515 BC), initiated the convention of dividing the globe we live on into five climatic zones: one torrid, one of moderation each side of it and — respectively beyond them — two chilly. Yet the earliest known example in the medieval West of a map of global climate zones comes in AD 1110 from Petrus Alfonsus, a converted Spanish Jew who had gleaned this knowledge from Arab sources.[2] Similarly, we are advised by a Hebrew University of Jerusalem study that "at the time of the First Crusade some Moslems had a more precise idea of Europe's geography than some of the finest map-makers in Europe itself".[3]

At that embryonic stage in the development of mapping, neat distinctions were never made between objective information and subjective impression, between prosaic materials to hand and the heady visions of politicians and their ilk. However, as cartography consolidated, especially from the seventeenth century, these two tendencies did separate into distinct genre.

Visionary essays readily became "mental maps", usually encapsulated in single sound bytes. Though many across the years have fostered aggressiveness, others have been circumspect and benign. Yet all could be dubbed two- as opposed to three-dimensional. Among the salient examples are the following: *Les limites naturelles;* Manifest Destiny; Great American Desert; wilderness; Cape to Cairo railway; the Raj; Home Counties; the Promised Land; *mare nostrum; lebensraum; Anschluss; drang Nach dem Osten;* Greater East Asia Co-Prosperity scheme; Atlantic Charter; *Festung Europa;* soft underbelly of the Axis; Iron Curtain; Free World; Near East; Middle East; South-West Asia; Enosis; Third World; Socialist Commonwealth; common European home; Evil Empire; Axis of Evil; Arc of Terrorism; High Frontier; Domino Theory; Universal Common; and Spaceship Earth.

Fascist dictators have, of course, been singularly addicted to mental mapping with sinister spin. Witness Hitler's confused endeavours to locate the USA, conceptually speaking. Until 1925 or so, he veritably overlooked its existence. Then he was captivated for some years by the American automobile boom, a phenomenon he attributed to boundless space and natural resources as utilized by aspiring Nordic immigrants and the zealously anti-Jewish Henry Ford. Some time after 1933, informed by what he understood to be the import of *The Grapes of Wrath* and of Orson Welles' Martian alarum, he plumped instead for the brash multi-ethnic decadence image.

Come 1942, the Führer was looking towards building a battle fleet ready for a final showdown between an Eastern Hemisphere overlorded by himself and a Western in hock to Roosevelt.[4] Meanwhile, those Americans apprehensive of this scenario were more insightfully persuaded that an ensuing "intercontinental" war could only be waged conclusively by intercontinental bombers. Hence the B-36 design.

Still Terra Incognita?

Evidently some of the expressions in the "mental maps" listed above are of recent origin and in current usage. The rest are in abeyance or treated as period pieces. May we continue to shed terms which substitute for thought more than they facilitate it. But are we overall gaining in sophistication to the extent required to address all the interactive problems bearing down on our global civilization?

What, for instance, is the level of comprehension in the democracies of the affluent West? Suppose one took a cross-section of voters to an outline map of the world and invited them to locate with adequate precision the following features or places, all of which have lately figured in our global discontents: Chechnya, Dharfur, East Timor,

El Niño, the Golan, Guantanamo, Kosovo, the Maldives, the Malvinas, the Sahel, Somaliland and Pyongyang. What might the mean success rate be? Not above 30 per cent?

Nor is it merely public awareness that may lag. Expert understanding sometimes remains surprisingly incomplete. Take the "indivisible world ocean", a fluid medium of immeasurable importance in virtually every way — transportation, food supply, military security, climate prediction and control, recreation, and even to site airports or building extensions.

Until the deployment, from 1920, of acoustic depth finders, the deep sea bed was widely presumed to be a "flat monotonous plain"[5] and the waters immediately above lifeless and still. That presumption is now discarded. So, too, are previous notions about the non-importance of sea surface currents. An influential rendition of weather prognosis first published in 1943 quoted approvingly this comment by an ex-President of Britain's Royal Meteorological Society: "If it were possible to divide the Atlantic into sections . . . thus preventing any flow of water, it would cause very little change in the climatic conditions of North-West Europe".[6] Since when, terrestrial energy emissions as measured by orbital satellites have revealed that ocean currents effect 40 per cent of all the heat transfer between the equator and 70° N.[7] Without this contribution, the mean gradient in surface air temperatures from the tropics to the polar regions might be twice as steep.

The situation today is that considerable uncertainty yet surrounds ocean behaviour in several respects germane to the looming ecological crisis. Will warming of the ocean surface inhibit the circulation of nutrients from the seabed ? At what stage may the Gulf Stream Drift crank into a less northerly circulation? How effective might iron filings dispersal be in stimulating a phytoplankton population explosion which could (a) restore fish stocks and (b) draw carbon dioxide back from the atmosphere? Towards the end of the twentieth century, it could still be said that we collected more geophysical data about the atmosphere every day than we ever have as yet about the ocean deeps.[8]

What is Humankind?

All the same, the most serious blind spot in our understanding of international unruliness is not the oceans, oilfields or whatever. It is our inner selves. Perhaps the best known precept in palaeobiology is the Phylogenetic Law: "The Ontogeny of the Individual tends to recapitulate the Phylogeny of the Species". In other words, the attributes successively revealed as any animal or plant matures are liable to reflect the way its species has evolved in response to changing environmental circumstances. Perhaps 97 per cent of the time which has passed since our emergence as a distinctive species (300,000 years ago?) has been spent in the Palaeolithic, the Old Stone Age. Spent, that is to say, within small and close-knit tribal groups itinerantly eking out a precarious existence and developing strongly exclusive group loyalties in order to do so. Moreover, the same could be said of pre-human ancestors across millions of years previously. So how might we and all else best survive in the shrinking global village we have created?

How positively that question has been answered by those who have asked it has fluctuated markedly since 1945. Obviously this has been in response to ever-changing specifics. But it may also betoken a more general feeling that humankind is poised as rarely before between resounding triumph and utter tragedy. Winston Churchill would have taken the point. So would H. G. Wells. So would many religious seers at different times.

Cuba and Beyond

Thus after the taut but peaceful resolution of the Cuban missile crisis of 1962, a mood near to euphoria prevailed awhile. The Cold War was sanguinely believed to be all but over, an era of peace was dawning.

Disillusion was therefore the greater when, in the middle sixties, the world descended into a veritable morass of violence: Vietnam; China's Cultural Revolution; the Middle East; Czechoslovakia; Nigeria; new dictatorships in Latin America . . . Duly, there was a strong *literati* reaction, one usually looking for explanations in terms of basic truths about human nature.

To Anthony Storr, a British consultant psychiatrist and social philosopher, the primordial within us came out in our ready recourse to pugnacious language: "We *attack* problems and *get our teeth* into them. We *master* a subject when we have *struggled with* and *overcome* its difficulties. We *sharpen* our wits, hoping our mind will develop a *keen edge* in order that we may better *dissect* a problem . . ." More basically, he invites us to admit that, apart from certain rodents, "no other vertebrate habitually destroys members of his own species. No other animal takes positive pleasure in the exercise of cruelty on another of his kind . . . The sombre fact is that we are the cruellest and most ruthless species that has ever walked the Earth."[9]

This statement was true to the deep sensibility of its author. Yet experientially and, indeed, logically, it went too far. We cannot measure or even define cruelty, ruthlessness or self-awareness with the exactitude direct comparisons would require. Nor do we know anything about the vast majority of the species who have "walked" this Earth at one time or another. As for ourselves, the twentieth century (that "of the Common Man" *sic*) saw several scores of millions die in wars or through subjugation. But even that harsh reality does not rule out the notion that our dispensation to violence is contingent upon given circumstances.

Konrad Lorenz, the Austrian ethologist and Nobel Laureate, demonstrated how constrained aggressiveness towards their own kind is a trait possessed by many other animal species and could, in principle, be beneficial. He was further at pains to emphasize that the innate mechanisms through which it operates can be sensitive to "small, apparently insignificant" environmental change.[10] Meanwhile, Claire and W. M. S. Russell showed how in Man and other primates as well as various other mammals violence was a threshold response to constraint of resources and especially of living space.[11]

Coming in from the theatre world, Robert Ardrey saw mainsprings for the appeal of war and territorial conquest in a deep need we higher animals feel for identity as opposed to anonymity, stimulation not boredom and security instead of anxiety.[12] To my mind, however, the most incisive insights were afforded by Arthur Koestler, donning a social philosopher mantle. Following on eclectically from Lorenz, he proposed that when the innate physical attributes of the members of an animal species are such that they could kill one another with single spontaneous strikes, their psyche will usually incorporate a mental block against acting thus. Ravens do not go for each other's eyes nor does one wolf rip another's jugular. Doves and chimpanzees are not comparably inhibited because no one bodily action will normally be that decisive. The same goes for ourselves. But we now deploy deadly weapons.[13]

Koestler envisaged a major disjuncture within the human brain. It was between an older part whence mainly comes our emotional drive to service and sacrifice; and the "all too rapidly grown" modern part which generates intellect and ingenuity, weapons development included. Each needs careful handling in relation to the social geography of modern mass society.

As the Cold War came to what proved to be its concluding climax, debate resurged albeit more along neo-Marxian lines. In a script originally due to be broadcast as one of the annual Dimbleby lectures on BBC radio, the late Edward Thompson, Marxian social historian and nuclear neutralist, wrote thus of a perceived "universal need for *the Other* as a means of defining the identity of any group and the individuals within it . . . The Cold War, by dividing this world into two opposing parts . . . has become necessary to provide both bonding and a means of regulation within each part".[14] Moreover, such attitudes are liable to translate into self-glorifying fanaticism more readily at group level than they would within a solitary individual: "the narcissistic image of one's own group is raised to its highest point while the devaluation of the opposing group sinks to the lowest".[15] Earlier, Thompson had depicted the arms rivalry between the two main blocs as betokening a mutual action–reaction commitment to "Exterminism". His depiction of "launch on warning" as still a strategic imperative was neither pertinent nor valid. **(See Chapter 9)**. But he was surely right in his insistence throughout that too narrow a logic can lead to supreme irrationality.[16]

In 1993, Sir Crispin Tickell, writing as retiring President of the Royal Geographical Society, warned that human progress could prove to be a "suicidal success".[17] But perhaps the most ominous admonition of all had come from Bertrand Russell in 1969, in a postscript to his autobiography. He was coming across too many "men in our dangerous age who seem to be in love with misery and death and who grow angry when hopes are suggested to them".[18]

A debate along similar anxiety-ridden lines may be reviving yet again, partly in reaction to "9/11" and its successive aftershocks but also in response to how a global ecology crisis is looming ever larger. In the following review of how we have come to this pass, climate change figures repeatedly. It will do so in the critical decades ahead though hopefully less exclusively relative to other environmental threats.

2 Geopolitical Concepts

Moving Frontiers

The use of the term "frontier" to connote the dynamic advance of settlement and control is traceable to twelfth-century Spain. Nor is that surprising. It has well been argued that "few periods can be better understood in the light of the frontier concept than western Europe between 800 and 1500 AD".[19] The author actually had in mind "external frontiers" from Finnmark round to Iberia but also "internal" since work and settlement spread upwards as well as outwards within districts long settled. But the point can still be taken.

All the same, the medieval understanding of how peoples might be bounded geographically contrasts with today's. Extended frontiers (military, economic, ethnic . . .) tended to be zones not lines. Big feudal lords often saw merit in spreading their estates around, maybe with varying enfeoffments. Sometimes, too, low population densities allowed of more or less peaceable interpenetration between peoples. And small and discrete settler communities might still retain close-knit identities. In the upper Elbe marshes, Flemish settlers who had arrived *c.* AD 1200 spoke their own language into this last century. You could say they had lived throughout inside their own micro-frontier.

A piece of received wisdom not well confirmed is that, in medieval times and often nowadays, districts peripheral to emergent nationhoods regularly fare less well than do the respective cores.[20] Surely too much has been made of how, in lately colonized districts, agricultural margins are prone to fluctuate under the influence of weather and other factors. Cores are similarly subject to variations in the agricultural internal margins. In medieval Europe, altitude limits could vary a hundred metres over several centuries, considerably on account of climate change.

Frederick Jackson Turner

Still, it was out of the post-bellum United States that the most celebrated frontier thesis to date emerged. In 1890, the US Census resolved that a "moving frontier" should no longer feature in its reports. In 1893, F. J. Turner of the University of Wisconsin proposed that any American claim to be exceptional compared with Europe must be set in the context of how "American social development has continually been beginning over again on the frontier".[21] After all, this process was tantamount to a "steady movement away from the influence of Europe,"[22] the steadiness giving way to a more radical attitude shift as the western mountains were crossed.

Not that Turner presented but a prettified view of the frontier experience. Through it, he told us, "a complex society is precipitated by the wilderness into a kind of primitive organization based on the family". Individualism and democracy were encouraged thereby. However, "individual liberty was sometimes confused with an absence of all effective government".[23]

One charge levelled against Turner has been that he preferred averration to solid comparison. Yet in a subject area so broad brush, others could be similarly indicted. Some have sought to draw a distinction between frontier societies which are inclusive and those which are exclusive, alternatives putatively relating to whether indigenous peoples are assimilated into the colonization process or whether they are perennially held at bay. Roman, Arab and Spanish colonizations are deemed to have favoured inclusion in the ultimate while American, Australian, Canadian and South African are seen as exclusive. However, such distinctions criss-cross and blur. Viking, Chinese and French settlement has conspicuously exhibited either trait or both according to circumstances.[24] Likewise in Central and South America, the pitch the Spanish adopted could be almost tantamount to spontaneous genocide to start with but then became a long process of top-down assimilation. Take, too, the early Afrikaners. There is a sizeable "Cape Coloured" — i.e. mixed race — minority in South Africa today.

The Turner vision of the homesteader (alias the yeoman or the kulak) is far from universally valid. As he himself conceded, the great ranges of Texas and the High Plains matched it ill. So, for that matter, did the ranches of Ohio.[25] Even in nineteenth-century Australia, with its robustly egalitarian ethos, the archetypical frontiersman "was a wage worker who did not, usually, expect to be anything else".[26] Similarly in

Brazil huge cattle ranches advanced the frontier in much of the north-east while in the south "sugar and coffee planters sought virgin forests to fell and transform into large agricultural estates worked by slaves".[27]

The latifundia of the Spain of the *reconquista* were toughly martial precursors of this genre.[28] So were the Junker estates formed during German eastward expansion. One thinks, too, of medieval England's marcher barons on its troubled borderland with Wales. Granted, Turner himself would never have suggested that the "frontier" *leitmotif* he perceived was bound to be replicated in all such circumstances. But latter day expositions of "New" and "High Frontiers" have seemed to imply such may be the case.

Frontiers of settlement still play a part in our terrestrial affairs, though often imprecisely defined. They would, in any case, remain worthy of study as expressive of lingering tribal aspirations. In modern parlance, the term "frontier" can further refer to the expanding limits of knowledge in a given sphere. More prosaically, it is a neat territorial divide between two states, a feature usually guarded each way against intrusion. The simplest form of delineation is a line of latitude or longitude. Its disregard of natural features can be symbolic. In the case of the 49th parallel of latitude between Canada and the USA, it indicates that the two states never expect to fight along their agreed border. In that of the 38th between the two Koreas (1945–50), the supposition was that the partition was seen all round as a temporary arrangement pending negotiated reunification. Least pragmatic of all, and now in abeyance, was the notion that control by a nation of a stretch of Antarctica's coastline gave it a claim, within the longitudinal bounds thus set, right down to the South Pole.

Borders and Barriers

A disposition long extant, certainly within Anglophone military cultures, has been to regard fixed defences as never impenetrable. An analogy long favoured is that you are trekking towards a range of hills. From ten miles out, it looks solidly impassable. Close to within a mile and ways through proliferate.[29]

The Great Wall of China in 1644 and especially the Maginot Line in 1940 are taken as instances. However, a material consideration is that, at their respective times of crisis, China and France were badly divided within. Moreover, the attacks respectively came in from an angle significantly different from how the defences were oriented.

May we consider further the Great Wall. Frontier walls against northern nomads were first built *c*. 300 BC. These had been developed into a single system by 214 BC, thanks to Qin: the death-obsessed genius seen as the first emperor of a united China. The Wall was a political statement, not just a military precaution. It said that, while China was always ready to barter with northern barbarians, it was the peoples to southwards she sought to integrate *pari passu*.[30]

Often in modern times, the statement transmitted by border control is ambiguous or equivocal. In 1946, Tito's Yugoslavia (then Moscow's willing vanguard in the Cold War) officially closed her border with Greece in order to facilitate keeping it open surreptitiously to benefit the Greek Communists. Yet when in 1949 Belgrade closed it again, it did so to offload this obligation in the wake of a bitter rift with Moscow.

In 1953 a Demilitarized Zone was established in Korea, its axis being the cease-fire line as accepted for the armistice. Across the ensuing years, this framework was to see a full measure of low level but vicious confrontation, aggressive tunnelling by the North Koreans being a prominent theme.

Come 2008, the boundary between Egypt and Gaza was marked by a wall of brieze blocks betokening the former's blockade of the latter until such time as Hamas recognized Israel. Yet serious though Cairo no doubt was about this basic purpose, it was keeping options open to a degree. Beneath those blocks ran multiple tunnels whereby arms reached Hamas via shadowy middlemen routinely bribing Eqyptian officials. Gaza's black market thrived on the blockade. Cairo would not have found a complete sealing of the border infeasible had it been so resolved. That it was not is shown by how, in January 2008, Palestinians surged across on a shopping spree after Hamas had blown several holes in the wall.[31]

The high barrier Israel has lately erected around itself has sharply reduced suicide strikes within its own territory. The wider ramifications are a moot point. Contrary to the hopes not a few Israelis raised on its commencement, its construction has not readily led to a more relaxed occupation regime on the West Bank. Instead more is being afforded in the way of continuous protection to the settlers out there.

In due course, however, the barrier may turn the Palestinian resistance more towards non-violence. Among those discussing this possibility is Avishai Margalit, a highly regarded philosopher at the Hebrew University. He endorses evidence that this approach could commend itself to a strong majority of Palestinians and elicit a comparable Israeli response.[32] Yet so positive an action–reaction could flourish only in the context of proactive bilateral and multilateral diplomacy.

A psychological aspect always to address is how far tightly held frontiers express a deeper rejection of the world without while masking internalized uncertainties. A high-profile case study from the recent past is the Iron Curtain across Europe. Its symbolism was underscored by barbed wire and minefields stretching not just down this, the main divide. They also extended in between the newly communized states.

Legend notwithstanding, the term "Iron Curtain" was not coined by Winston Churchill in March 1946. A year earlier, Joseph Goebbels had scorned how, every time the Bolsheviks took another territory over, they erected around it an "Iron Curtain" behind which their "dirty work" could proceed.[33]

From 1957, during the FLN rebellion, the French in Algeria created the electrified Morice Line eight feet high along the Tunisian border. A minefield 145 feet wide was laid each side. A civilian-free buffer zone 30 to 50 miles wide was cleared behind the line. Some 40,000 troops garrisoned the border area. The FLN incurred perhaps 6000 casualties trying to cross it; and succeeded several times by concentrated attack. The Morice Line was basically completed in September 1958.[34] But the French military–political crisis of that summer had indicated *algérie française* was already a lost cause.

From 1980 Morocco built a wall to seal off the Western Sahara from Mauritania, a territorial dispute yet unresolved. By 2006, Kazakhstan was building a wall eight feet high along 28 miles of its border with Uzbekistan.[35] Now India is erecting an iron fence eight feet high on its 2500-mile border with Bangladesh.

Much remarked of late has been the border between Mexico and the USA. The American side has become quite well fortified against illegal migration. A third of the 2000-mile border has been fenced so far. Some 18,000 guards man the main crossing points while another 18,000 patrol the sectors in between. Predator UAVs and "bullet proof" helicopters are the mainstays of mobility support.

The illegals' response has included ladders and tunnels but also trying to make it across unfenced desert. Since the turn of the millennium, over 1000 bodies have been found in the Tuscon area alone. The illegal flow has probably reduced of late for a mix of reasons. However, any diminution will have been in part offset by those who have got through being less keen to hazard returning home.

Neighbouring States

In discourse about geopolitics, several tenets regularly recur. Among them is spheres of influence. So is a disposition to align with the guy who lives next door but one. For some time prior to 1435, England and Burgundy were allied. Reciprocally France and Scotland were informally from late medieval times through the Renaissance and well beyond, the fabled "Auld Alliance".

Lately there has been the much derided "domino theory", a term which entered parlance in 1955 via the Eisenhower administration. It stems from the truism that if some party gains control of a given territory, they are the better placed then to absorb lands immediately beyond .. . In his geography of British history classic, Halford Mackinder put a positive gloss on somewhat the same idea. An on-going quest for secure borders had joined up the British Empire and its affiliates from Cape Town to Cairo.[36] Naturally, certain other countries (notable France and Germany) found this disconcerting. After 1945, the West as a whole felt threatened by a somewhat comparable prospect, a progressive extension of a seemingly monolithic Sino-Soviet bloc.

Visiting Vietnam in May 1961, US Vice President Lyndon Johnson put on it all a rather extravagant Mahanian twist (see below). He insisted that "the battle against Communism must be joined in South-East Asia . . . or the USA inevitably must surrender the Pacific and take up defences on our own shores". Periodically a less convulsive rendering of the domino effect surfaced in Washington and elsewhere. It was simply that if South Vietnam fell, Laos and Cambodia would perforce follow suit. One can add that the USSR's invasion of Hungary in 1956 was a classic application of domino thinking. So was the Warsaw Pact's invasion of Czechoslovakia in 1968.

The Landlocked Syndrome

Another recurrent theme has been the insecurity, sometimes the outright paranoia, engendered by being landlocked. Witness Woodrow Wilson's insistence in his otherwise very general Fourteen Points that Poland be given a corridor to the sea, this much to the discomforture of the Germans and especially the Danziggers. Witness, too, Serbia's *angst* as Yugoslavia broke up.

Salutary instances have presented themselves in Latin America, a continent still too disposed to see its geopolitics in trite two-dimensional terms.[37] Bolivia comes across as perennially aggrieved. In the 1860s, vast natural deposits of agricultural nitrates were discovered under the Atacama desert. At that time, the political sovereignty Bolivia claimed over that region could be exercised only loosely due to a paucity of transport links. Chile could therefore move to take commercial control of nitrate exploitation by building a railway from Antofagasta.

Duly, war was waged between 1879 and 1883 between Chile and a Bolivia backed by Peru. Chile won conclusively, obliging a spearhead Bolivian task force to surrender and capturing awhile Lima, the Peruvian capital. She secured thereby sovereignty over the Atacama.

More remarked because more recent, however, has been the Grand Chaco war (1932–5) between Bolivia and similarly landlocked Paraguay. This conflict, ostensibly about export access via the Paraguay river, ended inconclusively after 100,000 deaths. Futility had made it all the more bitter.

Subsequently, Latin American geopolitics acquired more of an ideological slant. Three populist state leaders — Juan Peron in Argentina, 1946–55, 1973–74; Fidel Castro in Cuba, 1958–2008; and Hugo Chavez in Venezuela since 1999 — were to confront the American hegemony implicit in the Monroe Doctrine. Evidently, none of these regimes were landlocked though Havana had to recognize that (a) its prime target for early subversion, Bolivia, was thus circumscribed and (b) Cuba itself was susceptible to the opposite hazard, maritime investment.

But old enmities smouldered on, aggravated awhile by Peru's arbitrary seizure of a solid tranche of Ecuador in 1941. During Britain's liberation of the Falklands from Argentina in 1982, Chile gave the UK's task force valuable assistance; and was collaborating militarily with Ecuador. Peru duly gave the Argentine dictatorship vocal diplomatic backing.

Since when, things have been calmer in this particular respect. The crisis which broke when, early in 2008, Colombia raided a guerrilla base in Ecuador proved readily containable, notwithstanding the acute strain on civil–military relations in Ecuador plus a strong though brief Venezuelan reaction. Meanwhile, however, the number of landlocked states worldwide has risen sharply. This has been due to decolonization in Africa and, more recently, the break-up of Czechoslovakia, Yugoslavia and the USSR. At the same time, Russia has become more hemmed in than the Soviet Union was. Should the world order ever collapse, being landlocked could render those concerned far less secure economically, militarily and psychologically. A classic European instance is Czechoslovakia in 1938.

Munich, a Case Study

This is not an attempt to assess Munich in the round. Could Britain and France have played their hand better? Should they have been there at all? Who gained most from a deferment of war? No, the aim is to test the proposition that landlocked Czechoslovakia could have been defended there and then by a pan-European alliance with Moscow as its "keystone" and Paris nodal too. Such pretty well became received wisdom on the broad Left in Britain through 1940 at least.

Take the Left Book Club text by G. E. D. Gedye. It went through seven reprints within twelve months of publication in February 1939. There is little need to concern ourselves with the author's simple faith in Stalin nor with his suggestion that the latter's hugely manic army purges were merely akin to Napoleon's sacking some top brass to hone things up. Our focus should be on absurdities more strictly geographic.

Gedye extolled the strength of the Czech fortifications along the Bohemian rimland, "their Maginot line". Yet this ignored how the Anschluss with Austria in March 1938 had outflanked them and how vulnerable Czechoslovakia was to Nazi blockade, before and after the Anschluss.

Disregarded, too, was the reluctance of the smaller nations of Eastern Europe to join a security coalition, especially one led by a Moscow whose clutches most of them had lately escaped from. Hungary and Poland were even to accept small portions of Czech territory during the Nazi occupation in March 1939, thereby giving aggression a slither of legitimacy.

Gedye further persuaded himself that Soviet bombers based in Lithuania could threaten Berlin more readily than Prague could be accessed by the Luftwaffe. Not merely did this judgement flout geography, it failed perhaps understandably to appreciate how low a priority Moscow was according strategic bombing. Flatly ludicrous was his averration that all Czech warplane losses could be replaced "from the first day" by the Soviets flying machines in. And what about pilot replacement?

He envisaged a "Russian army corps" (presumably just three divisions) transitting a corridor duly opened through Romania. He tacitly presumed the corps would integrate smoothly, linguistically and otherwise, with the Czechs, thereby tipping the balance of advantage. He adjudged the Romanians in no position to oppose transit. He said nothing about forced transits in principle.[38]

Similarly in 1937, the Oxford social scientist G. D. H. Cole, emergent *primus inter pares* of the Fabian Left, had argued British Conservatism was the sole obstacle to a pan-European security pact built around Moscow and Paris.[39] So how would the French army (then gripped by a totally defensive ethos) come to the aid of a beleaguered Prague in accordance with its 1924 treaty obligations? Through the Siegfried Line? Across the Rhine? Up and over the Black Forest? Or by levitation?

Landlocked Post-Cold War

For the European Union, the Strategic Revolution of 1986–91 has meant six more landlocked members, actual or prospective — the Czech Republic, Slovakia, Slovenia, Serbia, Kosovo and Macedonia. Were the EU to crumble in the context of a currency crisis or whatever, their situations could exacerbate tensions.

Then there is authoritarian Belarus, which boasts a booming economy by dint of cheap fuel imports from Russia. Moscow finds some profit or, at any rate, gratification in sustaining this rather direct legatee of the USSR.

Meanwhile, the former Soviet republics of central Asia gain security of a morbid kind from the obstacles to accessing them by sea or air. Kazakhstan is the prime instance, gaining leverage from (a) hosting a USAF base for support of operations in Afghanistan and (b) a Kiplingesque "great game" for access to her hydrocarbons and uranium oxide. This has made it easier for her governance to remain viciously dictatorial.

From the time of Peter the Great at least, Tzarist Russia was visibly exercised about being hemmed in by land borders, constricted seas and icefields. Stalin was to be too. In 1935 he waxed "almost obsessive" about acquiring the world's biggest navy; and shortly had some "superdreadnoughts" laid down albeit never completed.[40] In 1912 a very similar programme had been endorsed in principle. In 1906 an Imperial Commission had called for a Baltic to Black Sea waterway. One should now ask how far the new Russia is concerned to over-compensate; and enquire, too, whether Mackinder's "heartland thesis" deepens or stultifies our understanding of the real situation, past or future.

The Heartland Thesis

That Sir Halford Mackinder (1867–1947) was a towering personality and, in multiple directions, an achieving one is undeniable. His undergraduate training in History and Geology had been valuably bifocal. He was eloquent orally and in print. He helped develop Human Geography and other activities in British universities. His *Britain and the British Seas* was most refreshing in literary as well as educational terms. Yet none of this *ipso facto* confirms his Heartland thesis. Nor, indeed, does the pervasive interest it has long aroused, one way or the other.

First enunciated in 1904,[41] it survived little changed several recastings. Its key tenet was that the interior of Eurasia was a fastness defined by the watersheds of rivers draining into the icy Arctic or else into other waters not affording access to the world ocean. Once there had been due exploitation of the geometrical advantages this expanse had over world oceanic power, it would hold the key to global supremacy. To Mackinder, the Heartland was the centrepiece of an Afro-Eurasian "world island". Lands further out were designated "the rimland". Though the eighteenth-century construct, "interior versus exterior lines" seems not to have entered Mackinder's lexicon, it captures the very essence of his Heartland reasoning.

He duly inferred that a Russian heartland might launch "centrifugal raids" from her "central strategic position" much as the Mongols had.[42] The fact of the matter is though that Tzarist Russia repeatedly failed, in times of war, to demonstrate such facility: 1854–6, 1877–8, 1904–5, 1914–16 . . . And although the reinforcement of the Moscow front from the Far East late in 1941 was critically important, it was on a comparatively limited scale. Just 12 divisions were involved.

Mackinder did well to perceive in 1919 how much hung on competition between Germany and Russia/USSR for influence in Eastern Europe. But this particular insight cannot be said to have stemmed from a settled geopolitical perspective. Instead, that very year he had his Heartland extend westwards to include all territory draining into the Baltic and the Black Sea.[43]

At the global level, there is the deepest of contradictions. Notwithstanding his Kiplingesque apprehension of looming continental power, part of the Mackinder psyche was overtly hoping, from 1902 onwards, that the British imperial identity might survive more or less indefinitely: "The whole course of future history depends on whether the Old Britain . . . withstands all challenges to her naval supremacy until such time as the daughter nations shall have grown to maturity and the British Navy shall have expanded into the Navy of the Britons."[44]

Air Power and the Sea

Like so many of his generation, Mackinder found air power peculiarly hard to gauge. Witness his 1919 prediction that its advent would enable the Heartland to take "easy possession of Suez via Arabia".[45] How such a *coup de main* could ever have been easy went unexplained.

What could already have been said, in fact, was that aviation gave armed forces substantially more scope for striking deeply. Moreover, it became uncomfortably clear in 1940–1 that this especially threatened maritime assets. Comparisons of resource allocation to achieve a given result often point to the same conclusion. A yardstick for comparison could be sliced costing of all the investment involved in having aircraft available for deep strike. Working from common sense definitions and assumptions, it can be shown that making a strike aircraft available for operations was liable to involve several times more capital outlay were it based on a carrier as opposed to an air station. Worse, a sizeable warship could be either destroyed or turned into a major liability by a single well aimed shot far more readily than would be the case with comparable assets ashore.

However, maritime mobility might, in given circumstances, offset these considerations. In any case, the strategic implications of a tactical evolution are not tritely deducible. In 1941 Senator Robert Taft, then a leading isolationist, argued that the very fact that air power made it more difficult to advance an army by sea meant that the United States should base its security simply on naval supremacy in the Pacific and Atlantic.[46]

Mahan and Corbett

Otherwise the parameters of modern naval power were defined in the light of historical experience by Alfred Thayer Mahan (1840–1914) and Julian Corbett (1854–1922). Rear Admiral Mahan was at the US Naval War College, Newport, from 1886. Sir Julian Corbett was a lawyer and historian, Cambridge-trained. They have been jointly dubbed "the Clausewitz of naval warfare".[47]

Mahan was to exercise the more influence. He gleaned from naval history, especially British, the clutch of tenets well spelt out in his 1890 classic, *The Influence of Sea Power on History*. Among them were the following. Mercantile shipping affords a good basis for naval power albeit less crucially than once it did.[48] Naval dependence on harbours and ports has increased with the need for regular coaling. The absence of defensive topography at sea ineluctably places a premium on Defence through Offence. A *guerre de course* against maritime commerce can be invaluable but needed, he felt, the backing of a main battle fleet. He further thought the best way to keep an enemy away from one's shores was to blockade his closely.

Aspects of this Mahanian legacy surfaced in the International Seapower Symposium held at Newport in October 2007. A strategy paper then presented reaffirmed how "forward deployed" naval power could limit regional conflict. A further development of this approach can be seen in the elevation of humanitarian aid to be a core strategic principle.[49]

Sometimes Mahan seems to commend sea control as worth sustaining almost for its own sake: "Naval strategy has for its end to forward, support and increase as well in peace as in war the sea power of a country."[50] Something of the sort comes over, too, in his treatment of the British use of maritime blockade during the French Revolution and Napoleonic era. Convincing enough though he was when assessing its direct impact on the psychology and strategy of Napoleon as Emperor, he is jejune when considering its actual economic effect.[51] William Pitt the Younger (as Britain's Prime Minister, 1784–1801 and 1804–6) had spoken far more substantially on this score.[52] But the definitive verdict of history we still await.

Julian Corbett settled for a more moderate and contingent assessment of sea power.[53] In his classic work, he cited the famous averration by Francis Bacon (1561–1626) that whosoever "commands the sea is at great liberty and may take as much or as little of the war as he will". Nevertheless, a strong sense of the historical constraints on the use of naval power to affect situations on land led Corbett to seek the harmonization of the two domains. Correspondingly he laid emphasis on "limited war", the limits in question being set by perceived levels of importance or else by how readily a contested territory is conceptually isolatable. The Crimea and Korea are seen as locales "where sufficient isolation was attainable" (p. 47). The allusion to Korea was, of course, in relation to the Russo-Japanese War, 1904–5. But it seems even more apt as a prevision of 1950–3.

Being well disposed to amphibious operations, Julian Corbett stressed how close the much-castigated Walcheren expedition of 1809 came to being a resounding success. In any case, it induced Napoleon to divert 300,000 men to improving coastal defence. Corbett further endorsed a trend he perceived towards erasing the distinction between the battleships securing sea control and the cruisers exercising it. Meanwhile he was much concerned about the threat the torpedo posed to that "imperfect organism" (p. 108), the battleship. Neither he nor Mahan contributed at all evidently to the deliberations on naval arms control.

A Maritime Commonwealth?

It is not just Mackinder who has felt tempted to envisage a fraternity of maritime nations at peace with themselves and extending the benefits thereof. In August 1941, Roosevelt and Churchill enunciated the basic freedoms in an Atlantic Charter they signed together on board an American cruiser just off the Canadian coast. President Kennedy expressly saw the Atlantic Alliance as keeping the world free for diversity. And a well respected scholar of naval affairs has lately returned to the theme.[54]

Moreover, the sea can of its very nature be inspirational at a more primordial level. Forty years ago, a ranking American geophysicist averred that part of the "spell of the ocean came from half forgotten memories and images beyond imagining deep below the surface of consciousness".[55] In other words, it has survived since that distant time when our pre-human forebears left the relatively cosy sea for the land, an ambience richer in opportunities for triumph and disaster.

Such reflections bespeak a romanticism too mystical for the obsessively prosaic world of Strategic Studies. But does this not expose the latter's ingrained limitations? Maybe if we want to understand fully the attitude of Russia to constricted access to the Open Sea, mystic deprivation should be borne in mind.

The Interactive Superpower

Central, literally as well as figuratively, to any sea-based commonwealth must be the USA, set as she is between two very equivalent oceans. Admittedly the area of the Atlantic, North and South, is under half the Pacific's 70 million square miles. Naturally, however, the linear dimensions are more closely comparable. From Yokohama to San Francisco is 4525 miles; and from New York to Gibraltar, 3120.

Nothing would be gained from reintroducing here the term "Heartland". It could hardly displace the accolade "Superpower"; and if perchance it did, would diminish understanding, not enhance it. Mackinder ignored the existence of the USA to an extent tantamount to psychological denial.

What can better be focussed on is the perennial debate within the USA over striking a balance between inward looking and outgoing. During the agonies of disengagement from Vietnam, analysts[56] made appreciative reference to a synoptic review of United States history since 1776 conducted 20 plus years previously by Frank Klingberg, a professor at the University of California. He had concluded that the sequence divided quite neatly into alternating phases of extroversion and introversion, the former being characterized by "expansion and the extension of influence" and the latter by "consolidation and preparation — *plateaux* preceding the mountain climbs ahead". His reckoning was that the extrovert phases had seen 124 American annexations and expeditions as against 12 in the introvert. The mean duration was 21 years for the latter but 27 for the more dynamic former. Klingberg cautiously inferred that the extroversion which began in 1940 might give way during the late sixties to retraction. The transition will quite possibly "carry heavy moral implications . . . The aspirations of the people of Asia and Africa could well furnish the chief cause, along with repercussions from America's own racial problem".[57]

The Klingberg thesis was not without conceptual loose ends. One was how to differentiate "introvert" and "extrovert" in the age of the "moving frontier". Another was that he disregarded 1944–6, a time when United States policy had both extrovert and introvert features. Nevertheless, clear indications of retrenchment did appear in 1967, bang on time on a median extrapolation of the cycle as he defined it and in circumstances relating to Afro-Asia and to the domestic race issue. Race riots occurred in Detroit and other cities in the "long hot summer" of that year. Obversely, a saving grace of the Vietnam situation (as observed in 1966) was how black and white GIs were standing together to prosecute the war though also to vent scepticism. Other 1967 signs of retrenchment included Washington's disinclination to involve herself actively in breaking conflicts in Nigeria and the Congo or over Egypt's precipitate closure of the Straits of Tiran, a proximate cause of the June War.

A pitch towards "introversion" read as "isolationism" has long been recurrent in American attitudes to Europe. Not a few of those dubbed "isolationists" across the years have distanced themselves studiedly from that tormented continent yet remained willing to endorse forward strategies in the Pacific and the Caribbean.

On that interpretation, one can see a swing back to Atlanticist extroversion in the reining back of the Strategic Defense Initiative under way in Washington in 1987. This one says because the more heady visions of what SDI was about were much distrusted in Europe, especially in Britain and Scandinavia. Since when the "shrinking world" phenomenon has made the introversion/extroversion distinction hard to preserve except by treating "unilateral interventionism" as a prime manifestation of the former approach.

Global Asymmetry

There has been much talk of late about regional warlike conflicts being asymmetrical. Yet nothing could be more asymmetrical than how the two great spheres of influence formed up against one another in the 1940s, a situation Churchill depicted as the elephant versus the whale.

Among the inherent dangers was action being taken by one sphere in some part of the world to counter unwelcome initiatives by the other in another location entirely. During the 1962 Cuba crisis, the West subtly took precautions lest the Soviets retaliate against West Berlin. Raised to the level of actual hostilities, such a conflict could exemplify what some of us working in the Australian National University in the early eighties depicted as Limited World War.[58] (See 10).

Such a pattern of conflict was foreshadowed in some measure in the far-flung yet restrained European wars of the eighteenth century. Thus William Pitt the Elder, Britain's Prime Minister for much of the Seven Years' War (1756–63), observed that "I shall conquer Canada in Germany": that is to say, by encouraging Prussia to act as a continental distraction to France.

The seventies began on an optimistic note in world affairs. Come the turn of the decade, however, things turned rather fraught again. A number of crises suddenly broke — Afghanistan, Grenada, Iran, Nicaragua and Poland. A comparison was even drawn with 1914, a slide into hostilities few had sought.[59] Limited World War with a vengeance?

The Afghan Domino

Afghanistan looked the most immediately threatening. It is therefore instructive, and somewhat reassuring, to note two steps Moscow took to indicate that its major military incursion there was no precursor to wider conflict. One of the 20 divisions in Ground Forces Soviet Germany was withdrawn, the first and last time during the Cold War such an order of battle change was effected.

The other signal was the publication in New York of *The Coming Decline of the Chinese Empire* by Victor Louis. Son of a Frenchman, Louis had long since taken Soviet citizenship and become a roving ambassador for the KGB. Ostensibly, his thesis on this occasion was that separatist sentiment in the peripheral regions of China was becoming irresistible, due mainly to the alluring progress of Soviet Central Asia. Yet no deep insight was required to see this disquisition as a great allegory. It was not China's cohesion that was critically threatened, in Louis' real opinion, by centrifugal tendencies. It was the USSR's. The triumph of anti-Soviet factions in Afghanistan could trigger a secessionary domino effect in Soviet central Asia, an effect which could then spread to other constituent republics.

A revealing clue comes in the opening chapter, "Manchuria: A Country which is not". Louis acknowledged that the proportion of Han Chinese in its population has now reached 90 per cent. Yet he concurrently insisted that the native Manchu could no longer be denied their nationhood.[60] In short, he was resorting to a standard totalitarian ploy. Clinch your argument with a proposition that is arrant nonsense, logically or factually. This best betokens a determined pursuit of aims. In this instance, these were to subjugate Afghanistan but go no further.

This whole exercise can be seen as a lesson in the management of asymmetrical conflict. It can be seen, too, as a refutation of the Mackinder thesis. Taking the USSR as tantamount to a Soviet Heartland, one has to say that if it did pose a major threat to the outside world, the geopolitical part of the explanation would be not potential strength but structural weakness.

Successful Failure?

The plain truth about the fabled heartland thesis is that, in Sir Halford Mackinder's mind, it was strictly ancillary to his overriding aspiration which was to see a multicultural British Commonwealth "united by economic ties and a common stance on world affairs".[61] Had he given himself to geopolitical abstraction in 1906 instead of 1904, his emphases might well have been different. For in the meantime, Russia and Japan had made peace via the Treaty of Portsmouth, the former having been worsted on land as well as at sea.[62] This *bouleversement* would induce very soon a radical switch in Britain's external policy. Stop worrying about the threat the Tzar might pose and work with him to contain the one the Kaiser did.

Unfortunately, the fact that the Heartland thesis was so two-dimensional, in spirit as well as structure, meant it could be turned to advantage by people who by no means endorsed its author's actively Liberal politics. One has especially in mind Karl Haushofer, the German ex-soldier turned academic who as early as 1921 was seeking to apply Mackinder's disquisition to Friedrich Ratzel's notion of *Lebensraum*. At one stage removed, Haushofer early on influenced Hitler himself, this through a friendship with Rudolph Hess, the fledgling Nazi Party's deputy leader. Karl Haushofer and his similarly geopolitical son, Albrecht, were thoroughly self-interested in how they related to the Nazis and any likely successors.[63] This stern judgement is not gainsaid by Albrecht being shot for implication in the July 1944 bomb plot nor by his father's committing suicide in 1946.[64]

It remains a tragedy that Sir Halford Mackinder could not divert from his public duties enough to formulate a geopolitical hypothesis sufficiently robust in depth to resist predation. Yet for those duties posterity can still be grateful, not least three ranking British universities — Reading, London and Oxford. As a matter of interest, all three are currently contributing formidably to the debates about climate change.

3 Survival Studies

In 1964 Saul Cohen, a ranking Boston University geographer, drew a neat distinction between the world's "geostrategic regions" (i.e. its core areas) and its "crush zones" or "shatter belts". For understanding the recent (i.e. post-1945) past, this model was helpful.[65] Yet as a predictor for even the near future, it was too much within what can be seen as a two-dimensional geopolitical tradition. After all, 1964 was to see (in Berkeley, California) a sizeable campus protest against the Vietnam War. It heralded a good six years of youth revolt right across the West against "Vietnam" but almost everything else. The ambient stimuli had included a pronounced expansion of higher education and of the use of automobiles, television and the contraceptive pill.

Come 1968 this movement seemed to be putting at risk the very cohesion of society. It thereby showed once again how times of pronounced cultural change tend not to be times of political serenity. Earlier instances, in European experience, include Late Antiquity, the Renaissance and the follow through of the eighteenth-century Enlightenment. Concurrent with this latest was much violence worldwide including Indo-China, the "Cultural Revolution" (1966–9) in China, a savage pogrom against the Chinese minority in Indonesia (1966–7), the "Biafra" revolt in Nigeria (1967–8), the June War in the Middle East (1967), and the brutish if non-lethal crushing of the "Prague Spring" (1968).

Anglophone Insulation

Customarily, British strategic thinking placed little emphasis on the security implications of turbulence on the home front. The Pacification of the Highlands and the Irish troubles apart, the British Isles themselves had been remarkably free, for centuries past, from social unrest of such a scale and character as to threaten internal and, by extension, external security. Throughout the Reformation we experienced nothing approaching in horror the Thirty Years' War in Germany. As religious struggle gave way to secular, this habit of restraint survived. Pre-1939, the Fascists never became a force the way they did over so much of the continent. Nor in the forties did the Communists. Historically, the Straits of Dover have cut to a minimum the scope for foreign aggression or polarizing interference. And what has just been said about the British experience applies in general terms to Australia, New Zealand, Canada and the USA. Young men would set out from the home country to defend it on distant battlefields: "Fighting for New Zealand" in Crete and at Cassino.

Nuclear Dissuasion

Such thoughts as this author has lately had on the genesis of nuclear strategy have already been aired.[66] Suffice for now to insist that, while a nuclear preoccupation well identified the nascent subject of "strategic studies", it did so within too narrow confines. Not too surprising. After all, firepower had thereby "been transformed from being a resource perennially in short supply. Now it had become too abundant to release save in the tiniest fractions. Coming to terms with 'the Bomb' was duly seen as a task for intellectual heroes, paragons best excused more mundane concerns".[67]

The imbalances thus engendered were strikingly apparent awhile in the case of the Indian subcontinent, the Republic of India in particular. Early on analysts weighed the esoteric aspects of India's one day acquiring "the Bomb".[68] Yet they never cogitated the strategic reverberations of India experiencing a major developmental setback, a real possibility until the Green Revolution in agriculture got going from *c.* 1967. Nor did they much anticipate social change in Nepal, Sikkim or Bhutan, change liable to make the Himalayas less of a natural glacis than they have usually appeared histori- cally. Nor did they contemplate beforehand, nor much in retrospect, the geopolitical ramifications of the east wing of Pakistan seceding — the creation, in 1971, of Bangladesh.

Lately India has been registering an impressive economic lift-off. But, as with other fast-developing countries, this could lead either to a deeper "geography of anger" with the Hindu majority, in particular, being encouraged by its extremists to associate the perceived Pakistani threat with internal threats to Indian "security and . . . purity".[69] This latter tendency was encouraged by "9/11" and by the situation in South-West Asia generally. Nevertheless, in the 2004 election the rightist government coalition headed by the Bharatiya Janata Party (BJP) unexpectedly lost to one led by Congress. In office the BJP had sought to reconcile its predilections for free market liberalism, electronic technocracy and cultural fundamentalism by means of a hawkish foreign policy and heavier defence expenditure. On the other hand, in 2002 the BJP in govern- ment had stood back from all-out war with Pakistan.

Going Operational

What can be acknowledged, however, is that in the late forties two great American generals of World War Two showed operationally more awareness of the socio- economic dimension in peacebuilding than ever strategic doctrine did. One has in mind George Marshall in China then Europe; and Douglas MacArthur in Japan.

As he looked towards the White House in 1958–9, Senator John Kennedy did attempt to spell out a comprehensive doctrine for containment of the USSR. He spoke forcefully of "getting America moving" on a front ranging from civil rights to defence and foreign aid. Thus could one regain the initiative lost to the USSR, a state of affairs supposedly betokened by a "missile gap".

Once in place, his New Frontier administration (1961–3) sought interaction with the academic strategists, Robert McNamara (the Secretary of Defense) being especially keen. On matters military, the results were sometimes constructive. On other topics, each side was wont to be ill-prepared. Thus neither was ready for the windows of opportunity then presented for progress towards peace within the Middle East and Southern Africa. Nor had either been geared up to deal with Marxian insurgency, least of all in South-East Asia. On the other hand, the nuclear Partial Test Ban of 1963 did betoken a dawning awareness of the need for military policy makers to take more account of ecological concerns. That year also saw the publication of Rachel Carson's *The Silent Spring*, a book soon to be dubbed the Morning Star of the environmental movement.

Economics and Ecology

Now the quest for economic stability is interacting with that for biogeographical preservation. However, Ecology is often still regarded as a "soft" if not "soppy" approach to peace. Obversely, it is suddenly *à la mode* for those in high places to warn us that "climate change" rather than "terrorism" is what now poses the biggest global threat. Yet it surely requires little reflection to appreciate that, as and when any part of our planet experiences an acute Malthusian crisis occasioned by ecological stress, this is almost bound to find expression in insurgency and/or other patterns of organized violence.

Nor is the idea of some such linkage novel. Two centuries ago those who would defend British farms and woodlands against domination by pell mell industrialization had two objectives in mind. The one was to safeguard social stability, conservatively defined as a rule. The other was to preserve a material resource base balanced enough to meet warlike contingencies.

Through the troubled early decades of the twentieth century, this twinning episodically resurfaced. American exponents thereof included Theodore Roosevelt, Harold Ickes, and eventually Charles Lindbergh. Yet nobody anywhere in the Anglophone world demonstrated it as strikingly as did Field Marshal Jan Christian Smuts of South Africa (1870–1950). This renowned Boer commando leader went on to serve in the highest councils of the British Empire in both world wars. Prime Minister David Lloyd George credited him with being the father of an independent Royal Air Force. It dates from 1918.

His parallel connection with Ecology is evidenced via the etymology of "holistic", a term much in vogue in the New Left and counter-culture years in the sixties. It connotes a quasi-mystical all-embracing slant on the natural world, one those concerned contrast with the reductionist tunnel vision of scientific enquiry. We have it on good authority[70] that Smuts coined this term in 1926. A stimulating book he brought out that year was entitled *Holism and Evolution*.

Climate Change

Obviously climate change is now centre stage *vis-à-vis* discussion of the impact of ecological trends on global security. In fact, the sense of urgency has been made the more acute by the professional communities having dallied, on any reckoning, many decades.Penultimately, as one might say, the El Niño-driven (see 14 (1)) cool global interlude from 1940 to 1965 distracted unduly those of us with an active interest in the subject area.[71] Our failure to scrutinize closely enough the facts and theories led us too long to infer that the overriding prospect remained long-term cooling caused by a compounding of the Milankovitch "astronomical" cycles in the Earth's rotation and its revolution around the Sun.

Ecology in the Round

A deal of research is still needed on climate change. Nor is it just a question of refining further the global predictive models. More effort must go into relating world trends to regional ones, seeing that the said interplay can be far from obvious. Next, in logical progression, one must consider the impact of regional trends on the ambient ecology. Finally, one must routinely integrate this factor with all the others affecting the fortunes of a given region.

However, it is important to remember throughout that, even if the Earth's climate was entirely stable, there would still be a global problem of ecological impoverishment. Nor is it simply a question of pressure on our natural surroundings. In certain cardinal respects, our social environment has been deteriorating. Local communities and extended families have been dissolving markedly. So as the generation born the other side of 1940 progressively leaves the stage, our collective psyche could well experience a loneliness that more concern with the self or with transient interest groups can never offset. It is a rural problem but still more an urban one. We do well to heed a warning that a generalized dislike of urbanization cannot excuse neglect of the city as "one of the most important front lines of environmental issues".[72]

To apprehend this is not to discount the big advances made this last century, veritably the world over, in such fields as education, health, gender equality, social and geographic mobility. . . . Nor is it to deny that traditional backgrounds can be very constrictive, especially towards aspirant youth. Nor, one trusts, is it to pronounce on how individuals, young or old, should look on Mother Nature or on their own social settings. But it is to aver that it could be dangerous to smother attitudes rooted deep in our past at just the time when we are, for a medley of reasons, launched into cultural change at an unprecedented speed. For this could disconnect our reason and undermine our reasonableness.

Survival Studies

Accordingly, we needs make ready to move well beyond the corral that is Strategic Studies. Those of us who were in on its beginning got used all too easily to scenarios for crisis bargaining, hard-faced and narrowly rational, perhaps between adversaries by then responsible for the deaths of millions. Since when, its confines have too often remained too arbitrary and narrow. The widening of horizons lately under way has been altogether too episodic, random and disjointed.

We dare not renounce Geopolitics, nuclear or otherwise, *en bloc*. Nor can we not address matters military. But neither can we afford not to embrace such Human Ecology issues as how far may even the post-industrial Western democracies contain within themselves latent threats to general peace. Might we not propound Survival Studies, this as a field of enquiry broader, more participatory, more planetary, less exclusivist, and — in the final analysis — less utopian than its precursors?

Geography is seen herein as having concerns singularly apposite to this end. Granted, distance and direction tend to matter less in warlike crises than once they did. But in a world shrinking by the week, such area-related concepts as coverage, density and distribution often count for more. This can apply to détente as well as to conflict, to cold wars as well as to hot ones . . .

The outright conquest of territory has all but lost, for the time being at least, the kind of quasi-legitimacy it could sometimes be graced with in the past. Meanwhile, the time dimension concerns us more. Our concern to achieve for planetary purposes a more functional world order does not allow us to be relaxed about regional conflicts staying deadlocked decade after decade or even year after year. We needs also be mindful of how readily, in this age of information, internal tensions can reverberate externally and sometimes *vice versa*.

Among the imperatives especially to bear in mind is a radical reform of the existing world order in good enough time to head off what not a few of us see as the biggest security threat civilization has ever faced, that of biological warfare. On a conservative reckoning that surely means by 2025 whatever one imagines may be the circumstances.

It will further be argued herein that an important strand in a duly joined up survival strategy will be promotion of regional groupings actively committed to mutual development and security. As is evident from endeavours to date, however, it will not be easy to establish affiliations that relate to the world scene in any very consistent way. Our planetary home is too diversified. Geography never repeats itself.

Action–Reaction

The impact of the current information explosion cannot be anticipated at all definitively. It is altogether too novel in character and scale. What can be said, however, is that it will radically affect the collective psychology of defined groups: familial, ethnic, occupational, national, religious . . . In a nutshell, it could make any of them looser, more open and less self-obsessed. Alternatively, it could cause a group to batten down, withdraw unto itself, and generally be a sight less open to the outside world than it could have been had the whole ambience not been so information saturated from all directions. In a shrinking world, any mind that is to remain closed must perforce become more closed.

As and when the latter course is taken, it may well be in reaction to similar groups the other side of a confrontation. The hawks on each side will have common interests. Over against them, the respective doves will as well. Unfortunately, two considerations may work in the hawks' favour. The one is instinctual tribal loyalty. The other is that recourse to violence is liable to uglify a conflictual situation in a very dramatic way. Action–reaction can be traced at every level from a private contretemps to nuclear confrontation. Even in military campaigns, however, an escalatory effect is often constrained or masked by doctrine, geography or force levels. The outcome may be back-down, not back-lash. An instructive example of informal de-escalation comes from when battalions faced each other in the trenches of the Western Front, 1914–18. Sometimes one side would abstain from firing, perhaps for hours and days, provided the other did likewise. The same has applied elsewhere.

The concept of action–reaction cannot proffer a unilinear interpretation of crisis any more than that of deterrence can. Not infrequently, indeed, any backlash against a hostile action may be vented on a third party. Even so, strategists should not neglect this factor the way they have done, strategic arms races sometimes excepted.[73]

War and Social Bonding

The truism that action–reaction operates at all levels of social organization does not connote a consistent relationship between internal discord and response to external threats. In 1935, Poland and Greece, each deeply divided politically, set strong liberal proclivities aside in favour of Rightist authoritarianism. But that did not stop both peoples effecting great solidarity soon thereafter in the face of aggression: German and Soviet against Poland in 1939; Italian then German against Greece, 1940–1. Then again, the fact that Israel has perennially been riven by philosophic and social divisions has never compromised unity critically in the face of armed attack. Contrast all that with how France in 1940 was paralyzed by internal dissent, political and religious.

Still, the most awesome instance to date of the masses on all sides being galvanized by a call to arms has to be Europe in 1914. Twenty years later, George Dangerfield reflected on how Britain was wracked ahead of this dreadful *démarche* by a raft of problems: industrial disputes, House of Lords reform, Irish home rule, women's suffrage . . . The upsurges of anger these gave rise to could not be accommodated by the Respectability which was "one of the chief articles in the Liberal creed, unwritten deep in the heart". Dangerfield was wanly mystified by how, as soon as war broke out, "a single nation" was instantly reforged.[74] But need he have been so surprised? After all, anger had thus been afforded a legitimated conduit. In every country turning to belligerence, young men were flocking to enlist, to find fulfilment in war.[75]

Religion and Peacefulness

The role of religion in either consolidating the prospects of planetary peace or devastating them may seem less urgent a question today than it did immediately post-9/11. All the same, it still assumes more importance than seemed likely, say, 30 years ago. What we must do is avoid mental mapping of the kind that would have us depict the Earth's religious landscape as a surface on which well defined belief systems are mosaiced together either neatly or abrasively. We need geographic expression of the nuances and ambiguities within belief and commitment, especially these days. We also need to allow not just for collaboration but for confluence, a process that will usually involve coming to terms with the Natural Sciences as well.

The sharpest philosophic divide within modern youth, in particular, is not between one religion and another. It is between those who see little value in any religion and those who do feel they derive something from one or, quite possibly, more than one faith. The issue is another Survival Studies could hardly ignore. Still, the first thing to gauge is not how we see our place in Infinity and Eternity. It is how we are shaped by our own evolutionary background, by the history of human geography.

Notes

1 David Chandler, *The Campaigns of Napoleon* (London: Weidenfeld and Nicolson, 1967), p. 1067.

2 Evelyn Edson, *Mapping Time and Space* (London: British Library, 1999), p. 121.

3 Benjamin Zadar in John H. Pryor (ed.), *Logistics of Warfare at the Time of the Crusades* (Aldershot: Ashgate, 2006), p. 176.

4 Gerhard L. Weinberg, "Hitler's Image of the United States, *American Historical Review* 4, LXXIX (July 1964): 1006–21.

5 H. U. Sverdrup *et al.*, *The Oceans: Their Physics, Chemistry and General Biology* (Englewood Cliffs: Prentice-Hall, 1942), p. 2.

6 George Kimble, *The Weather* (Harmondsworth, Penguin, 1951), pp. 216–17.

7 John G. Harvey, *Atmosphere and Ocean: Our Fluid Environments* (London: Artemis, 1985), p. 45.

8 *HCC*, p. 45.

9 Anthony Storr, *Human Aggression* (London: Allen Lane, 1968), Introduction.

10 Konrad Lorenz, *On Aggression* (London: Methuen, 1966), Chapters III and IV.

11 Claire and W. M. S. Russell, *Violence, Monkeys and Men* (London: Macmillan, 1968), Chapter 3.

12 Robert Ardrey, *The Territorial Imperative* (London: Collins, 1967), Chapter 9.

13 Arthur Koestler, *The Ghost in the Machine* (London: Hutchinson, 1967), p. 307.

14 Quoted in Dorothy Rowe, *Living with the Bomb. Can we live without Enemies?* (London: Routledge and Kegan Paul, 1985), p. 146.

15 Eric Fromm in Ralph K. White (ed.), *Psychology and the Prevention of Nuclear War* (New York: Doubleday, 1968), p. 146.

16 Edward Thompson, "Notes on Exterminism, the Last Stage of Civilisation", *New Left Review* 121 (May–June 1980): 3–31.

17 Crispin Tickell, "The human species: a suicidal success", *Geographical Journal* 159, 2 (July 1993): 226–29.

18 *The Autobiography of Bertrand Russell* (3 Vols.) (London: George Allen and Unwin, 1969), Vol. III, p. 221.

19 Archibald Lewis, "The closing of the medieval frontier, 1250–1350", *Speculum* 33 (1958): 475–83.

20 For a rounded discussion of this matter see David Abulafia in David Abulafia and Nora Berend (ed.), *Medieval Frontiers: Concepts and Practices* (London: Ashgate, 2002), pp. 6–9.

21 Frederick Jackson Turner, *The Frontier in American History* (New York: Holt, Rinehart and Winston, 1920), p. 2.

22 *Ibid.*, p. 4.

23 *Ibid.*, p. 30.

24 Martin W. Mikesell, "Comparative Studies in Frontier History", *Annals of the Association of American Geographers*, 50, 1 (March 1960): 62–74.

25 Turner, *The Frontier in American History*, p. 16.

26 Mikesell, "Comparative Studies in Frontier History", p. 71.

27 Alida C. Metcalf, *Family and Frontier, Colonial Brazil* (Austin: University of Texas, 2005), p. 4.

28 Elena Lourie, "A Society Organized for War, Medieval Spain", *Past and Present* 35 (December 1966): 54–76.

29 E. S. May, *Geography in Relation to War* (London: Hugh Rees, 1907), p. 21.

30 C. P. Fitzgerald, *The Chinese View of their Place in the World* (London: Oxford University Press for Chatham House, 1966, Chapter 1.

31 "A riddle of rockets", *The Economist*, 385, 8550 (13 October 2007): 71.

32 Avishai Margalit, "A Moral Witness to the Intricate Machine", *New York Review of Books* XIV, 19 (6 December 2007): 34–37.

33 Patrick Wright, *Iron Curtain* (Oxford: Oxford University Press, 2007), p. 351.
34 Alexander Alderson, "Iraq and its Borders", *RUSI Journal* 153, 2 (April 2008): 18–22.
35 *New York Times*, 20 October 2006.
36 Halford Mackinder, *Britain and the British Seas* (London: William Heinemann, 1902), p. 334.
37 "Talk fraternally but carry a big stick", *The Economist* 387, 8582 (31 May 2008): 61
38 G. E. D. Gedye, *Fallen Bastions* (London: Victor Gollancz, February 1939), pp. 383–7.
39 G. D. H. Cole, *The People's Front* (London: Victor Gollancz, 1937), p. 175.
40 "Stalin's Big Fleet Program", *Naval War College Review* LVII, 2 (Spring 2004): 87–120.
41 H. J. Mackinder, "The Geographical Pivot of History", *The Geographical Journal* XXIII, 4 (April 1904): 421–44.
42 W. H. Parker, *Mackinder. Geography as an Aid to Statecraft* (Oxford: The Clarendon Press, 1982), Chapter 6.
43 H. J. Mackinder, *Democratic Ideals and Reality* (London: Constable, 1919), pp. 34 and 140–1.
44 Mackinder, *Britain and the British Seas*, p. 358.
45 H. J. Mackinder, *Democratic Ideals and Reality* (London: Constable, 1919), p. 143.
46 N. Schoonmaker and D. Reid, *We Testify* (New York: Smith and Durrell, 1941), p. 215.
47 John Reeve in Hugh Smith (ed.), *The Strategists* (Canberra: Australian Defence Studies Centre, 2001), Chapter 4.
48 A. T. Mahan, *The Influence of Sea Power on History, 1660–1783* (London: Sampson, Low, Marston, 1890), p. 46.
49 Professor Geoffrey Till, "The New American Strategy – What's New? What's Next?", *The Naval Review* 96, 1 (February 2008): 7–17.
50 Mahan, *The Influence of Sea Power on History, 1660–1783*, p. 89.
51 A. T. Mahan, *The Influence of Sea Power on the French Revolution and Empire* (London: Sampson, Low, Marston, 1893), 2 vols., Vol. 2, Chapter 19.
52 Ernest Rhys (ed.), *Orations on the French War to the Peace of Amiens by William Pitt* (London: J. M. Dent, undated), Chapters I, IV, XIII, XV, XVI and XXI.
53 Julian Corbett, *Some Principles of Maritime Strategy* (London: Longmans, Green and Co., 1919), p. 48.
54 Dr Norman Friedman, "The Sea-based Commonwealth", *Naval Review* 95, 4 (November 2007): 370–2.
55 Roger Revelle, "The Ocean", *Scientific American* 221, 3 (September 1969): 55–65.
56 Zbigniew Brzezinski, "US Foreign Policy: The Search for Focus", *Foreign Affairs* 51, 4 (July 1973): 708–27.
57 *Ibid.*
58 See the author's *Limited World War?*, Canberra Papers on Strategy and Defence 32, Canberra, Australian National University, 1984.
59 Miles Kahler, "Rumours of War: The 1914 Analogy", *Foreign Affairs* 58, 2 (Winter 1979/80): 376–96.
60 V. E. Louis, *The Coming Decline of the Chinese Empire* (New York: Times Books, 1979), p. 11.
61 Brian W. Blouett, "The political career of Sir Halford Mackinder", *Political Geography Quarterly* 1, 4 (October 1987): 355–67.
62 N. C. Hayes, "The Russo-Japanese War, 1904–5", *Naval Review* 95, 4 (February 2008): 379–86.
63 Henning Heske, "Karl Haushofer: his role in German geopolitics and Nazi politics", *Political Geography Quarterly* 6, 2 (April 1987): 135–44.
64 For a succinct but most informative summation of the mechanics of geopolitical influence on the Nazis and the limits thereof see G. R. Sloan, *Geopolitics in United States Strategic Policy, 1890–1987* (Brighton: Wheatsheaf, 1988), pp. 232–5.

65 Saul Cohen, *Geography and Politics in a Divided World* (London: Methuen, 1964), pp. 62–87.
66 *GISC*, pp. 38–42.
67 *Ibid.*, p. 39.
68 Leonard Beaton and John Maddox, *The Spread of Nuclear Weapons* (London: Chatto and Windus for the Institute for Strategic Studies, 1962).
69 Arjun Appadurai, *Fear of Small Numbers, An Essay on the Geography of Anger* (Durham: Duke University Press, 2006), p. 94.
70 *Supplement to the Oxford English Dictionary* (4 Vols.) (Oxford: Oxford University Press, 1976), Vol. 2, p. 120.
71 S. Brönnimann, "The global climate anomaly, 1940–42", *Weather* 60, 12 (December 2005): 336–42.
72 Andrew Light, "The Urban Blind Spot in Environmental Ethics", *Environmental Politics* 10, 1 (Spring 2001): 7–35.
73 Lawrence Freedman, *Nuclear Strategy* (Basingstoke: Macmillan, 1981), pp. 254–6, 335–8.
74 George Dangerfield, *The Strange Death of Liberal England* (London: Constable, 1936), pp. 355–6 and 409.
75 Hew Strachan, *The First World War* (Oxford: Oxford University Press, 2000 (5 Vols.), Vol. 1, pp. 111 and 133–62.

2 | Acquiring the Globe

1 Miocene to Iron Age

How our pre-human forebears then ourselves evolved in the face of endless stress can instructively be traced back to the Miocene geological epoch within the Tertiary period, an epoch defined as ending five million years (myr) Before Present (BP). The Miocene marked the zenith of a great era of mountain building. The uplifting of the Rockies and the Andes progressed decisively then; and a particular manifestation of this Western Hemisphere orogenesis was the formation of the Columbia and Snake lava plains which cover 300,000 square miles and remain up to 10,000 feet thick. The Deccan of India likewise witnessed huge lava extrusion. It was collateral with the rise of the Himalayas and the Tibetan plateau.

Human beings (*Homo sapiens*) belong to the Primate order as do apes, monkeys, lemurs etc. The Hominid family comprises just ourselves and certain extinct relatives (above all, the Neanderthals) closer to us genetically than any extant apes are. Today there are five ape genera, these limited to a few niches in Africa and South East Asia. Through the Miocene and well into the succeeding Pliocene, there were dozens. This narrowing down owed much to climate change.

For 50 myr or so, our global climate has been subject to an underlying cooling trend. In the early to middle Miocene (*c.* 15000000 BP) this underwent a step-like acceleration, one result apparently being the establishment of the pan-oceanic circulations, the gyres. During the Pliocene (five to two myr ago) there were longish intervals of abnormal warmth. Overall it was an epoch of marked climate alternation in tropical Africa though mainly warmish and wet through the first half, cooler and drier in the second.[1]

In 1924, Raymond Dart of the University of Witwatersrand discovered *Australopithecus*, a Pliocene genus ancestral to ourselves. A modest five feet tall, australopithecines could walk upright; and by 2500000 BP some hunted in packs. These attributes were adaptations to being squeezed out of native rain forests by prolonged and searing droughts. Australopithecine remains occur extensively across South and East Africa. An acquired ability to walk upright will have enhanced vision, diminished solar exposure at high noon, freed the forepaws for other duties, and kept the brain cooler. Skeletal remains often include fractured skulls indicative of harsh social discipline and/or intraspecific warfare. There are indications of some speech development by 2000000 BP and of laid camp fires from nearly 1500000 BP.[2]

Australopithecus had evolved by 1750000 BP to another African species, *Homo ergaster* — a biped with a 900 to 1200 cc. brain. However, this divided relatively soon into one branch leading to *Homo erectus* and *Homo floriensis*; and another leading to *Homo neanderthalis* and *Homo sapiens*. Especial interest currently attaches to *Homo floriensis* in that it was first identified in 2004 on Flores in Indonesia. All the same, the Neanderthal interaction with humankind is critical to any evaluation of our basic nature.

Human versus Neanderthals

In the accepted chronology, the last two million years are the Quaternary period; and the most recent ten millennia of this span constitute the Holocene epoch, the rest the Pleistoocene. The Quaternary is remarkable for a high incidence of vulcanism with temperatures depressed accordingly. The contrast between glacial spells and warmer interludes was often accentuated by (a) methane-producing peat growth,[3] and (b) the ice-albedo feedback. **(See 14 (I)).**

Despite a tendency for biological evidence to become corrupted over time, DNA is already invaluable for evolution mapping. An early milestone was the Berkeley report in 1987 concerning mitochondrial DNA gathered from five disparate human populations. It indicated all had derived from a female who lived in Africa something over 200,000 years ago. The inference was that *Homo sapiens* as a whole is descended mainly from this "African Eve", rather than the Neanderthals then spreading across Eurasia to reach eventually Java and Siberia. The first human sally into Eurasia was into the Levant via the Sinai rather before 100000 BP. It proved unsustainable. The next one, which was decisive, took place *c.* 85000 BP via what were then somewhat narrower straits at the Red Sea entrance. The latest evidence suggests Eve lived more like 300000 BP. Otherwise it confirms the "out of Africa" thesis.

For long there was no conclusive evidence of fruitful miscegenation between humans and Neanderthals in the 40,000 years or so the two species lived in Eurasia in loose or sometimes very direct proximity. Then in late 2006 some findings were reported from work ongoing at the Howard Hughes Medical Institute and the University of Chicago. These were that in 37000 BP a Neanderthal mutation of the gene microcephalin entered the human gene pool. It is now carried by 70 per cent of us.[4]

Homo sapiens had evolved a brain nearly three times the volume of an *australopithecine*, partly by dint of sustaining cerebral growth throughout the first year after birth.[5] Yet at 1450 cc the median brain size of a Neanderthal was somewhat larger. Moreover, this five-foot hominid was well endowed in sundry other ways. Large-boned with strong muscle attachments, it was physically robust. The long gestations it probably underwent were likely beneficial.

The Neanderthals used stone tools for butchering and often lived in caves. A stone and bone face mask (from 32000 BP) has been attributed to them. The varying elaboration of the graves in which they buried their dead betokened social structuring. Sometimes flowers were placed on graves, perhaps in emulation of human practice.

Most remarkably, they appear to have had much the same "vocal anatomy as we have. The possession of a similar hyoid bone" in their voice box, "an enlarged thoracic spinal cord and an enlarged orifice to carry the hypoglossal nerve to the tongue are consistent with Neanderthals speaking".[6] Opinion is still divided as to whether they did achieve what could fairly be termed 'speech'. At all events, their culture advanced but slowly.

Per Ardua

The social and territorial progress of *Homo sapiens* was repeatedly punctuated by stressful climate shifts. The account that follows is perforce confined to the major turning points as eurocentrically perceived. But it will throughout be needful to bear in mind that shorter term and/or more localized erraticism could be very impacting. The Potsdam Institute for Climate Impact Research has identified 20 episodes between 120,000 and 10,000 BP when the mean temperature in a transition zone shot up 5 to 10°C within a decade.[7]

A valuable pointer to an ever-shifting climate has been a seabed core extracted from mid-Atlantic at 53°N. Taking the broad sweep, it reveals a summer sea-surface temperature peaking at 16°C in 124,000 BP. Next came an irregular downward progression which culminated in an abrupt plunge to 6° around 72000 BP. This was due to a supervolcano explosion at Toba in Sumatra, by far the worst Quaternary eruption anywhere. DNA indications are that the human total worldwide may have fallen to several thousand. Depending on geographical spread, this must have been perilously close to the threshold of irrevocable decline.

A post-Toba recovery gave way to another peak-to-trough fall (from 12 to 7°C) between 54,000 and 46,000 BP. During this cooling phase, the Cro-Magnon people entered Europe from Anatolia. These representatives of *Homo sapiens* introduced their dynamic Aurignacian culture which "quickly soared far above that of the Neanderthals"[8] or, indeed, of previous humans. Their craft skills ranged from rudimentary hearths to sewn clothes. Portable artefacts were fashioned not just from stone but from bones, antlers and ivory. Elegant spearheads betoken a certain martial sophistication. An hierarchical social structure finds expression in how certain of their dead were buried in elaborately beaded costumes. It further seems that a subtle jaw mutation was facilitating progress with speech and therefore thought. May one add that Cro-Magnon males typically stood six feet high.

Spirituality found expression in the cave paintings discussed below but also in the Venus figurines which gradually became abundant. All in all, one may conclude that Cro-Magnon ascendancy over the Neanderthals was assured.

Nevertheless, mystery still surrounds just how the latter were driven to total extinction, this finally in south-west Europe *c.* 28,000 BP. Some of us suspect that, being so strategically mobile, the Cro-Magnons proved morbidly efficacious at spreading among their Neanderthal neighbours (by accident or design) diseases they themselves had gained a goodly measure of immunity from. Alternatively, should we speculate that, in the presence of creatures so similar yet demonstrably more accomplished, the Neanderthals effectively died of a collective broken heart?

Annihilating Large Animals

The definition of a large animal, alias megafauna, is for this purpose one appreciably bigger individually than its human predator. A simple view of the human contribution to large animals (e.g. mammoths, woolly rhinosceroses, great elks, kangaroos . . .) becoming extinct during the late Pleistocene would have it that, within one or two thousand years of humankind arriving in a given region, many such species will perforce have been eliminated.

Nonetheless, debate continues about how far hunting to extinction was driven by threshold effects induced by the Malthusian pressure of human populations on resources. Computer-driven studies of mammoths in North America have indicated how their numbers could have crashed once the human population has passed a critical threshold. In this rendering, the immediate effects of climate vicissitudes may be secondary.[9] However, in thinly peopled Siberia extinctions were mainly due to the paucity of escape routes southwards in times of glacial advance.

Also to be considered is our deliberately burning landscapes, especially woodland, one purpose being to flush out prey. In south-west Germany forest clearance by burning, primarily to gain more land for cultivation, diminished markedly from c. 4000 BP with the introduction for hunting of bronze blades.[10] In Australia, the aboriginals burnt grasslands and woods very extensively and recurrently. Among the motives were removing dead grass and reducing the populations of mosquitos and snakes. But it is to be observed that all land mammals, reptiles and birds weighing more than 100 kg. were eliminated not so long after the aboriginals arrived. Students of early humankind seem generally persuaded that a need for teamwork in pursuit of more powerful animals or for forest clearance spurred our development of syntactical language.

Language and Religion

The many languages still extant (lately reckoned at 6000) can usefully be grouped into "families", their respective commonalities going back scores of thousands of years. Indeed, the syntactical structuring of language is a universal human trait stretching back to "the beginning of time" as we perceive it. There are no "primitive" human languages. It is surely this more than anything else which sets us apart, even though not a few other species have means of communication more sophisticated than once we realized.

Over time, the coherent evolution allowed by syntax produced extreme diversification, a distinct language or dialect often being a defining attribute of a particular tribe. It will be powerfully cohesive in peace and war. Yet by virtually the same token it will be liable dangerously to exacerbate misunderstandings between tribal neighbours — the Tower of Babel effect.

2 Planetary Tribalism

With the emergence of structured language came also a burgeoning capacity for joined up thought: an attribute basic to knowledge accumulation, decision taking and, indeed, the formulation of religious perspective. Workaday experience interacted with language structuring to favour a dualist view of reality: good or bad; male or female; light or dark, us or them . . . That the Sun and our solitary Moon appear almost the same size in the sky (thanks to a fluke of astral geography[11]) reinforced this imprinting of our psychology.

Dualism was also perceived in what we would understand to be more metaphysical terms. On the one hand, Creation was orderly and rhythmic to an extent expressive of an immanent authority. On the other, this sublime order was punctuated frequently though irregularly by acute criticalities: weather extremes, climate shifts, earthquakes and volcanoes, meteorites, epidemics . . . Prominent in many legends was the Noahic flood, its factual foundation apparently being the formation of the Black Sea *c.* 7550 BP as water massively overflowed what became the Turkish Straits.[12] The human dispersion thus necessitated much encouraged linguistic variegation. Not quite four millennia later (in 3635 BP), this same region was to experience the eruption of the Thira volcano on the Aegean island of Santorini, a cataclysm which caused *tsunamis* around the Eastern Mediterranean. It was likely the inspiration a millennium later for Plato's *Atlantis*, his utopian critique of city-state strife.

Together these two events were a background to how the sea was depicted in the early literature of the Eastern Mediterranean and Near East, the Bible included. This was as the bellicose "Other", a vast entity that can be harnessed to God's service only after it has been subdued.[13]

Long before and ever since, more subtle signs have been discerned in the sky. Witness the planetary conjunction the "Star of Bethlehem" likely represented. More often though a sign has been read as connoting divine displeasure though what about might be unclear. A lively display of aurora borealis use to cause consternation. The usually prosaic *Anglo-Saxon Chronicle* tells how, shortly before Vikings sacked Lindisfarne monastery in AD 793, the people of Northumbria had been "wretchedly terrified" by "excessive whirlwinds, lightning storms and *fiery dragons . . . flying in the sky*".

Three-tiered Creation

A belief that interaction with the Supernatural took place via the netherworld as well as the skies above was encouraged by volcanic extrusions and by earthquakes. A singularity was the Apollo oracle at Delphi. There a psychogenic cocktail of gases from bituminous limestone effuses through a cave mouth.[14]

Of wider import have been the deep limestone caves of France and Spain where *c.* 35,000 years ago Palaeolithic artists began to engrave and paint. Often working in barely accessible passages, chambers and nooks, sometimes a good kilometre down, they depicted people but also larger mammals (e.g. lions, mammoths and woolly rhinosceri) or maybe mythical half-human creatures. They have commanded the respect of modern viewers ever since their rediscovery a century or so ago. Surmise that they fed into deep religious experience has lately received corroboration from similar imagery on the southern African plains being studied at Witwatersrand in relation to folk myths lately still extant.[15] The Palaeolithic art will have served to underline a sense of human exceptionalism, a sense which must have waxed the stronger with the final vanquishing of the Neanderthals in that same region not so long afterwards.

An approach to conflict resolution has not infrequently been to postulate a pantheon in which all the lesser gods have specific responsibilities, communal or thematic, yet are themselves answerable to a paramount God. The Celtic cult of Lug, which extended in pre-Roman times from Iberia to Iceland, would have overlorded quite a few lesser immortals. So, a millennium or two earlier and within a more confined region, would have whoever laid claim to Stonehenge.

The Neolithic Revolution

Well before then, however, the multilateral transition to the Neolithic material culture — the New Stone Age — had been effected. Deriving its name from the polishing and refinement of stone tools, it involved, too, the development of pottery; and a partial switch from nomadism to sedentism — the latter considerably in small villages. The domestication of animals and cultivation of plants also figured.

Strictly speaking, one should think of "evolution" not "revolution". For though these several advances were considerably related, their timings could vary markedly. Take domestication, a practice sometimes hard to delineate. The dog–man relationship is well confirmed for the last 10,000 years but may sometimes reach back 100,000. The earliest travel on horseback confirmed, this east of the Dnepr river, was *c.* 6000 BP.

Nor should one exaggerate how far the Nile to Mesopotamia arc spearheaded the Neolithic advent. Take pottery. The earliest fired-clay figurines known are 27,000 years old and from central Europe. Then again, the earliest clay pot was till recently from the middle Yangtse, 14000 BP. But now one from northern Japan has been dated by Kokogakuin and Nagoya universities at 16500 BP.

Curiously though, the regional time lag between ceramics and crop cultivation can be some millennia. Similarly delayed may be the proliferation of cultivation. Rice growing began in the middle Yangtse (whence wild rice had already spread from the south) around 9000 BP.[16] But it may not have overtaken the wild rice, in terms of volume consumed, for 1500 more years. It arrived in Korea around 4500 BP, and thence diffused to Japan.[17] Likewise, data recovered in Mexico indicate that there "the transition from hunting and gathering to food production occupied over eight millennia". There, too, the "beginnings of agriculture were not associated with sedentary village life, pottery, polished stone artefacts and other traits" customarily regarded as integral to Neolithic change.[18]

From 11000 to 10300 BP, the Younger Dryas interlude turned much of South West Asia cooler but also much drier. One extrinsic input early on was a huge overflow of glacial meltwater down the St Lawrence valley into the Atlantic. Another was a peaking in seasonal temperature contrasts due to the Milankovitch effect. (See 14 (I)). Across much of the South-West Asia to South Mediterranean zone, cooler winters tended to mean less rainfall while hotter summers always caused more water loss.[19] Life in the axial river valleys therefore became more difficult; yet at the same time, the option of dispersal into steppic hinterlands beyond was undermined.

A response was found in more insistent innovation. Cultivated grainstuff became a staple, this apparently happening first in the Karacadag mountains, south-east Turkey. There wild einkorn grass is still to be found which has a genetic fingerprint closely akin to that of modern wheat. Thenceforward, something of a momentum gradually built. Written languages decipherable by modern scholars were extant in the Middle East five millennia ago. By then, too, Mesopotamia could boast several cities with well over 10,000 inhabitants.[20]

Metals as Materials

The smelting and working of various metals developed on a limited scale. By 5000 BP the merits of the alloy of copper and tin we know as "bronze" were appreciated in

Mesopotamia, India, Asia Minor and Greece; and by 3000 BP were so in China too. Before so very long, Egypt and Mesopotamia would be generating "the first full chapter in the history of science and technology. The early history of chemistry, agriculture, stockbreeding, medicine, hawking horses (*sic.*), astronomy, not to mention law, begins here."[21] Then during the last millennium BC, more efficient furnaces allowed of iron's considerably displacing bronze. It is a matter of respective melt temperatures.

A study of 350 British skulls dated between 6000 and 5200 BC showed 20 with depressed fractures consistent with deliberately inflicted blows. In six cases, these would have been fatal.[22] But in Britain, the advent of iron smelters encouraged a political consolidation into something like a dozen main kingdoms, overlording many lesser communities. The centre of any community was what we habitually speak of as a hill fort. At this stage, however, physical security was probably a less active factor in its location than religious and political symbolism.

General Inferences

Such interpretation of our cultural history and psychological evolution does not show *Homo sapiens* to be a fighting animal in some kind of all-enveloping sense. Nevertheless, it confirms how prone we are to intra-specific violence when pushed beyond thresholds of discomfiture often not identifiable in advance. Our capacity to be rational and reasonable can be similarly threshold delimited; and is, in any case, subject to a yearning to devote ourselves to communities ideally several score or hundreds of persons strong, these likely defined linguistically or metaphysically. Along with all of which goes (a) a special sensibility in respect of the heavens above and (b) a proclivity to delay adaptation to changing circumstances until it is almost or absolutely too late.

Yet one needs recognize, too, how the exploitation of metals in particular eventually led on to the formation of large and enduring empires. These had a propensity to acquire remarkably similar basic features even when seldom or never in contact: towns, metropoli, temples, agriculture, irrigation, crafts, coinage, slavery, writing . . . How far are these entities to be explained in terms of a widespread longing for law and order? How often do they develop as a quest for immortality on the part of death-obsessed individuals? Are the most longlasting realms those that best accommodate local community identities?

Underlying such driving forces has been the presumption that we have to subjugate part of Nature to prevent the whole overwhelming us. Accommodating these legacies in such a way as to restore the world we live in is what Survival Studies has essentially to be about.

3 Hydraulic Despotisms?

Since the collapse of Soviet Communism, there has been a surprising absence of works aspiring to comprehend the course of history, recent and future. But this cannot continue. Texts are bound to appear which claim to demonstrate either that democratic capitalism cannot save the planet or else that only it can. Likely, indeed, the banking credit crisis of 2008 will call such demonstration forth.

One must expect, too, that these hypotheses will be endorsed all too uncritically by their respective disciples in academe. The result will be to lock those concerned into closed-minded actions and reactions, instead of leaving them to engage in free-ranging debate. The threat which could thus be posed to opinion forming and policy making that is rational and reasonable can be illustrated by reference to one text that has been treated as an authoritative scholarship on many a Social Science course, though not methinks Geography ones.

In 1957, Karl Wittfogel (1896–1988), a sinologist of German descent, brought out *Oriental Despotism*. Though ostensibly geographical, it evinced a cavalier disinterest in natural ecology as a factor in historical causation. This failing it shared with Marxian utopians but also the hard Right in academe and elsewhere. Hardly surprising, given that its author had sojourned awhile in each camp. The gist of the text was that broad valleys and plains within which agriculture required large-scale irrigation were strongly conducive to unitary despotic government. Such hydraulic societies were deemed especially characteristic of south and east Asia. Curiously, Wittfogel early in the text professed unhappiness with the title he himself had opted for.[23] Yet several pages later he could say, "I have no hesitancy in employing the traditional designations *oriental society* and *Asiatic society* as synonyms for *hydraulic society* and *agromanagerial society*; and using the terms *hydraulic, agrobureaucratic* and *Oriental despotism* interchangeably".[24]

So was it all just Cold War spin? A revamp of "heartland" thinking? Its author's peregrinations encourage such a reading. In the twenties, he had been prominent in the ultra-Left Frankfurt school. By 1940, he had swung well Right, impelled by disillusionment with Stalin. Correspondingly, he stretched his thesis impossibly by seeking to apply it to Soviet Communism via Byzantium. He himself duly admitted that neither in Muscovy nor in middle Byzantium did agro-hydraulics play a determining part.[25] No matter. The book was acclaimed on the American Right as "flawless scholarship", a veritable "watershed".

A basic indictment of Wittfogel must be his not addressing the physical aspects. There are no indexed references in *Oriental Despotism* to soil quality, groundwater, salination, topography, silting or climate change. Yet he had once studied inscriptive evidence of warming in North China *c*. 3100 BP.[26]

In any case, endeavours to make a region able to support more people is liable to encourage population growth and raise expectations. Yet this may leave those concerned all the more vulnerable to climatic vagaries. Therefore neither the Orient, however defined, nor the world beyond separates neatly into water regimes conducive to overarching despotism and those more favourable to devolution. Usually the biggest factor in imperial ascension has been the nearby emergence of a similar polity which needs be counterbalanced. More generally, flattish open landscapes have favoured imperial development since they have facilitated the flow of data and the march of armies. Then again, good local stone has allowed of monumental building.

These radical reservations were given vent in a critique of Wittfogel delivered at the 2001 Bergen conference of the International Water History Association.[27] Wittfogel's exposition still seems to me too exclusive and determinist; and so organized as to be hard to assess in detail. Nevertheless, any study which lays emphasis on human influence on a water cycle as a key factor in holistic survival studies can hardly ignore "hydraulic despotism" as a theme to evaluate.

Korea

A paper presented at Bergen by Dr Bong Kang of Kyongju University warned that all judgements about Korea were bound to be tentative, given the weight of evidence still to be processed. Still, his provisional judgement was that pressures for wet-rice irrigation were not critical in the formation of the Baekje kingdom (emergent by AD 260) nor the Silla (operative by AD 356–402). The relevant schemes were operative either too long before or too long after. Chronicles mention irrigation far less than they do war. The irrigation works in question tend to be well removed from the national capitals.

Latin America

Here pre-Columban empire-building tests Wittfogel stringently. Crucial to any interpretation thereof will be the climatic differences between the "Mexican" Maya/Aztec realm and the "Peruvian" Inca one. The latter has a strikingly even spread of rainfall seasonally. Therefore, while large-scale projects were needed to water the desertic Peruvian coast, inter-seasonal storage was little needed. The socio-political structure of the Andean region cannot be "hydraulically" explained because its "irrigation systems were small-scale and few communities depended on them".[28] Nor, of course, can it account for the scale of Inca empire-building, this without wheeled vehicles.

Conversely, in Mexico City (formerly Tenochtitlan, the capital the Aztecs founded *c.* AD 1325), 90 per cent of the rain falls from May through October which has posed a storage requirement. Especially impressive historically is the Tehuacán valley network 200 km south-east of the basin of Mexico. It comprises 1200 km of canals; dams, large and small; covered aqueducts crossing watersheds; and extensive terracing. Some canals can be dated to 800 BC; and development of them continued until the early 1500s. There and elsewhere are shallow though wide local wells, some dateable from 9500 BC. Furthermore there are indications, archaeological and anthropological, that the communal *sociedades de agua* (water management societies) active today have deep roots.[29] *A priori*, Wittfogel seems of some relevance.

Sudanic Civilization

Also of interest in this context is the medieval development of quasi-bureaucratic Sudanic kingdoms in the transition zone of jungle, parkland and savannah on each flank of the African rain forest. These polities tended to evolve out of clusters of small theocracies, an interpretation corroborated by the prevalence along the western reaches of the Niger of urban sites dating back, in some cases, over two millennia. It may also be significant that across "large stretches of the central and western Sudan, traditions ascribe the foundation of early dynasties to pre-Islamic immigrants from the Yemen".[30] Germane may be the climate crisis that smote the Arabian peninsula's south-west (*Arabia Felix* or *Arabia Odorifera)* from AD 535 (see below). Pertinent, too, is the then recent introduction into Black Africa of the camel.

On the southern flank of the rain forest, these exogenous influences were absent. Nor did any great tradition of town-building develop. Nevertheless, in Zimbabwe and the Transvaal good building stone did encourage prestige architecture.

In West Africa, Kanem and Ghana were the first statehoods the Arab chroniclers recorded, each strongly emerging by AD 1000. Kanem, located around Lake Chad in a steppic landscape, never really urbanized. Conversely, Kumbi Saleh, the capital of Ghana, and set firmly within what today is parkland (i.e. savannah with clumps of trees) achieved considerable urbanization reliant on fossilized groundwater from deep wells.[31] Describing it in 1067–8, the Córdoban geographer Al-Bakri credited Ghana with an army 200,000 strong. This could connote a population of two to three million or 20 to 30 per square mile, close to what the region was supporting when the Sahelian drought cycle struck from 1960.

Ghana had by 1000 assumed nodal importance in a trans-Saharan trade dominated by alluvial gold from West Africa and salt from the Saharan seas of long ago. Yet barely a decade after al-Bakri's account, the kingdom was overrun by the Almoravid confederation of desert tribes, a group lately strengthened by incorporation of the Berbers and the general adoption of Islam.

With Ghana's demise came wanton mayhem. Yet that was not what had gained for the Almoravids control over the west Maghreb and, even awhile, Muslim Spain. This uglier denouement rather suggests an abrupt passage through a negative ecological threshold. Admittedly, evidence about climate change then around the Sahelian fringe of the Sahara is skimpy, indirect and rather contradictory. What can be proposed, none the less, is well depletion and overgrazing.[32]

Arabia Odorifera

Around 950 BC, the Queen of Sheba (that is, most probably, Saba) led a resplendent caravan to visit King Solomon. Two centuries later, a stone-faced dam was built close by Marib, Saba's capital. Quite exceptional at two thousand feet wide, its main section impounded flash floods sufficient to irrigate 24,000 acres. Early in the Christian era, Greek and Roman scholars were celebrating *Arabia Odorifera* (alias *Arabia Felix*) for the exotic life style it derived from a grip on the Indian Ocean cosmetics trade and the indigenous production of perfumes/medicines in the frankincense and myrrh range of tree resins.

Having assumed this centrality, the region lacked any security of obscurity, an exposure which political unification *c.* AD 300 could not offset. Cultural impacts proved the harbingers of political intrusion. Nature, too, intervened. For some time, it has been evident that *c.* 540 the world's weather underwent acute perturbation; and in 1999, David Keys demonstrated that 535 witnessed the worst volcanic explosion (this by Krakatau) in 72,000 years.[33] In erstwhile *Arabia Felix*, the upshot was decades of rainfall erraticism. In the 540s, a flood breached the Marib dam for the first time since 450. A decade later this happened again, with silting so severe that half the irrigation was lost.

After a third breach in 590, the Marib was abandoned for good. Tens of thousands emigrated, many to Africa. The economic underpinning the dam proffered had proved irreplaceable. But its construction had followed rather than effected the initial surge of prosperity. Nor had it led forthwith to political unification. *Arabia Odorifera* was not therefore a convincing "hydraulic society".

The Nile

The circulation of the Indian summer monsoon regularly extends over the southern

Red Sea and northern half of the horn of Africa. So since Nile flood levels are determined largely by fluctuations in the Blue Nile and the Atbara, which rise in the Ethiopian highlands, there will be a positive correlation between the flood's abundance and that of the monsoon rains across the northwest of the Indian subcontinent.

The criticality of the Blue Nile derives most basically from the fact that it traverses but six degrees of latitude whereas the White Nile flows across 15. Moreover the latter crosses the swamps of El Sudd. These have exercised a modulating influence ever since their evolution in the aftermath of the radical adjustment the Nilotic drainage system went through *c.*12000 BP as the equatorial climate turned moister within that longitude zone. The river further down stream ceased to be subject to erratic flow and extensive braiding. It became what it has since remained, a well-incised feature with strong though seasonally accented flow. This adjustment coincided with the Older Dryas, a defined cool interlude — lasting two centuries — in the post glacial warming.[34]

What interdisciplinary research on-going suggests is that *c.* 5500 BP (a few centuries ahead of the first pharaonic dynasty) the middle Nile valley dried sufficiently to favour agriculture. At the same time, desertic hill country to the west became too dry for existing settlements. This encouraged the migration of the hill people (with their superior culture) to the valley. Admittedly, the capital of the first dynasty was at Memphis, much further downstream. But that does not exclude a causative link.[35]

Since the Blue Nile will have been a powerful scouring agent, the Nilotic flow was little compromised by silting. The big hazard was weak monsoons, individual or clustered. Water shortage would be registered throughout the valley, save that those in the upper reaches might hold back too much. The inference could be that the Nile lent itself to synoptic centralized governance, to autocratic despotism.

What became customary, however, was that the Pharaohs left the life-giving river to local control, perhaps for fear of their mandate from Heaven being undermined by weak seasons. This disposition continued through the Roman era (58 BC to AD 616). The Roman digging of many deep wells was consonant with it.[36]

Mesopotamia

In contrast with the Nile Valley, Mesopotamia was beset with environmental problems. So much so that one can only dismiss as otiose Wittfogel's sole contribution on this score, his casual suggestion that the presence of two axial rivers, the Tigris and Euphrates, was an enviable basis for a network of hydraulic canals.[37] Otherwise one could have said that Mesopotamia fits his thesis better than almost anywhere else.

In fact, these "twin rivers" have always been prone to silting. Furthermore, a notorious earthquake zone bestrides the whole territory. Besides which, the climatic vicissitudes are complex. Mesopotamia is less firmly under monsoonal influence than is the Nile. On the other hand, it is considerably reliant on winter depressions off the Mediterranean advancing across the Anatolian plateau, then sometimes curving down the Persian Gulf and beyond. However, this pattern is sensitive to climate change.

These circumstances, plus geopolitical exposure, help explain why the Mesopotamia of Antiquity experienced some radical political transitions. Yet maybe these factors also account in part for its prominent role in Antiquity as a pacemaker, not least in science and technology.[38]

Four millennia ago, the Akkadian empire arose in Mesopotamia. It was enjoying unprecedented prosperity in 2200 BC. Then came a sudden collapse centred in Subir province by the Tigris head waters. The desolation lasted three centuries with the Subir situation having knock on effects throughout. Coherent government effectively ceased. The population overall fell more than a half. A paper published in 1993 indicated that acute aridity, coupled with stronger winds, had suddenly set in then persisted the while.[39] This was quality research. However, the authors then implausibly presumed that this episode was down to a local volcanic eruption. Surely, the only explanation that could make sense (albeit none too easily) is that the wintertime depression flow was medianally displaced southwards several degrees.

Among other such denouements was one two-and-a-half millennia later. Under the 5th Abbasid caliph, Harun ar Rashid (ruled AD 786 to 809), Baghdad "was the richest city in the world. . . . Arab merchants did business in China, Indonesia . . . a businessman could cash a cheque in Canton on his bank account in Baghdad".[40] Yet by Harun's death, unrest was becoming endemic on the caliphate's peripheries (see 3); and, through the tenth and eleventh centuries, factional violence was endemic in Baghdad itself. Against this background, the then caliph invited the Seljuk Turks into the city in 1055 to prevent disintegration.

The caliph is a Shi'ite concept compromised for Sunnis by the dispute over the succession to the Prophet. That apart, however, this Abbasid caliphate was beset by ecological disaster. Between a peak in AD 500 and a troughing out *c.* AD 1150 rainfall means about the headwaters of the "twin rivers" decreased as pronouncedly as at any time the last 3500 years.[41] Silting compounded the problem. Witness the decline to crisis point by 1150 of the brilliantly designed irrigation network the Persians had installed *c.* AD 500 in the Nahrwan district near Baghdad.[42] To cap everything, Syria and Mesopotamia endured from 1094 to 1204 a whole series of earthquakes. All of which strengthened a Seljuk disposition to focus on Anatolia and beyond. There the will of Allah might better prevail. Herein lay the seeds of Ottoman ascendancy.

Persia

Karl Wittfogel placed Achaemenian Persia in the third of the four stages he claimed he had delineated on a descending scale from "compact" to "loose" central hydraulic control, historically speaking. He thus rated it among several within that category which exhibited compact control just within limited areas.[43] These four stages he related to a notion about "hydraulic density" that was neither self-explanatory nor explained.

The steppic/desertic plateau comprising the Persian heartland long relied for water on *qandts*, man-made underground channels maintained by local not imperial arrangement. The political centre served mainly to protect the whole territory against belligerency. The wide horizons so often in view engendered a sense of strategy. Chess — the original war game — was a Persian adaptation of an Indian pastime.

The way the landscape allowed of military mobility guided by rapid data transmission favoured both reactive defence and pre-emptive attack. Cyrus created the Achaemenian empire(from India to the Aegean; from the Black Sea to the Nile) in barely half a century. Its noble state architecture (Persepolis, Bihistun . . .) bespoke a despotism enlightened enough to accommodate a spread of local cultures. Herodotus said, "There is no nation which so readily adopts foreign customs as the Persians." The Old Testament praises Cyrus for his magnanimity towards the Jews.

But the end result was the tight yet fragile unity of a jigsaw. Alexander the Great took the Persian Empire apart with surprising ease.

An instructive interpretation of the subsequent Persian experience was advanced in 1968 by Xavier de Planhol. He noted how, within Persia, nomadism had long retained a singular dynamic. There were many instances of switches from agriculture back into nomadism, these by no means always due to alien incursions.[44] Granted, De Planhol attached little or no importance to climate flux. Yet its possible impact on Persia was actively being debated a good century ago, notably by a persuaded Ellsworth Huntington and a dismissive Lord George Curzon.[45] Not that anything was ever said on that score to counter the view that hydraulic management there remained devolutionary, not empire-generating *per se*.

The Indian Subcontinent

The subcontinent is less rich in processed materials on this subject than might have been expected. Currently this shortcoming is being addressed by paying attention to the small town and village sites of the Indus Valley of the third millennium BC. It has long been admired for its city water management. Many houses had shower rooms and flush toilets which discharged via pipes into roofed sewers. But also remarked is how suddenly this way of life collapsed *c.* 2000 BC. With our present awareness of catastrophe theory we are less surprised than previously. Even so, elucidation remains difficult. Perhaps one must look to a conjunction of causes, including the intra-annual variability of the mighty Himalayan thaw.

A smallish scale yet illuminating case study is afforded by Sri Lanka. It goes unmentioned in *Oriental Despotism*. However, a year or so after the book appeared, an analysis in a well-respected Marxian journal showed how, from the third century BC to the twelfth AD, the Sinhala regime flourished in the northern dry zone of the island on the basis of water control largely devolved to village level.[46] What could be worth exploring is how far this relates to the annual import by the Arab world (later on within this time span) of up to 10,000 tons of sword-grade iron produced by thousands of small Sinhalan blast furnaces.

The Temple States

Also ignored in *Oriental Despotism* are the "temple states" which thrived in South-East Asia between the seventh and fourteenth centuries AD. Quite the most remarked elsewhere is the Khmer Empire in the lower Mekong valley. Initially its capital was Angkor Wat. Later it was Angkor Thom, a city which grew to 250,000. One can surmise that this genre of statelets emerged against a background of diminished mean rainfalls. However, the theoretical justification for saying as much is flimsier than one might wish.[47]

Khmer temple architecture appears in the seventh century. Then between the eighth and tenth, surplus wealth was directed towards two macro-exercises in social engineering. The one was to switch most of the agricultural base to tens of millions of plots newly created on forest slopes above the flood plains. The other involved investment in "theocratic hydraulics". The Angkor cities (and, less convincingly, the rural hamlets) were turned into replicas of Heaven as per Indian cosmology. Not one instance was known to a ranking researcher "where a temple pond was equipped with a distribution system to water the fields".[48] In other words, this hydraulic society was binary not unitary. On which basis it flourished four centuries, despite the Khmers' failure to build dams properly to cope with the strongly seasonal rainfall.

China

Wittfogel is exceptionally confused *vis-à-vis* the territory his early career should have left him most at home with — namely, China. In his descending four-fold ranking, he rates the Qin "probably" in the second bracket from the top and Zhou, a preceding dynasty "perhaps" in the third. Behind it all is his failure to mention the German word "loess". It refers to deposits of loosely assorted earth by winds blowing from the vicinity of ice sheets. Over north-west China loess is very widespread, often 150 to 300 metres thick. It seems clear that, through the late Quaternary, porosity left loess virtually devoid of trees and shrubs, therefore easily accessible. Varieties of sorghum and, more especially, millet were sown there way back into the Neolithic. Accordingly, the "cradle of Chinese civilization is the south-eastern part of the loess highland, an area that has little in common with the great flood plain of the Yangtse".[49]

The oldest recognized dynasties are the Xia (1900 to 1356 BC); the Shang (1742 to 1050 BC) which eventually made Anyang its capital; and the Zhou (1122 to 256 BC) somewhat to westward of Shang. All extended from the loess uplands towards the North Chinese flood plain. By the Shang, social stratification was more in evidence, as was artistic sophistication. Military defence was proffered by aristocratic clans at long last driving chariots when on the river plain. During the Zhou, the rudiments of bureaucracy appear.

The next defined period, that of the "Warring States", is characterized less by time actually spent at war than by mass mobilization whenever war loomed. Infantry armies began to appear. Combatant numbers rose sharply, one nominal roll telling of 600,000. In agriculture, iron implements, fertilizer and — again at long last — irrigation became more common. It is also the time of Confucius; of Sun Tzu (the presumed author of *The Art of War*); and, in due course, Mencius, another philosopher.

Archaeology no longer endorses the view that better inventories of iron weapons were decisive in the emergence of the state with Xian its capital under Qin, the first Emperor (reigned 221–206 BC). But a strong economy counted. So did location to westward, encompassing the Hwang ho-Wei confluence.

As Qin effected the first unification of China at near to the present shape of China Proper, he dourly consolidated the unity achieved along personalist lines. Major administrative and infrastructural programmes were launched. He imposed a tough penal regime. He killed scholars and burned books, Confucian texts included. Obsessed with lonely death, he eventually built for himself an enormously iconic tomb complete with the terracotta army.

Water control was extended by regional schemes distinct one from another. Following lesser precedents, a military "magic transport canal" was built, linking the Wei and Yangtse diagonally. The tradition thus established would culminate in the imperial transport canal (linking the Hwang-ho and Yangtse) built by the progressive Sui dynasty (AD 581–618). Such work under Qin was of importance but did little to determine the extent and character of his new Chinese state.

The Han dynasty (206 BC to AD 220) was closely comparable with contemporary

Rome in cities, roads and so on. Well houses with pulley systems were brought into service. So, apparently, were waterwheels.[50] But canalization apart, water management at macro level was by no means to Roman standards (see below). In AD 11, the Hwang-ho switched direction to pass to the south of the Shantung peninsula. This disaster, causing huge mortality, might have been forestalled.

Rome

What then of Wittfogel's dubbing Rome a "loose" hydraulic polity, one of variable but mainly low "hydraulic density"? The search for an answer might begin with an interpretation of what Rome felt it was about.

From AD 208, under Emperor Severus, a campaign to conquer all Scotland was launched only to stop with his death in 211. The next year, Emperor Caracalla extended Roman citizenship to all freemen within the Empire. Philosophically, these turning points were linked. Henceforward progress is measured not by territorial expansion but by "concentration of resources in order to protect the solidarity of culture . . . ".[51] This solidarity particularly found expression in the architecture, sculptures and lay-out of the cities.

However, the *Pax Romana* still needed defence. Folk memories of Hannibal's dreadful invasion (218–201 BC) were too bitterly ingrained for that to be forgotten.[52] Somewhat averse to dense woodland, the Romans came to accept the Rhine and Danube as their "natural limits". Still, a long and complicated front had to be held by an army amounting to well under one per cent of the Empire's population. In Britain, the only land frontier in 211 was Hadrian's Wall, a mere 120 km. Even so, the British garrison was built around three or four legions of *c.* 5000 men apiece. But the whole army only boasted 33 legions. Yet the North Sea to Black Sea frontier was 18,000 km. Moreover, those two big rivers were crossable, being braided extensively and sometimes frozen hard enough to march over. The fact remains this frontier line was the soundest militarily.

In the provision of clean water, metropolitan Rome especially favoured itself. By AD 52, its aqueduct network (400 km, five-sixths underground) stood almost complete. The city would soon number a million free people plus 150,000 slaves. The capacity flow to them has been put at 600 litres per capita per day,[53] five times what a modern Londoner averagely consumes. But mean inflow may have been but a minor fraction of this. Nor do we know much about equity in distribution.

What is clear is that neither water nor *panem et circenses* ensured metropolitan stability. In a benighted third century, riots became endemic. During the crisis of AD 238, indeed, urban civil war raged savagely for days. Next century, as things worsened, Constantinople was made the new capital.

About 150 cities around the Empire received aqueduct support, the structure most celebrated by posterity being the Pont du Gard in the lower Rhone valley. But these locally focussed adaptations were never going to determine imperial bounds. The same applies to Rome's refining considerably the waterwheel-cum-revolving millstone invented in the Mediterranean several centuries before Christ.

Rainfall variation will perforce impact on any imperium with Mediterranean shores. Additionally, a less direct effect can perhaps be seen in the pressure the Huns so exerted on other tribes or directly on the legions between AD 370 and 422. Their belligerence pattern across that particular span resembles remarkably a 12- to 14-year drought cycle which emerges, raggedly yet insistently, from rainfall trends in Russian/Soviet grain-producing regions, 1890 to 1970 — after which point it breaks down, likely on account of global warming. The proposal would be that this cycle had deep historical roots; the Huns were especially liable to turn predatory when drought-stricken thus.[54] However, these climate influences cannot be accommodated within Wittfogel's "hydraulic" model.

In terms of what we know of Classical Rome and, indeed, China or Pharaonic Egypt, it is unhistorical to say that "The bureaucratic density of an agromanagerial society varies with its hydraulic density".[55] So far as Rome's singular political culture as well as her territorial limits are concerned, some would attribute much to Hannibal's great intrusion. But many of the specific linkages proposed are tenuous and unproven.[56] It is best to acknowledge most basically the nodal importance of the city's location.

The utility of this central placing was underpinned by the Imperial Post network started by Augustus. Its very consistent delivery at 50 miles a day in all seasons and places did much to consolidate imperial unity.[57] But there was no hydraulic imperative requiring such territorial extension.

All in all, one is left to conclude that only with the Aztecs and Mesopotamia does the thesis enunciated in *Oriental Despotism* stand up to critical examination — just about. Otherwise it totally lacks the intellectual rigour our civilization must foster if it is to surmount the problems it has brought upon itself.

Notes

1 Lloyd H. Burckle in Elisabeth S. Verba *et al.* (ed.), *Palaeoclimate and Evolution. With Emphasis on Human Origin* (New Haven: Yale University Press, 1995), p. 3.
2 Richard Dawkins, *The Ancestor's Tale* (London: Weidenfeld and Nicolson, 2004), p. 60.
3 Lars G. Franzen, "Are Wetlands the key to the Ice Age cycle enigma?", *Ambio* 23, 4–5 (July 1994): 300–8.
4 *New Scientist*, "The Neanderthal Within" 193, 2593 (3 March 2007): 28–32.
5 Steven M. Stanley, *Children of the Ice Age* (New York: Harmony Books, 1996), Figure 7.1.
6 Stephen Oppenheimer, *Out of Eden* (London: Constable, 2003), p. 30.
7 "Chaotic warnings from the last Ice Age", *Discover* 23, 6 (June 2002): 14.
8 Steven M. Stanley, *Children of the Ice Age*, p. 211.
9 "How humanoids massacred the mammoths", *New Scientist* 134, 1819 (2 May 1992): 14.
10 J. S. Clark *et al.*, "Post-glacial Fire, Vegetation and Human History on the Northern Alpine Forelands", *Journal of Ecology* 77, 4 (December 1989): 897–925.
11 See *EC*, p. 5.
12 Richard A. Kerr, "Black Sea deluge may have helped spread farming", *Science* 279, 5354 (20 February 1998): 1132.
13 Christopher Connery, "There was no more sea: the supersession of the ocean, from the Bible to cyberspace", *Journal of Historical Geography* 32, (2006): 494–511.
14 J. Z. Boer *et al.*, "New evidence for the geological origins of the ancient Delphic oracle", *Geology* 29, 8 (2001): 707–11.
15 David Lewis-Williams, "Explaining the Inexplicable: Upper Palaeolithic Cave Art", *Minerva* 17, 3 (May–June 2006): 165.
16 C. F. W. Higham in Barry Cunliffe, Wendy Davies and Colin Penfrew (eds.), *Archaeology. The Widening Debate* (Oxford: Oxford University Press for British Academy, 2002), p. 342.
17 Gary W. Crawford and Chen Shen, "The Origins of Rice Agriculture: recent progress in East Asia", *Antiquity*, 72, 278 (December 1998): 858–66.

18 David R. Harris, "The Origins of Agriculture in the Tropics", *American Scientist* 60 (March–April 1972): 180–93.
19 Joy McCorriston and Frank Hole, "The Ecology of Seasonal Stress and the Origins of Agriculture in the Near East, *American Anthropoligist* 93, 1 (March 1991): 16–69.
20 V. Gordon Childe, *What Happened in History* (London: Max Parrish, 1960), p. 73.
21 Nicholas Postgate in Barry Cunliffe *et al.*, *Archaeology. The Widening Debate*, p. 108.
22 *New Scientist* 190, 2550 (13 May 2006): 16.
23 Karl A. Wittfogel, *Oriental Despotism* (New Haven: Yale University Press, 1957), pp. 2–3.
24 *Ibid.*, p. 8.
25 *Ibid.*, p. 182.
26 Karl Wittfogel, "Meterological Records from the Divination Inscriptions of Shang", *Geographical Review* XXX (1940): 10–33.
27 *The History of Water* 4 Vols. (London: I. B. Tauris, 2006), Vol. II.
28 Gustavo Politis in Barry Cunliffe, Wendy Davies and Colin Renfrew (ed.), *Archaeology. The Widening Debate* (Oxford: Oxford University Press for British Academy, 2002), p. 218.
29 S. Christopher Caran and James A. Neely, "Hydraulic Engineering in Prehistoric Mexico", *Scientific American* 295, 4 (October 2006): 56–63.
30 Roland Oliver and J. D. Fage, *A Short History of Africa* (New York: Facts on File, 1989), p. 32.
31 "Fossilized" means derivative from a moister historical past.
32 *HCC*, pp. 170–1.
33 David Keys, *Catastrophe* (London: Century Books, 1999), Chapters 32 and 33.
34 D. A. Adamson et al., "The late Quarternary History of the Nile", *Nature*, 288, 5786 (6 November 1980): 50–5.
35 "Pharaohs from the Stone Age", *New Scientist*, 193, 2586 (13 January 2007): 34–38.
36 A. B. Lloyd, "The Late Period, 664–323" in B. G. Trigger *et al.* (eds.), *Ancient Egypt: A Social History* (Cambridge: Cambridge University Press, 1983), pp. 279–304.
37 Wittfogel, *Oriental Despotism*, Chapter 6B.
38 *EC*, pp. 18–19.
39 H. Weiss *et al.*, "The Genesis and Collapse of Third Millennium North Mesopotamian civilisation", *Science* 261 (20 August 1993): 995–1004.
40 Sir John Glubb, *A Short History of the Arab Peoples* (London: Hodder and Stoughton, 1969), pp. 104–5.
41 Arie S. Issar, *The Impact of Climatic Variations on Water Management Systems* (Sede Boker: Jacob Blaustein Institute, 1992), Fig. 5.
42 Thorskild Jacobson and Robert M. Adams, "Salt and silt in Ancient Mesopotamia", *Science* 128, 3334 (21 November 1958): 1251–8.
43 Wittfogel, *Oriental Despotism*, pp. 166–7.
44 X. de Planhol in W. B. Fisher (ed.), *The Cambridge History of Iran*, 7 Vols. (Cambridge, Cambridge University Press, 1968), Vol. 1, Chapter 13.
45 *HCC*, pp. 187–91.
46 E. R. Leach, "Hydraulic Society in Ceylon", *Past and Present* 15 (April 1959): 2–26.
47 *HCC*, pp. 144–6.
48 W. J. Van Liere, "Traditional Water Management in the Lower Mekong Basin", *World Archaeology* 11, 3 (February 1980): 265–80.
49 Ping-ti Ho, "The Loess and the Origin of Chinese Agriculture", *American Historical Review* LXXV, 1 (October 1969): 1–36.
50 Michael Loewe, *Every Day Life in Imperial China* (London: Batsford, 1968), pp. 175 and 198.
51 Norman H. Baynes, "The Decline of Roman Power in Western Europe", *Journal of Roman Studies* xxxiii (1943): 29–35.
52 J. F. Lazenby, *Hannibal's War* (Warminster: Aris and Phillips 1978), pp. 248–57.

53 Susanna Van Rose, "Water Problems from the Past", *Geographical Magazine* LXI, 10 (October 1989): 36.
54 *HCC*, p. 268.
55 Wittfogel, *Oriental Despotism*, p. 167.
56 Lazenby, *Hannibal's War*, Chapter 8.
57 A. M. Ramsay, "The Spread of the Roman Imperial Post", *Journal of Roman Studies* XV (1925): 60–74.

3 | Vikings and Mongols

I Northmen Abroad[1]

Viking forays began in the 750s with Danish raids on the Frisians. But big leaps outwards came with a Norwegian probing of the Dorset coast of England in 786; and, then, seven years later, the sacking of Lindisfarne monastery. In 836, the Norwegians founded Dublin. In 874, they appeared before Lisbon and Seville. From 865, the east centre of England was heavily settled by Danes. By 878 only the southerly English kingdom of Wessex stood against them; and this by dint of its monarch, Alfred the Great, holding out in Athelney — an islet enveloped by marshes. Yet the next year Alfred gained at the ferocious downland battle of Eddington a compromise peace betokened by the conversion to Christianity of the Danish leader, Guthrum.

Early next century, Vikings mainly of Danish extraction seized a sizeable enclave about the lower Seine where they built up with some expedition the Duchy of Normandy, a polity ahead of its time in terms of medieval martial statehood. This bold initiative pretty much set the seal in Western Europe on predatory maritime nomadism.

Iceland

Around 865, during a coldish interlude (840–70) hemispherically, Norwegian farmer Flóki Vilgeröarson sought to settle in "Iceland". He named it thus, having lost his wherewithal and seen "a fjord filled up with sea ice" (Landsman Saga c. AD 1200). Then in 874, Ingolf Arnarson gained a solid footing. By 930, 20,000 Norwegian immigrants had distributed themselves in workable lodgements under quasi-priestly chieftains or *godi* (see below). That year, the leading *godi* came together at Thingellir, the Plain of the Thing: a flattish rift valley of heroic aspect inland from Reykjavik. They launched the Althing, a national assembly (legislative and judicial) destined to meet there every midsummer till 1798. The Althing's endorsement in AD 1000 of Christianity was a decisive step towards working consensus.

By 1100 slavery had died out in Iceland and Greenland, a century or two ahead of Scandinavia proper. Throughout, it had been subject to a relatively benign code of conduct. Until the thirteenth century, indeed, Icelandic experience as a whole was that of a reasonably open though clannish frontier society with highly accessible governance.[2] An absence of external threats allowed of this. So did longitudinal variations in climate change. Over Greenland, medieval warming had begun in the sixth century and continued to the tenth.[3] Though the piecemeal evidence available for Western Europe is hard to summate, one should probably speak of warming setting in there *c.* 750 and peaking *c.* 1275. Iceland will have been well forward. The benefits will have been considerable for its rich stocks of cod, a fish which finds sea temperatures below 4°C uncomfortable and below 2°C impossible. The mean sea temperature off the southern Icelandic coast today is *c.* 6°C.

On land, however, the prospects were compromised by poor ecological management. Critical was the destruction of woodland to smelt iron. Prior to the Viking advent, two-thirds of the landscape was vegetated — over a third of this being birch woodland. Soon, however, the situation drew near to what it is today — a quarter of the area vegetated and but one per cent wooded.[4] An early result was inability to cope with what therefore became the great famine winter of 975/6. As another saga puts it, this was the terrible dearth "in heathen days, the severest there has ever been in Iceland. Men ate ravens then and foxes, and many abominable things were eaten which should not have been eaten; and some had the old and helpless killed and thrown over the cliffs". It all served to undermine the pre-Christian gods.

Greenland and Beyond

In 981 or thereabouts, Erik the Red went into northerly exile from Iceland for three years for blood-feuding, a sequence he had been through once already in his native Norway. He and his fellows returned to advocate the colonization of what he, with studied optimism, dubbed "Greenland". About 985, he led an expedition back which founded two outposts for what became 5000 Viking people. Termed the Eastern and Western Settlements, both lodgements were actually on Greenland's western seaboard with its fjords reminiscent of Norway's and its less storm-swept open coasts.

Many cod bones in middens suggest a salubrious climatic phase. However, this may not have meant the open sea was less stormy. On the contrary, northwards ridging of the Azores anticyclone in association with the general warming could well have steepened barometric gradients. At all events, of the fleet of 25 ships with which Erik the Red returned, only 14 made it. Several are known to have foundered.

A further question is how deeply the Vikings probed into North America. For a few years in the late tenth century, they did maintain an eight-building outpost at L'Anse aux Meadows in Newfoundland. Also extensive trade with indigenes is confirmed by Norse artefacts discovered in many sites in Labrador, in Baffin and Ellesmere islands, and around Hudson Bay. A primary aim early on was procuring ivory for export to Europe. There is no solid evidence of sojourn within what we know as the USA unless it be one Norse penny (dated 1065–80) found on the coast of Maine.[5]

The Rus

Yet quite the most remarkable dimension in the Viking phenomenon was the Swedish. Having refined their boats and boat handling on Baltic herring grounds, these eastern

Vikings (alias the "Rus" or "Vararingians") projected themselves along lakes and rivers. A number of major streams rise in or near the Valday hills (roughly half-way between Moscow and St Petersburg), then flow to several well separated coasts. This allowed of continent-wide travel, given some taxing but not impossible portering between the upper reaches. Initially Moscow's nodality stemmed from this circumstance. The place first figured in the Russian chronicles in 1147.

Around 840, a Rus protostate arose about Kiev; and raiding along the Black Sea's southern shores soon commenced. Then in 860, 911, 941 and 944, successive Viking fleets (reportedly of 200 to 2000 ships apiece) appeared before a somewhat weakened Constantinople. Secular economic recession had reduced the population within its mural bounds from perhaps 800,000 in 650[6] to maybe fewer than 250,000.[7] Juxtaposed, those respective figures may exaggerate the decline. But there were, by 850, wide areas of urban dereliction. Even so, the city could still beat off predators with its formidable walls,[8] quality ships, martial traditions and "Greek Fire" — an incendiary mix based on Black Sea naptha and discharged through tubes, sometimes from large rowing boats. Byzantium and the Rus therefore moved unsteadily but insistently towards accommodation. The acceptance of Orthodox Christianity by Vladimir, Grand Prince of Kiev, signified this in 988.

Root Causes

Being so well defined in time, space and character, the Viking experience invites explanation. A salient theme could be Malthusianism aggravated by primogeniture. Material advances coupled with medieval warming encouraged a rise in population and consumption which eventually became excessive. But before examining further this ambient circumstance, one should ask what the Viking eruption owed to technical innovation. Regarding weaponry, there is little sign. Norse nautical science, on the other hand, had lately caught up with itself.

By the Bronze Age there had evolved the high end posts and freeboard planking now regarded as characteristic of the Vikings' longships, their famous assault craft. By AD 200, rowlocks were being fitted. Then by AD 700, longship development had been effectively completed with the installation of rudders and a single large sail. The salient features of the overall design were slender profile and shallow draught; light yet sturdy construction, and ready reversing.

That sails were not mounted in longships earlier is surprising, seeing how some communities around the southern North Sea had been using them since Roman times. Already, too, the broad-beamed *knörr* the Norse used in long distance commerce were sail-reliant. Perhaps a mast and sail had been thought too cumbersome for tactical use inshore. Also imaginable, given Norse ardour, is that sails were seen as somewhat sissy. Such thinking could have seemed pertinent midst Baltic ice floes and in the Norwegian "leads": the stretches of often remarkably calm water between the successive islands and the mainland.

On the open sea, however, sails would have been mandatory regardless. And sea control became for the Vikings their overriding advantage. Charlemagne and Alfred the Great each made progress with naval development, the latter reputedly building super-longships for proactive costal defence. But neither their navies nor anybody else's hit back at the Viking homelands.[9] In 810, an ageing Charlemagne did contemplate a land–sea campaign against Denmark to curb her support for the Saxons he had

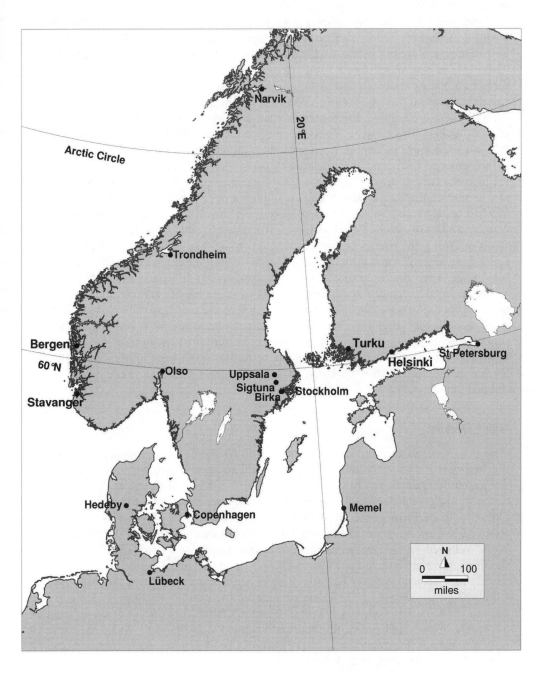

The Baltic

been converting to Christianity so forcibly. Then changes in the Danish political scene made such riposte unnecessary.

It has well been said that, whereas the sea divided their opponents, it united the Vikings themselves. Moreover, this unity was reinforced by cultural affinities including much commonality — *dönsk tunga* — between the respective languages. This explains how an individual from one of the three Norse home nations might well serve in a Viking foray organized by another. More striking, it helps explain how closely concurrent the onset of the Viking era was in the three cases.

Still, there were subtle differences as regards origination. In Denmark, the initial sorties against Frisia were intended to constrain Charlemagne's expansion. Afterwards, however, a strong rise in agrarian productivity (as from the sixth century) led to demographic overspill aggravated by greater social stratification.

In Norway, topography was an acute obstacle to sustained agrarian growth but also to political unification. Operating out of his ancestral hearth by Oslo fjord, Harald Fairhair sought acceptance as the first Norwegian king. This ambition he realized by gaining control of much of the south-west through a big victory over the petty kingdoms at Hafs fjord near Stavanger in 872. Of those Norwegians who migrated to Iceland between 874 and 930, about half came from the Bergen area. Their exodus was in part a reaction to kingly authoritarianism. By AD 1000, the Svear tribal area around Lake Maloren was emerging as the core area, politically and commercially, of the embryonic Swedish state. The parallels with Rome-cum-Ostia were striking. Both were near coastal waters and close to the centre of a large peninsula running longitudinally into an elongated sea with more of a latitudinal alignment. Stockholm, ultimately the centrepiece of the Swedish node, has often been spoken of as "the Venice of the North", this by virtue of its archipelagic setting. It could as well be dubbed "the Rome of the North".

Viking Decline

One way Sweden expressed its singularity was in resistance to Christianity. The faith reached there early in the ninth century but paganism would long remain entrenched, not least among the Svear. In 1066, the incumbent of a new bishopric at Stiguna was driven out. Several decades later, the Svear banished King Inge for refusing as a Christian to perform pagan rites. Only in the 1130s, when Uppsala symbolically replaced Stiguna as a see, could you say Christian ascendancy was confirmed. In Norway, the conversion phase was less drawn out, having started in 995. In Denmark, the first Christian King, Harald Bluetooth, had ruled 935–85.

We have seen above how Christianity was early on accepted on some prominent Viking frontiers, very early in Danelaw and by the new millennium in Kiev and Iceland. Not just accepted but affirmed. Besides, Christianity could anywhere make a sizeable impact well before its final endorsement. Charlemagne and the Saxons notwithstanding, this impact would very generally be to dampen down martial exclusivism, its Viking variant included.

A major motivation for Rus adventurism, not immediately but soon, was access via the Volga to silver mined in various uplands of the Abbasid caliphate centred in Mesopotamia. Before long, however, the Abbasids were in general crisis. Archaeological surveys of the Mesopotamian flood plain reveal an 80 per cent reduction in the rate of settlement expansion during the 840 to 870 interlude, already cited as a time of hemispheric climatic anomaly. Yet even before that there were problems. Witness the movement of the capital from Baghdad to Samarra in 836.

In due course, there was an 80 per cent collapse in Abbasid bullion exports. Denmark was much affected since its port of Hedeby had become a key bullion transhipment point as between Sweden and Western Europe. A recovery in this Baltic trade early next century was sharp but also brief. Meantime, however, Denmark was acquiring more of the infrastructure, institutional and material, of an integrative kingdom. It was embarked on modal change. The pressures which, in 1013–14, obliged England to acknowledge formally the suzerainty of Denmark's King Sweyn and his son, Canute, belonged more to the world of medieval dynasties than to the opportunistic forays of Viking berserkers.

Acting against expansionism from whatever quarter was the Holy Roman Empire. Charlemagne had been anointed its first Emperor in Rome on Christmas Day 800. Originally its formation had been engendered by the Islamic threat. Shortly before his death in 1935, the great Belgian medievalist, Henri Pirenne, enunciated the thesis that "Without Islam the Frankish Empire would never have existed and Charlemagne without Mohammed would have been inconceivable." As things transpired, the Empire came under sustained attacks in the ninth century from Saracen Moslems but also Magyars and Danish Vikings. "Far more tenacious than either Magyars or Saracens were the Vikings . . . their attacks were also far more extended in time".[10] Being extended also in space, they had a very differential impact, thereby exacerbating geographical discordances. By 850, the Empire was fissiparous. Nevertheless, outright invasion had been avoided.

Greenland's Decline and Fall

The later history of the Greenland Norse unfolds against a background of continual climatic decline in a region always marginal for human existence in sundry ways. Other considerations also come into play. Pasture productivity decreased by 75 per cent partly through the erosion of ill-protected soil. Therefore there was a dietary shift towards sealing and fishing, this quite early on. There was no propensity to learn from the Inuit, with whom these Norse were widely in contact by the twelfth century. No Norse skin-covered boats (as per the Inuit kayaks and umiaks) have yet been found, invaluable though they would have been inshore. Nor have Norse harpoons or barbed spears (for use through ice holes) been discovered. There may also have been undue reliance on cows as opposed to sheep and goats. Wood for construction and charcoal became desperately scarce.

Religion likely played just a moderate part in the early life of the Norse settlements. But in 1082 the first regular bishop was consecrated.[11] Tom McGovern has shown how the profile of the Church rose subsequently. The next two centuries, some elegant (and overly large) churches were built; and, by 1300, the Church owned two-thirds of the better grazing land. He sees it all as part of a Greenlander inclination to lock into a preset pattern, "elaborating their churches rather than their hunting skills".[12] But is not some such reaction common among all who see their problems as insurmountable?

It is needful to feed into any interpretation international factors beyond the control of the Greenlanders if not beyond their ken. In 1261, Greenland (and the next year, Iceland) accepted Norwegian sovereignty, hoping thus to consolidate trade links. But in 1294, the year a monopoly of dependency trade was given to Bergen, not more than one voyage annually was made to Greenland. Yet Bergen's majestic harbour area could accommodate upwards of 200 ships at a time.

In the fourteenth century, this nexus weakened further. The Black Death badly ravaged Bergen. The European demand for Greenland ivory lessened. Scandinavian interests re-oriented more towards the South and East with the growth of Hansa influence and after Norway, Sweden and Denmark had come together with the Kolmar Union of 1380. Meanwhile Iceland's booming fish trade was becoming for her seamen an attractive alternative to Greenland trips.[13]

The last recorded visit from Norse Greenland to Europe was in 1408. Several decades later the Greenlanders died out. But why? Starvation, disease or despair? Armed conflict? The absence of genetic signs of miscegenation rules out the assimilation no one envisages in any case. But does it not also render less likely rape in the aftermath of battle? And what should we make of the large-scale Inuit movement into southern Greenland (and Labrador) in the 1350–1450 frame? Does this bespeak a quest for a more sustainable resource base or simply worsening conditions to northwards? Does significance, in this context, attach to indications that, by the late fifteenth century, other Western Europeans were fishing the Grand Banks? The data is so scanty that surmise can proceed unrestrained.

The Viking Legacy

With the Viking expansion elsewhere, even more than with the earlier *völkerwanderung*, one is struck by how readily aggressive warfare could give way to convergence with indigenous people: to intermarriage, the adoption of Christianity and trade. The Whiggish historian, G. M. Trevelyan, said the Vikings "combined the pride of the merchant with the very different pride of the warrior as few people have done".[14] Halford Mackinder is among the commentators who have averred that Anglophone democracy is still sustained by the Danelaw "sokeman" tradition.[15] More equivocally, a leading American legal historian wrote of medieval Iceland as "an almost unique instance of a community whose culture and creative power flourished independently of any favouring material conditions . . . having produced a Constitution unlike any other whereof records remain, and a body of law so elaborate and complex that it is hard to believe it existed among men whose chief occupation was to kill one another".[16]

Granted, the Viking world was never utopia. It has been suggested that a fifth of the population of eleventh-century Sweden may have been slaves.[17] Moreover, the Vikings were continentally the "greatest slave traders of their day", many of their victims being sold into the Abbasid caliphate in exchange for silver.[18]

What Norse political culture in the round can be said to have rested on is the basic rights of freemen. On these were superimposed the leadership of *godis* locally chosen, more or less consensually, to govern and represent within a devolved structure. A persuasive model will surely have been the longship. So may one move forward near to a millennium to consider a letter sent to the Naval Committee of the US Congress in November 1775 by Paul Jones, a sailor of the "rough tarpaulin" ilk destined to become a folk hero of the Revolutionary War. His missive's "main burthen" is summated in the sentence: "Whilst the ships sent forth by the Congress may and must fight for the principles of human rights and Republican freedom, the ships themselves must be sailed and commanded under a system of absolute despotism." This harsh rendition from a harsh era by a former slave-trader did not say the last word on the nautical ethos, now nor then. But it assuredly said the first. Disciplined democracy is what longships were about.

Alfred's engendering of a naval tradition was in emulation. Indeed one can even see in Athelney a miniature of Viking maritime strategy. The marshes, suitably water-logged in winter, shielded his base area from attack. Yet they afforded boat access to all territories around. More encompassing though was to be his response to democracy, which was to make his monarchy the most accessible in Europe. Twice a year the Witan (i.e. elective folk assembly) met to voice concerns and to witness the dispensation of royal justice. Also shires were defined, self governing while answering directly to the king.

Russia

The subject comes up again *vis-à-vis* the interaction with what became "Russia" of the Viking and Mongol expansions in turn. The most immediately obvious difference between them is, of course, that the one expansion pattern was basically waterborne while the other was essentially equestrian.

George Vernadsky, writing as Professor of Russian History at Yale, stressed Moscow's Tsardom emerging to embrace "an entirely new concept of society and its relation to the state. All classes . . . were bound to the service of the state".[19] One may question the notion of "entirely new" but the most important point for now is that direct comparison was drawn with what had been the much freer polity of Kiev.

By according the Mongols significance in disorienting the whole course of Russian development, Vernadsky was following in the steps of that renowned Russophobe, Karl Marx: "in the terrible and abject school of Mongolian slavery . . . Muscovy was nursed and grew up. It gathered strength only by becoming a virtuoso in the craft of serfdom". This famous passage was cited with sardonic approval by Tibor Szamuely, the Hungarian dissident and one-time inmate of Stalin's *gulags* who later worked in exile as a political philosopher at the University of Reading. He likewise averred that "the Muscovy that emerged from the fragments of the old Rus and the break-up of the Mongol Empire bore hardly any resemblance to the free society of Kiev".[20]

These expositions were forceful and compelling. But one must ask whether they were just identifying a tendency rather than an iron law of causation. This question becomes all the more germane when it is argued that, because Mongol suzerainty over Russia was only indirect, it was all the more enduring and its legacy the more profound.[21]

2 The Mongol Blitzkriegs[22]

An analysis at the University of Michigan in 1982 sought to enunciate guidelines about nomadic behaviour, particularly in times of territorial expansion. It acknowledged that nomads have to progress beyond elementary kinship links when they seek to establish a new imperium. One could perhaps cite Genghiz Khan's insistence on Mongolian becoming a written language as recognition of this. But the Michigan study also

stressed how regularly the efficiency of kinship bonding in effecting the initial mobilization is underrated by adversaries. In this and manifold other respects, there is often
a big intelligence gap between the two sides.[23]

Unfortunately, historians, too, may have a knowledge deficit. They certainly do
regarding the Mongols. In his creditable 1986 study of their great expansion, David
Morgan reviews a range of Chinese, Persian and European sources. Two invite immediate comment. Morgan is not entirely dismissive the way some distinguished scholars
have been of *The Secret History of the Mongols*. Rather he accepts it as the one and only
"substantial surviving Mongol work about the Mongol Empire". True, its authorship
is unknown; its coverage stops in 1240; and it comes down to us via a fourteenth-
century Chinese transliteration. Still, he makes a good case for drawing on it, subject
to the pinch of salt so often needed with pre-modern sources.[24]

David Morgan further presumes that the *Travels of Marco Polo* are an authentically first-hand account of how China was in the reign of Kublai Khan. However, in
1995 a book by the Curator of Chinese at the British Library put forward cogent
reasons for doubting whether Marco Polo ever set foot there, as opposed to hearing
embroidered accounts from some who may have.[25]

Climate Modulation

Indirect evidence is what we must turn to when judging how climate change may have
affected the situation the Mongols found themselves in. Still, this evidence is quite
variegated and comes together tolerably well. The gist is that, during the first half of
the thirteenth century, Mongolia passed through a decade or two of cool droughtiness
between two eras of warmth and moisture.[26] This could mean that Genghis Khan was
bestirred to start his great offensive by acute aridity. Obversely, the preparations had
been, and the follow through would be, facilitated by more benign weather.

Two commentaries by non-climatologists merit mention here. Owen Lattimore
drew a distinction between (a) the "neo-classical" view that heavy nomadic migration
came about through exogenous changes in ecology, especially climate, and (b) a
"semi-classical" view that undue cultivation or grazing could affect desertification plus
(and this arouses scepticism) basic climate change. However, he went on to acknowledge how interactive the two patterns of causation were. Moreover, tribal responses
were modulated by cultural diversity between neighbouring tribes, a diversity clearly
manifested in tent design.[27]

Rather more recently, G. F. Hudson, an Oxford historian of the Far East, argued
that worsening aridity could have directly favoured desertic nomads (as the Mongols
more or less were) as against those more concerned to grow crops. He presented a
model of three concentric zones. The innermost, the desert, would expand with
aridity. Yet the *taiga* (the Siberian coniferous forest) would stoutly resist displacement;
and would, in any case, have generated acidic soils ill-suited to grasses. Therefore the
grassy/partly cultivated middle zone would be squeezed.[28]

There is, however, one climatic parameter which tends to be neglected in historical contexts yet may have figured in the Mongol situation. It is windiness. Medieval
warming probably weakened awhile the huge anticyclone observed in recent times to
extend regularly over Siberia in the winter half; and to draw round itself very strong
and persistent winds. Witness a participant's account of a November storm during the
Sino-Swedish expedition to north-west China from 1927. Come one morning, "it

roared and howled and raged still worse than on the day before". Millions of "tiny flying particles lashed one's skin". Estimated windspeed reached 60 mph.[29]

No doubt centuries of these and other extremes of weather had inured the Mongols to suffering, theirs and others'. But what if medieval secular warming had lately led to fewer sandstorms of such intensity? Would not that have left these equestrian warriors more attuned to the wide horizons about them and the inviting prospects beyond?

Objective China?

Genghis Khan and his entourage were by no means the first among those who "lived in tents of felt" to have quasi-imperial aspirations. Three centuries earlier, Mongolia and much of north China were conquered by the Khitans, a semi-nomadic community from north of the Great Wall who were subsequently, as a Chinese dynasty, to assume the name of Liao. Then, in the 1120s, the Khitans were displaced by the Jürchen of Manchuria. They were ethnically related to the Manchus, the people that (in the seventeenth century) were to provide China with its last imperial dynasty.

Genghis was bound to see a decision about whether or not to attempt the conquest of China as quite the most defining he would ever face. That was heavily underscored by the transformation China was effecting. Demographic estimates for that time and region are hard to make with assurance. But the total for China in the early thirteenth century might have been 110 million as against 60 million in 850. The population of the rest of the Mongol empire at its zenith was probably well below 15,000,000 — Persia and Mesopotamia excepted. And those two territories were located eccentrically in relation to the rest of what some modern histories, written with a broad brush, have termed *Pax Mongolica*.

How China had been developing demographically and more generally was remarkable. In the authoritative opinion of Albert Kolb of the University of Hamburg, "No other people have achieved a comparable transformation or shown themselves to be so adaptable".[30]

China Emergent

This phase in China's long and often too colourful history emerged out of the empire's breaking into ten statelets in 907, at the end of the illustrious Tang dynasty. This episode affords us a classic study in the relationship between physical and political geography. Evidently, China's topography hardly favoured imperial unity in that the three great rivers effectively flowed West to East whereas North to South could have more readily invited expansion and control from the North. Besides, a map of these statelets shows their boundaries lying across the contours quite incongruently.[31] So perhaps a single all-encompassing polity was the best way through, regardless. By 979, the imperium had been restored, save that part of the north had fallen to depredatory nomads.

The revolution Kolb was referring to began in the eleventh century with the introduction from Fukien (in accordance with an imperial edict of 1011) of rice originating in Champa, a temple statelet in the lower Mekong. Influential in this process had been secular dryness on the Yellow River/Yangtse flood plain. Though not high in standard yield, Champa rice possessed, in one variety or another, two most positive traits: pronounced resistance to water shortage and early ripening. The area thus planted gradually doubled.[32] This made it that much easier for the Song to cope with a drought incidence that apparently was higher in the 1127–1279 era.

Thanks to this demand-led innovation, the whole Chinese economy concurrently experienced a market-led lift-off. William McNeill of the University of Chicago found as follows. Though the imperial government kept "the capitalist spirit firmly under control, the rise of a massive market economy during the eleventh may have sufficed to change the world balance between command and market behaviour. . . . China swiftly became by far the richest, most skilled and most populous country on Earth".[33]

A truly remarkable aspect of this economic revolution is the long experiment with paper money. By the eleventh century, this enabled successive governments to sustain non-convertible currencies while printing extra to head off budget deficits.

Not that monetary management was all coherence. The semi-isolated "red basin" of Szechwan maintained its own paper money till the thirteenth century. Then having assumed the imperial throne in 1264 (see below), Kublai Khan standardized all the paper currencies in circulation. However, the large gold and silver reserves he took over were not actively used for monetary stabilization. By 1330, the silver value of the note issue had risen 4000 times.[34]

Mongol Imperialism

In 1211, Genghis Khan had launched an offensive against the Song; and in 1215 had stormed and sacked a well-fortified Beijing. Next, campaigns towards south-west Asia led to the fall of various cities including Bokhara in 1219 and Merv in 1221. Chronicles record 700,000 being massacred in the latter instance, undoubtedly a wild exaggeration but on that account all the more expressive of the Mongol ability to paralyse through fear. Two years later, a Slav combined force under the Prince of Kiev was defeated.

North China finally fell in 1235 but the surviving Southern Song realm held out till 1279. The victor then was Kublai Khan, grandson of Genghis who, after several years of internal strife, had emerged as the first Yuan Emperor in 1264. Apparently a major bone of contention with his rivals had been his disposition towards enlightened moderation in occupied territories.

In China he took ready heed of advice that "The Empire was won on horseback but it won't be governed on horseback". As intimated above, his prowess as a currency manager left something to be desired. But his endeavours to foster Chinese art and literature evoked quite a positive response. A fine tradition of nature painting was upheld. Vernacular drama and fiction likewise underwent something of a boom. The pre-existing imperial infrastructure (granaries, road, canals . . .) was renovated. Not least, Kublai transformed Khanbalik (his name for Beijing) into a truly imperial capital. He also promoted diplomacy and foreign trade. The whole represented a process of accommodation singular in Mongol terms.

From 1236, some terrible devastation had been visited on Christendom, especially Russia. Thus Riazan, Moscow and Vladimir were sacked as Mongol horse advanced up frozen waterways. Yet how crippling to Russia as a whole these atrocities were has been questioned. Alexander Nevsky was still able to defeat the Swedes and German knights in 1241–2 though aided, no doubt, by prior attrition of the latter by the Mongols.

However, a general resumption of offensive action in 1240 had culminated in a comprehensive sacking of Kiev and left the Mongols poised to cross the Vistula. Then April 1241 witnessed a two-pronged offensive into Catholic Europe. An army of Poles and Teutonic knights was defeated at Liegnitz in Silesia. Two days later, another Christian army was savaged in Hungary. But then contingent circumstances checked the Mongols at what could fairly be deemed, in any case, their natural limits. Even with a most efficient pony express, it would have taken the best part of a fortnight to get a message from Danubia to the Gobi. Still more germane would have been the difficulty of absorbing Catholic Europe, another peripheral node not so far behind China in terms of population size and development level while much more variegated geographically.

The death in December 1241 of Ogedai Khan (the son of, and successor to, Genghis) distracted the Mongols, albeit not that drastically. More directly to the point, a brilliantly innovative research paper by Andrea Kiss of the Central European University in Budapest has drawn on a range of Hungarian and Dalmatian primary sources to demonstrate that, while the 1241–2 winter may not have been too abnormal across Catholic Europe as a whole, January was extraordinarily cold in Hungary — enough so to freeze the Danube solid (under normal circumstances) within a fortnight. Since the Mongols were poised for a Danubian crossing in force, its freezing thickly would have been welcome to them. However, this outcome was postponed a fortnight by repeated Hungarian shatterings of the ice which presented the Mongols with the worst of alternatives, cascades of ice flows. A firm freeze came only towards the month's end. A few days later a thaw began.[35] What with all this and the succession question, the Mongol army turned homeward in the Spring. North Italy and beyond had likely been saved.

In south-west Asia, a campaign waged by the Mongols in 1253 effectively wiped out the Persian wing of the Assassins, a heretic fraternity of Ismaili Moslems who, for two centuries past, had terrorized the Islamic mainstream from their base area in the Elburz mountains. In Abbasid Baghdad, this was acclaimed. But then in 1260, that city (still rated the most resplendent in all Islam) was ravaged by the Mongols.

After which, they turned towards Egypt. However, in September 1260, a depleted Mongol army was defeated by a larger Mameluke one at Ain Jalut near Nazareth. Mongol strength had lately undergone sharp reduction, thanks to the difficulty of keeping so pony-dependent a force in the Fertile Crescent through long hot summers[36] though also because the Mongol world was riven with tension following the death in 1259 of Great Khan Möngke, the eldest son of Genghis. By 1261, there was intermittent warfare between the Great Khanate and the considerably Turkic Khanate of the Golden Horde, as it was later dubbed. By then, indeed, the Golden Horde was talking with the Mamelukes about regional stability long-term. Mongol forays into the Fertile Crescent continued into the fourteenth century but consistency of purpose had been lost.

The Kamikazes

However much Kublai Khan could be said to have gone native sophisticate as Yuan Emperor, the historical assessment has been that he extolled the while his Mongol identity. Above all, he aspired to regenerate pan-Mongol unity by launching two major expeditions, in 1274 and 1281, to conquer Japan. Both were disasters from the outset, most basically on account of the weather but also because of Japanese stout hearts.

Regarding these, it was emphasized in *History and Climate Change* that, lying in middle latitudes just eastwards of a large continent, Japan has perennially been subject to acute temperature contrasts between summer and winter, contrasts aggravated by monsoonal summer humidities.[37] Faced with this situation, the early medieval Japanese felt themselves confronted with a stark choice in respect of architecture. With it, they could mitigate either the sticky summers or the freezing winters but not properly both. They resolved to give the former priority, coping with the rigours of winter "by stoicism and the wearing of additional clothes".[38] **(See 15 (3))**. Physical toughness initially engendered by architectural disdain for winter thus became a national trait. Witness the Japanese troops who, each time round, fought often to the death defending island outposts and foreshores against Kublai Khan's amphibious forces.

What then of the part a *kamikaze* (literally, "wind of God") played in thwarting each invasion attempt? The 1274 armada comprised 40,000 soldiers and seamen, many of them impressed Koreans and Chinese. On 19 November its troops came ashore in Hakatā bay, Honshu, to engage the Kyushu *samurai*. The latter fell back in good order, pending heavy reinforcements. Then the cautious Mongol commander ordered re-embarkation. As the ships reached open sea, a mighty storm struck. Korean sources tell of 13,000 fatalities, ashore and afloat.

Hakatā was again the chosen landing point in 1381. An initial assault in June involved 40,000, mostly Koreans. They abandoned their beachhead after six weeks largely because Japanese infantry and light naval forces had them hemmed in. So come August, they were joined at the offshore islet of Takashima by a second armada, its complement perhaps 160,000. But as the combined force turned again to offensive action, the typhoon supervened. More than half those embarked perished. By invoking such a *kamikaze*, the Shinto priests on the lusciously beautiful island of Ise could claim to have saved the nation.

As the present debate about global warming has progressed, a consensus has emerged to the effect that a sea-surface temperature of 27°C represents a threshold above which one is much more liable to witness the generation of the tropical revolving storms respectively known in different oceans as "typhoons" or "hurricanes" or "cyclones". Moreover, through the late twentieth century, the said threshold was regularly reached in the late summer/autumn in waters arching round from the tropics to Kyushu, waters over which typhoons may travel *en route* for the Japanese archipelago. The inference can be that seas even slightly warmer seasonally, if that they were, through the climax of medieval warming could well have induced more and stronger typhoons surviving with vigour to Kyushu and beyond. There could also have been a broader annual spread. The six most severe typhoons (as judged from human fatalities) to hit Japan between 1945 and 1960 all occurred between 15 and 26 September.[39] Both the *kamikazes* were well outside those dates.

One must add, however, that the susceptibility to storm damage of the second armada had been much accentuated by the haste with which it had been created — in one year not the six really needed. Chinese shipbuilding was then quite the best anywhere. Yet this time round, many masts were ill finished, joints rough, nails rotten . . . There are even indications of blatant sabotage. Worse still, many of the vessels used were impressed river craft.

It is hard to see, in any case, what Kublai really stood to gain from conquering Japan, other than conquest for its own sake. Had it been effected, it would surely have exacerbated the crisis of pan-Mongol unity, not resolved it. Much the same must be said of his subsequent forays southwards from China. An expedition to Java in 1293 withdrew when the monsoon broke. An incursion into India in 1296 "clogged the roads with refugees but was turned back with heavy losses". From Champa and Vietnam, "tribute was raised at a rate too low to cover the cost of the campaign". Sumatra was held but briefly.[40] All in all, this strategic lunge went well beyond "natural limits".

The Mongol Armies

Military thinking in the West throughout most of the twentieth century was suffused with the merits of movement, tactical and strategic. Granted, Britain's Royal Artillery cherished the adage that those in command looked for movement on peacetime exercises but for firepower in real war. Yes, but the literature was dominated by the apostles of mobility.

Their *primus inter pares* was Basil Liddell Hart. To him, a Mongol army was a "machine" which worked like clockwork; and this very mobility made it "irresistible to troops far more strongly armed and numerous". Once again, he may have been too disposed to isolate "mobility" as a distinct attribute. As he noted himself, the Mongols were also rich in firepower, their armies very considerably comprising mounted archers bearing compound bows of great range and penetrative force intended for firing from horseback just as your mount was turning from the enemy — a tactic the Parthians had used against the Romans to terrible effect in 53 BC. But he acclaimed, above all, how much their dexterity depended on their ultimate sense of oneness.[41]

A decimal command-and-control system was devised to instil this sense. The smallest formation had ten men (*cf.* the squad in a modern army). Above were those of a 100 (*cf.* a company) or a 1000 (*cf.* a battalion). The largest unit, a *tumen*, was 10,000 strong (*cf.* a division). Such a pyramidal structure could facilitate a Mongol army's switching between dispersal for foraging and concentration for decisive encounter. It might also squeeze out tribal differences persisting in the wake of the drive by Genghis to unite via a written Mongolian language "all who dwelt in tents of felt". But Liddell Hart was surely wrong simply to presume that perfect oneness was achieved even within these tribal levies from immediately around the Gobi. No residual disjunctions in language or dialect, not to mention tent design?

In any case, this structuring could be considerably modified or even displaced as circumstances required. Early on, men from territories occupied might be obliged to form siege trains for the reduction of investing cities. In China, the Yuan imperial forces, rank-and-file were largely Chinese and Korean, mainly infanteers.

The steppe ponies on which the Mongol hordes rode into battle were well able to withstand temperature extremes. Yet as smallish animals living on pasturage as opposed to grainstuff they may have tired easily, a state of affairs which could explain why (according to our understanding) a Mongol cavalryman averagely had five of them. The consequent demands on pasturage imposed, not only in the Fertile Crescent but in Danubia and elsewhere, limits on strategic mobility. One should add that, initially at least, a shortage of weapon-grade iron may well have been a critical constraint on manpower build-ups. Otherwise the Mongols were able to raise largish armies and commit them without forewarning.

The Mongol Legacy

Like the Vikings (and especially, perhaps, the Mediterranean Normans),[42] the Mongols had a strongly pragmatic and eclectic side to their character. They rebuilt Beijing (as Khanbaliq or Tartu) along classical Chinese lines to be the Yuan imperial capital. They learned to write their own language in a borrowed Uighur script because, otherwise, the natural limits of their imperial compass could have been an order of magnitude less. They involved the Persians fully in the management of their own country, somewhere they — the Mongols — had despoiled badly during the takeover. Conversely, they involved few Chinese in the central government of China.

Their eclecticism will have been rooted in a need to survive Mongolia year in and year out; in their shamanist religious tradition with its emphasis on innumerable local spirits; in the yearning of all tribes dwelling in "tents of felt" in or by the Gobi to preserve identities via cultural diversity; and in the cosmopolitan life of the nearby Silk Road.

All of which might conceivably have led to the Mongol imperium's facilitating a bringing together of Europe and East Asia via the Silk Road. Yet this was hardly how things worked out. Few missionaries went East. The "first definite papal interest" in China was expressed in 1289 with the dispatch of six Franciscan missionaries.[43] Imports into Europe of silks, muslins, embroidery and so on were heavily from or via the Middle East. The term "silk road" was not in contemporary use.[44]

In fact, of course, there never was a road, instead intermingled tracks which formed something of an axis for trackways spread broadly across Asia. Between AD 300 and 1850 camel caravans between the taiga and the monsoon lands played their part in the movement of merchandise around the region.[45] As regards the spread of disease, *Pax Mongolica* was a very negative development. Singularly awful was the 1347 event which brought the Black Death to Europe. (See 13 (3)).

One contention has been that the Mongols inflicted on a once splendid "Arab civilisation a blow from which it has never fully recovered". The upshot was the establishment across the Arab world of a dreary Ottoman Empire. So for "nearly 500 years" this world stagnated. "No creative worker or writer or thinker appeared among the Arabs; and they were not awakened from the slumber until the western nations appeared in the East."[46]

The sector bounded by Mesopotamia and the Nile Valley was the nearest the Arab world had got to having a heartland. Of the three main foci of Arab culture therein — Baghdad, Damascus and Cairo — the first two were sacked by the Mongols. Nor should we forget that, not even there did the Arab world have the geographical strength in depth Russia or China or Europe had.

There is a further aspect to the Mongol saga worth considering here. It is that it figures in a tradition of romanticizing the East which is even more callow and insidious than the glib talk about Oriental Despotism evaluated above. In the twentieth century, this perspective was often adopted by Western progressives, particularly in their perceptions of the regime of Mao Tse-Tung. It led to a whole succession of dangerous distortions.

An *Annales* appreciation has been that a supposedly enclosed Indian Ocean once served as a "mental horizon, the exotic fantasy of the medieval West, the place where its dreams freed themselves from repression".[47] Then with the dubious expositions of Marco Polo, this role was switched more towards China. But things reach their giddy limit with Edward Gibbon talking about Genghis Khan. This normally most insightful of historians opines that his religious outlook "best deserves our wonder and applause" because he "anticipated the lessons of philosophy and established by his laws a system of pure theism and perfect toleration".[48] The absurdity of identifying the Mongol supremo with the eighteenth-century Enlightenment in Western Europe and America (and this so dogmatically and tritely) can be gauged by one simple test. The Enlightenment military philosophy included a taboo against the wanton devastation of captured areas especially urban ones. Moreover, this taboo was quite strictly adhered to.[49]

Likewise the Vikings had behaved far less badly than Genghis or his confreres in this regard. Could this comparison owe anything to an enduring difference between the continental and the maritime approaches to war?

Notes

1 See also the author's "Vikings and Mongols, Part 1", *The Naval Review* 96, 1 (February 2008).

2 Gunnar Karlsson, *Iceland's 1100 Years* (London: Hurst, 2005), Map 1–12.1.

3 Willi Dansgaard *et al.*, "Climate changes, Norsemen and Modern Men", *Nature* 235, 5503 (1 May 1975): 24–8.

4 Kevin P. Smith, "Landnám: The Settlement of Iceland in — Archaeological and Historical Perspective", *World Archaeology* 26, 3 (February 1995): 319–47.

5 William W. Fitzburgh in Barry Cunliffe, Wendy Davis and Colin Renfrew (eds.), *Archaeology. The Widening Debate* (Oxford: Oxford University Press for British Academy), 2002, pp. 130–2.

6 Steven Runciman in M.; Postan and E. E. Rich (eds.), *The Cambridge Economic History of Europe* (8 Vols.) (Cambridge: Cambridge University Press, 1952), Vol. II, p. 51.

7 Geoffrey Barraclough (ed.), *The Times Atlas of World History* (London: Times Books, 1979), p. 108.

8 *HCC*, Map 4.2.

9 A few sallies by Wends and North Franks are alluded to in Paddy Griffiths, *The Viking Art of War* (London: Greenhill, 1995), p. 210.

10 Geoffrey Barraclough, *The Crucible of Europe* (London, Thames and Hudson, 1976), p. 77.

11 Gunnar Karlsson, *Iceland's 1100 Years* (London: Hurst, 2005), p. 38.

12 Thomas H. McGovern, "The Economics of Extinction in Norse Greenland" in T. M. L. Wigley *et al.* (eds.), *Climate and History* (Cambridge: Cambridge University Press, 1981), Chapter 17.

13 Jón Th. Thór, "Why was Greenland 'Lost'?", *Scandinavian Economic History Review* XLVIII, 1 (2000): 28–39.

14 G. M. Trevelyan, *History of England* (London: Longmans Green, 1942), p. 74.

15 H. J. Mackinder, *Britain and the British Seas* (Oxford: Clarendon Press, 1907), pp. 357–8.

16 James Bryce, *Studies in History and Jurisprudence* (2 Vols.) (Freeport: Books for Libraries Press, 1968), Vol. 1, p. 263.

17 Irene Scobbie, *Sweden* (London: Ernest Benn, 1972), p. 21.

18 James Graham-Campbell, *The Viking World* (London: Frances Lincoln, 1980), p. 18.

19 George Vernadsky, *A History of Russia* (3 Vols.) (New Haven: Yale University Press, 1953), Vol. III, p. 337.

20 Tibor Szamuely, *The Russian Tradition* (London: Secker and Warburg, 1974), pp. 19–20.

21 Charles J. Halperin, "Russia in the Mongol Empire in Comparative Perspective", *Harvard Journal of Asiatic Studies* 43, 1 (1983): 239–61.

22 See also the author's "Vikings and Mongols, Part 2", *The Naval Review* 96, 2 (May 2008): 165–71.

23 Rudi Paulo Lindner, "What is a nomadic tribe?", *Comparative Studies in Society and History* 24 (1982): 689–711.

24 David Morgan, *The Mongols* (Oxford: Basil Blackwell, 1986), pp. 9–14.

25 Frances Wood, *Did Marco Polo go to China?* (London: Secker and Warburg, 1995).

26 *HCC*, pp. 216–17.

27 Owen Lattimore, "The Geographical Factor in Mongol History", *Geographical Journal* XCI, 1 (January 1938): 1–20.

28 Note by G. F. Hudson in A. J. Toynbee, *A Study of History* (New York: Oxford University Press, 1962), Vol. III, Annex II, p. 453.

29 Sven Hedin, *Across the Gobi Desert* (London: George Routledge, 1931), pp. 250–2.

30 Albert Kolb (C. A. M. Sym, trans.), *East Asia. Geography of a Cultural Region* (London: Methuen, 1972), p. 41.

31 *Ibid.*, p. 126, Fig. 3.

32 Ping-ti Ho, "Early Ripening Rice in Chinese History", *Economic History Review* 18, 2, (1956): 200–18.

33 William McNeill, *The Pursuit of Peace* (Oxford: Basil Blackwell, 1983), p. 50.

34 Gordon Tullock, "Paper Money — A Cycle in Cathay", *Economic History Review* IX, 3 (1957): 393–407.

35 Andrea Kiss, "Weather Events during the first Tartar invasion, 1241–2", Department of Medieval Studies, Central European University, Budapest. Unpublished as of 2001.

36 Reuven Amitai, "The logistics of the Mongol-Mamlūk war . . . " in John H. Pryor (ed.), *The Logistics of Warfare in the Age of the Crusades* (Aldershot: Ashgate, 2006), pp. 25–42.

37 *H.C.C.*, p. 155.

38 Andrew Gordon in H. R. Hitchcock *et al.*, *World Architecture* (London: Paul Hamlyn, 1963), p. 113.

39 *The Kodansha Encyclopedia of Japan* (9 Vols.) (Tokyo: Kodansha, 1983), Vol. 8, p. 122.

40 Felipe Fernández-Arnesto, *Civilisations* (New York: Free Press, 2002), p. 112.

41 B. H. Liddell Hart, *Great Captains Unveiled* (London: Greenhill Books, 1989), Chapter 1.

42 Trevor Rowley, *The Normans* (Stroud: Tempus, 2004), p. 13.

43 J. R. S. Phillips, *The Medieval Expansion of Europe* (Oxford: Oxford University Press, 1988), p. 87.

44 L. Boulnois (Dennis Chamberlin, trans.), *The Silk Road* (London: George Allen, 1966), Chapter XV.

45 William H. McNeill, "The Eccentricity of Wheels or Eurasian Transportation in Historical Perspective", *American Historical Review* 92, 5 (December 1987): 1111–26.

46 Alfred Guillaume, *Islam* (Harmondsworth: Penguin, 1954), pp. 66–7.

47 "The Medieval West and the Indian Ocean: An Oneiric Horizon" in Jacques Le Goff (Arthur Goldhammer, trans.), *Time, Work and Culture in the Middle Ages* (Chicago: University of Chicago, 1980), pp. 189–200.

48 Edward Gibbon, *The Decline and Fall of the Roman Empire* (6 Vols.), (London: David Campbell, 1994), (Vol. 6, p. 311).

49 Christopher Duffy, *The Military Experience in the Age of Reason* (London: Routledge and Kegan Paul, 1987), pp. 12–13.

World Strategy

4 | A Eurocentric Era, 1492–1942

1 Renaissance and Reconnaissance

An approach to what Survival Studies should broadly cover might start with reflection on how Eurocentric the world became from 1492 to 1942: the former year seeing Columbus' first visit to the West Indies and the latter decisive victories at Stalingrad and Midway respectively for the USSR and the USA. *The* two Superpowers, as they were soon to be, thus finally arrived. But each continued to concern itself with Europe Proper in a very special way.

Europe's assumption of a high profile had been rooted in location. A French bifocal map of 1798 presented the world as two hemispheres. The one, antipodally centred several hundred miles north-east of New Zealand, is sea very preponderantly. The other, centred in western France, covers rather more land than sea. Europe therefore became pivotal to global communications.

A complementary reality was neatly observed by David Hume (1711–1776), the lead philosopher of the Scottish Enlightenment. He averred that "of all parts of the Earth, Europe is the most broken by sea, rivers and mountains . . . and most naturally divided into several distinct governments". Especial note was taken of how its bounds are penetrated latitudinally by two articulated arms of the sea, the Baltic and the Mediterranean. Its ridges, peaks, troughs and indented coastlines bespeak past orogenesis (i.e. mountain building), this most characteristically effected by extremely gradual but truly mighty collisions between very slowly drifting continental blocks. A connection can thus be divined between Europe's intricate geography today and its situation well within the assemblage of blocks which had been, till 225 million years ago, the one continental mass — Pangaea.

Poised to Conquer

Europe covers a wide arc of latitude, 36° from Gibraltar to North Cape or 46° if Svalbard is encompassed. This could connote marked differentials in the impact of climate change — e.g. on rainfall. On the other hand, the climate of much of Europe is under the moderating influence of ice-free sea.

An interesting, though rough hewn, endeavour has been made to show how liberal values flourish best where temperature differences between high summer and midwinter are relatively small as per Western Europe.[1] It is hard to believe there has never been anything in such correlation. Yet it is no less hard to relate much of Europe's history (especially in the twentieth century) to Voltairean smiles of reason. Likewise the fractured geography Hume extolled has sometimes favoured local freedom but otherwise called forth tyranny to impose cohesion.

This then was the continent that came to dominate the world, arguably too suddenly for anybody's good. Its preliminary reconnoitres were completed in remarkably short order. Forward location invited Iberia (and, firstly, Portugal) to lead the way. The Baltic and Nordic lands did not feature a long while, nautical traditions notwithstanding.

Prince Henry the Navigator of Portugal (1394–1460) seized Ceuta in Morocco in 1415. Two of the expeditions he later sponsored respectively gained Madeira in 1420 and passed Cape Verde in 1444, by which time gold and slave imports had begun. Then in 1486 another Portuguese, Bartholomew Dias (d. 1500) rounded the Cape of Good Hope.

Christopher Columbus, a businessman apparently born in Genoa and perhaps of Jewish extraction,[2] spent eight years urging several European governments in turn to finance his westabout Enterprise to the Indies. The Spanish agreed to sponsor in 1492, the year the *reconquista* of Islamic Spain, Al-Andalus, as well as the dynastic union of Aragon and Castile were finally effected.

In 1519, Ferdinand Magellan of Portugal broke with his compatriots and, under Spanish auspices, set off westabout round the world with 265 men in five ships. Three years later, just the one ship completed. Magellan himself had perished. But, for good or ill, he can fairly be said to have clinched the Eurocentric era.

Location

A question to ponder is what kind of interaction there was between the launching of this great Western European "reconnaissance" and the Renaissance, the indigenous cultural flowering with which it closely coincided.

Part of the answer is that the two phenomena are centred in adjacent regions, Iberia and Italy respectively. However, the Iberian initiation of the age of reconnaissance must be explained without reference to the Renaissance in that, until 1492, the Christians there were too geared up to forceful recovery to be much open to philosophic rediscovery.[3] Italy emerged early on as the Renaissance epicentre first and foremost because it was uniquely rich in classical remains.

Yet the Italians had also been stimulated by close proximity to the remarkable civilization the Moslems had created in Iberia. Roman irrigation works had been restored and then much extended. Over a hundred kinds of fruit and vegetable were cultivated, including (for the first time regionally) rice and sugar. Emphasis was placed on grapes for wine, the Qu'ran notwithstanding. Woodland reached perhaps its greatest extent there the last two millennia.[4] Córdoba may have had close to a million inhabitants prior to its falling to the Christians in 1236; and Granada nearly half a million in 1491. For comparison, the key Baltic mart, Lübeck, had 22,000 post-1350.

Upon Al-Andalus prosperity was built a strong intellectual tradition which veritably climaxed with the life of Ibn Rushd, alias Averroës (1126–98). His influence on

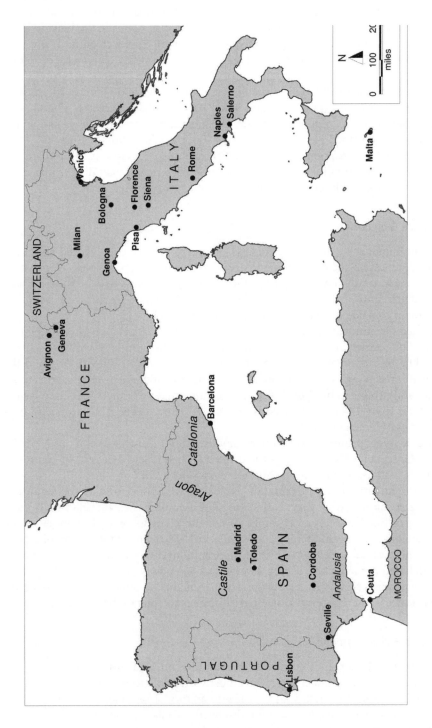

The Renaissance Heartland

Islam elsewhere was to be minimal, until recently at least. But he had "a far more successful afterlife among the Jewish communities in the medieval world, and a widespread influence on the Christian world".[5] Latin Averröists active in the University of Paris in the thirteenth century largely moved to Padua in the fourteenth. There they fostered scientific enquiry. Copernicus spent three years at the University of Padua shortly before he first drafted his heliocentric thesis.[6]

To which one could add that, particularly between 1300 and 1450, northern Italy experienced an economic transformation rather Andalusian in character. The allusion is to the evolution then "of a new agricultural landscape of irrigated grass and arable fields which was destined to become, before 1500, the admiration of Europe".[7] The Italians were also pre-eminent within Western Europe in the cloth trade and a cash-cum-credit money economy, some banking collapses notwithstanding. This most specifically applied to Florence, nodally located in the upper Arno valley.

Renaissance Singularity?

The Italian humanists who led the Renaissance seem to have been agreed about the following tenets. The decline of Antiquity had led to centuries of darkness from which Christendom was only now recovering. Furthering this process depended, especially in architecture and sculpture, on the inspiration proffered by Classical remains. Developments outside of Italy were usually well ignored.[8]

Two decisive innovations emanating from northern Europe were oil painting and printing. The former involved the use of ground pigments mixed with oil, usually linseed. The big advantages as against tempera or fresco were slow drying and the suitability for receiving extra layers once dried. These attributes allowed the artist more flexibility and produced brighter finish. Oil painting may well have been used by Jan van Eyck (*c.* 1390–1441) at the Burgundy court. Within decades it had been introduced through Venice into Italy.

Movable type printing proliferated rapidly after its introduction in 1455 by Johannes Gutenberg. By the turn of the century, it had reached dozens of centres as far afield as Lisbon, Seville, Constantinople and Stockholm.[9]

One consequence of this information explosion was bound to be a broader awareness of the voyages of discovery. But it was slow a-coming. Machiavelli briefly cited the finding of the Americas as symbolic of new horizons in general. Thomas More established an oft-followed literary convention in 1516 by using the fiction of a New World island, *Utopia*, to develop a humanistic alternative to how things are. His envisagings included garden cities, a craft economy, a less adorned religion, and rejection of aggressive war.[10] Nevertheless, the impact on thought and outlook of Europe's great reconnaissance is not registered in strength till the second half of the sixteenth century.

Accelerated Follow Through?

In the main, the Renaissance culture of the late fifteenth and early sixteenth century effected the acceleration of certain trends mostly traceable to the High Middle Ages. Interest in individual personality and in the human form is one such. More naturalism in general is another. Meanwhile, the rediscovery of Classical literature highlighted Plato dramatically. And Renaissance *virtuosi* at least affected a *joie de vivre* considerably male, sensual and bacchanalian, not too guilt-ridden.

This said, the notion Jules Michelet expounded from 1838 of a sudden "renaissance" throughout European culture can be nailed once and for all. Eighty years ago, Charles Haskins introduced the concept of a twelfth-century Renaissance starting *c.* 1070 and climaxing *c.* 1250. Come 1204, Europe had universities at Paris, Montpellier, Bologna, Salerno and Oxford. Paris was pre-eminent in vigour, scale and diversity. France as a whole was a pacemaker with its philosophers and vernacular poets and, above all, its Gothic architecture.[11] Literacy was the hallmark of a restless new non-clerical elite geared up for government.[12]

This era initiated the translation into Latin from Arabic or Greek of many texts from Classical Antiquity otherwise no longer accessible. Aristotle as interpreted by Averroës was very much a centrepiece. Two contributions to climatology exemplify the range and quality of the former's thought as then received. Two editions of *Meteorologica*, a text reliably ascribed to him, were published pre-1163 and in 1260 respectively. It tells *inter alia* how, since the Trojan War, some Greek valleys had become moist enough for cultivation whilst others had turned too dry.[13] Aristotle also gave a succinct account of how dew and hoar frost form which remains hard to fault.[14]

Two new orders of Friars, the Dominicans (founded 1216) and the Franciscans (founded 1210), had been especially instrumental in getting universities going. Among the Franciscan schoolmen of the 1200 to 1350 era at Oxford were three of considerable import for the emergence of natural science: Roger Bacon, Robert Grosseteste and William of Occam. Meanwhile, what we may term a Culture of Calculation was spreading across Western Europe more generally.[15] Witness the beginnings of double-entry book-keeping.

Then during what so often comes across as the forlorn fourteenth century, independent colleges of higher education were founded abundantly across Europe, above all in war-torn France. In one study, 25 new colleges were listed for the thirteenth century of which 20 were in France, while 87 were for the fourteenth with 54 of them in France.[16] An arresting claim made for the fourteenth century was that "no power — king, pope, bishop, ecclesiastical or lay authority — ever attempted to press its own candidates for college fellowships".[17]

Emergent Statehoods

A main aim of this higher educational drive was to provide for the emergent sovereign statehoods enough personnel to sit on representative councils (advisory, legislative or executive) and to staff professional bureaucracies. Correspondingly, a primary manifestation of how the roots of the Renaissance can be traced back to or, indeed, through the High Middle Ages concerns the emergence of statehoods, nation states and city states.

A verdict sanguinely delivered in a 1980 Princeton study was that "Europeans had created their state system only in the nick of time" in terms of coping with a drawn out general crisis under way by 1275,[18] the date some of us would take as also marking the onset of climate decline. A less positive view would have to be that the European state system as it was by 1500, say, by no means proffered a framework firm enough for the generation of philosophies, religious and political, soundly adapted to changing circumstance. This has surely to be a cardinal reason why the sixteenth century lapsed so deeply into organized violence. Fine paintings were no substitutes for social thought.

The physical intricacy of Europe means many locales were able to serve as foci for emergent polities. In several early instances, these locales were sea areas. A prototype was the North Sea empire of King Canute (1014–35). Among other examples were the Norwegian kingdom; the late medieval Danish kingdom; Aragon and also Venice *c.* 1350; and, indeed, Byzantium and the Aegean, especially with the big retraction in Anatolia hard upon severe military defeat by the Seljuk Turks on the enclosed plain of Manzikert in 1071.

Later, however, new statehoods were essentially land based, a response due in considerable part to better overland transportation, restored Roman roads included. The opening of the St Gotthard pass to normal through traffic *c.* 1230 led to armed conflict between the Hapsburgs and the free mountain peasantry. Out of it was born the Swiss Confederation.

The English Question

The biggest wars and rumours of wars in that era were to be those concerning whether England could embrace all the British Isles and/or whether it might retain territory in France. Wales had fallen to Edward I in 1284, albeit on terms. But in 1297, at the battle of Stirling Bridge, William Wallace struck a first blow against the idea that the English crown *ipso facto* overlorded Scotland.[19] Though alternating fortunes in raids and pitched battles hammered out the eventual border, that outcome matched closely the geological divide between austere Silurian rocks to northwards and the somewhat more moderate relief of carboniferous limestone to southwards.

One stark reality was that the monarchs of that day and age found it easier to extract dues from the downtrodden than to constrain the better off. They also found it easier to spend on the prospects for war than on social purposes. The ill consequences bore heavily on France.

In 1328, the year the French crown passed to Philip VI, that monarchy governed half the territory between what were already vaguely perceived as *les limites naturelles*, the Rhine and the Pyrenees. But nowhere was secure. Defeat in battle by the Flemings at Courtrai in 1302 had underlined the lesson taught at Bouvines in 1214. France could never conquer Flanders at a stroke. It was already too urban, insufficiently feudal. Meanwhile, a French seizure of Gascony from the English in 1294 would be reversed in 1303. Come 1337, Edward III of England (reigned, 1327–77) flatly accused Philip VI of harbouring the Scots, an accusation that triggered the "Hundred Years' War". Le Roy Ladurie was to stress how much that dragging out of misery on Gallic soil was aggravated by the soldiery carelessly spreading disease and destroying rural capital.[20] Asymmetric conflict was the problem. The French lost the three great battles (Crécy, 1346; Poitiers, 1356; and Agincourt, 1415), in each case because the English responded better to wet conditions on the day. Between times, the intermittent campaigning was considerably *chevauchée* — skirmishes and raids, a conflict mode the French were at home with.

What strikes one is how well France still progressed on the civil side, conspicuously so in higher education as noted above. A headstart had some bearing on this. As early as 1277, faculty at the University of Paris had formally challenged Aristotle's view that the precept of natural symmetry precluded life's existence outside of this Earth. Here can be seen the genesis of a long Gallic tradition of assertiveness within the Roman obedience.

Its geography makes Ireland look destined to have generated a cohesive identity early and then reconnoitred to advantage. Led by St Patrick (c. 375–461), Ireland initially appeared to be doing just that through a recharging — under druidic and bardic influence — of Christianity. The most remarkable, if least productive, of missionary ventures overseas were the Atlantic forays made in tiny coracles by St Brendan (d. 577) and brethren. They reached Iceland and maybe Greenland. What the summer weather was like then in those waters is hard to elucidate. But it is unlikely to have been better for such sailing than today's.[21]

Come 1171, however, advance towards a higher political organization had not progressed sufficiently to preclude the king of England's landing, with papal approval. He did so to underpin an accelerated colonization in line with the moving frontier trend continent-wise. He was keen to keep on better terms with Rome after the knightly murder of the Archbishop of Canterbury, Thomas à Beckett. Besides which, he saw in the Roman Church a sheet anchor of Irish stability.[22] Under the Treaty of Windsor a deep littoral area from Dublin round to Waterford came under English suzerainty. For the next 750 years Ireland would effectively be controlled from over the water. The developing of trans-Atlantic sea lanes was to make the English very persuaded this was a strategic necessity.

Iberia and Germany

Portugal might be seen as having had little prospect *a priori* of achieving statehood independently of Spain. However, her identity derived from a strong and exclusive sense of pre-Roman heritage, plus a coastal plain exceptionally broad and deep in Iberian terms. She did lose her independence to the Spain of Philip II in 1580 but regained it 60 years later through Madrid's being distracted by a Catalan revolt. Since the high Middle Ages Portugal had been pivotal to the maritime trade round Europe.

Two major regions effectively remained outside the process of nation state building as yet. These were those occupied by the German and Italian peoples. The German case mattered less at this stage. For one thing, what had been the eastern moving frontier had afforded new opportunities. For another, the German Renaissance though important — and not only for the printed word — was hardly focal.

Italian City States

Italy was another question. The physical geography of its backbone peninsula, elongated and with much indentation and high relief, and much ready access to the sea, encouraged a measure of cultural diversity. Yet so did it political divisiveness. Rome, however defined, no longer seemed able or willing to impose its nodality in this regard.

The Classical revival from the twelfth century had done something to revivify old aspirations for free cities. But free from what? Aristocratic oligarchs? Mercantile oligarchs? A fusion of the two? Enraged artisans? Embittered paupers? Florence? Venice? The next door city state? The Hapsburgs? The Vatican? Godlessness? In reality, oligarchy would very generally remain the kind of government extant the next couple of centuries.

There was similarly a lack of political direction north of the Appennines. Any prospect of the Po valley's becoming, through its economic progress, the launch pad for an Italian "universal league" finally died in 1495–8.[23] Well before, the signs had largely been discouraging. In 1174, the Lombard League (an association of north Italian cities founded in 1167) had severely defeated imperial forces at Legnano. Yet it could struggle on fractiously only until 1237.

Through the fifteenth century, a Venetian sense of exceptionalism acquired a dimension veritably ideological. It hinged on how the lagoon, protected "eternally" by the sea and St Mark, had been in Late Antiquity a haven for those resolved to retain their liberty. Now, the argument ran, this blessedness was being upheld by the balance and harmony inherent in a "mixed government" which combined monarchy in the person of the Doge, aristocracy in the 300 senators, and popular republican rule in the broad membership of the 2000–strong Great Council.[24] Also in the fifteenth century Venice pioneered a modern diplomatic service, oriented to the city's role as a commercial hub of the known world. But she remained vulnerable to the Straits of Otranto passing under hostile control, not to mention Adriatic weather.

Archetypal Florence

How Italy's political fragmentation impacted on Renaissance society at large is demonstrated by the Florence experience. In 1406, the city annexed Pisa to ensure itself good access to the sea. From 1434, its own affairs were dominated by the Medici banking family. In 1490, the Renaissance savant, Lorenzo de' Medici (1449–92), invited to Florence the apocalyptically-minded ex-monk, Girolamo Savonarola (b. 1452). He effectively took over after the Medici were ousted in 1494 against the background of a French invasion threat. Then a backlash against his morbid fervour culminated in his torture and burning at the stake in 1498.

The sacking of Rome by imperial forces in 1527 left Clement VII, the second Medici to be Pope, stranded. Yet in 1530, a restored Emperor–Papacy alliance reimposed the Medici on Florence. In 1553 Cosimo de' Medici occupied strife-torn Siena, a city home to a school of Renaissance painting of more conservative hue. Through most of this era, actual warfare was liable to be conducted by *condottiere*, the captains of very venal mercenary bands. However, by 1550 much of Italy, Florence included, was well under the influence of the Spain of the Counter-Reformation.

A Flawed Renaissance

Granted, this febrile ambience did not rule out individual artists and literati moving from one city state to the next. Indeed, it could almost be said to have encouraged this. What was precluded, however, was the development of stable communities of scholars sufficiently well founded to explore properly the novel and complex issues then arising in fields such as nationhood, accountable government, law, secular schools, church reform, rules of war, business management and the world renaissance.

A demonstration of what this could mean is afforded by the Florentine celebrity, Niccolo Machiavelli (1469–1527). He held high public appointments before being ousted on the Medici's return in 1512. The next year he was tortured for alleged involvement in an attempted counter-coup. Thenceforward he was overly concerned to ingratiate himself with the Medici regime. Accordingly, he never carried forward the discussion of statecraft in *The Prince* (1513), the best known of his clutch of public works. Its rather trite overview came to be seen in the decades after his death as not just "amoral" (a judgement that would still command support) but downright "immoral".

A rather different creature of the times is Desiderius Erasmus (c. 1469–1536), alias Erasmus of Rotterdam. He was and likely remains the most respected and loved of all the High Renaissance figures. His overriding concern, in politics and religion, was to have change take place without catastrophic schisms. In his long and peripatetic career, only a couple of years were spent in Italy. He thought Classicism had taken the Renaissance scene over too completely.[25]

The view that, however adventurous and brilliant artists were individually, the Italian Renaissance ultimately remained too monolithically classical has been aired recurrently since. John Ruskin (1819–1900), a lodestar of architectural taste in Victorian England, subscribed to it.[26] Then again, the first half of the twentieth century saw the emergence in the Low Countries of a distinguished coterie of historians of the Late Medieval/Early Modern period. *Primus inter pares* was Johan Huizinga. He depicted the Renaissance as but another Culture of Authority "much less modern than one is constantly inclined to believe". After all, it exhibits a "purblind reverence for the perpetual authority and exemplary quality of the ancient".[27]

Accordingly, the big contributions to our understanding were too few and too late and, may one add, from outside Italy. Leonardo da Vinci was undeniably a great artist. He was, too, amazingly polymathic and imaginative in his understanding of straight-down-the-line Science, pure and applied. Yet he contributed nothing to wider thought. Nicolaus Copernicus finally brought out his revolutionary heliocentric study at Wittenberg only in 1543. A quantity theory of money to explain the sixteenth-century price inflation had to wait upon the French philosopher, Jean Bodin, in 1576.

England's Francis Bacon (1561–1626) was a highly dubious personality but one who in certain respects did set wide bounds. He felt printing, the compass and gunpowder had "changed the appearance and state of the world". They presented the opportunity of progressing beyond the relative stagnation of the last two thousand years in regard to the "prudent investigation of Nature" and would thereby advance "the happiness of mankind". He did not feel too tied to a Classical age which "had not a thousand years of history worthy of that name but mere fables and ancient traditions". Even so, he nowhere proffers a systematic account of what Progress means. Paradoxically, he intimates that the "modern age" (in which he includes the Middle Ages) might well be the last in human experience. His cosmological understanding remains obstinately pre-Copernican.[28]

The trouble with this whole era is that too little was done in its early decades to engender a spirit of toleration open-ended enough to make the world reconnaissance more benign while precluding Church reform splitting Christendom so savagely. Erasmus might helpfully have been more proactive in word and deed. However, the big impediment to a more urgent and broader development was structural. It was that the geography of Italy, being so disparate, induced inter-city strife which foreign interventionists could turn to their advantage. When ancient Rome was first welding together the concept of Italy, this last factor had not applied, not post-Hannibal at any rate.

Geoeconomics

Received wisdom used to be that what set reconnaissance going was the seizure by the

Turks in 1453 of a Constantinople deeply in decline. For this obliged the Europeans urgently to seek alternative routes for the spice trade from the East. In fact, however, spice exports had already been moving predominantly through Cairo and the Levant to Venice. Yet neither would it be right to assume that the conquest by the Ottomans in 1517 of the Mameluke empire altered the situation materially. Clearly it brought the Levant and Egypt formally under their suzerainty. Even so, the Mameluke administration was left pretty much in place. You could add that the people who did attempt to interfere with spice shipments were the Portuguese. This was at one with the secrecy surrounding their early forays. Western Europe's globalization drive was doomed to proceed on the basis of bitter rivalry between the states involved.

By 1550, spice exports through Alexandria were at least back to their level 80 years before. Yet it cannot be said that Ottoman interruption thereof was inconceivable. Take the Genoese entrepôt of Kaffa in the Crimea, the port through which the Black Death had been casually transmitted to Europe. **(See 13 (3))**. In 1453 it was a regional commercial hub with a cosmopolitan population nearing 100,000; and was so by virtue of Italian mercantile strength. Come 1486–90, however, only 10 per cent of the ships calling there were Italian.[29]

Through the middle decades of the fifteenth century, Europe was tangibly in recession. It is a judgement not gainsayed by Italy's moving then into the High Renaissance — e.g. Alberti and Raphael. To cite a celebrated essay informed by the late medieval/early Renaissance era: "There is no heap of riches and no depth of poverty that will automatically ensure or forbid artistic achievement. Intellectual developments must be traced primarily to intellectual roots".[30] Investment in the fine arts or uplifting architecture have tended to ride high when more profitable avenues are hard to find. Closely akin is the Byzantine experience in the two centuries before 1453. They were "politically decadent and disastrous; but intellectually and artistically they formed a period of exceptional brilliance". Indeed, they influenced the High Renaissance, especially through migration upon Constantiniple's fall.[31]

Although secular weather tendencies had again turned adverse, bullion shortage had been the immediate cause of the said European recession. By 1400, political changes in west Africa had effectively curtailed a centuries-old trans-Saharan trade in alluvial gold. Meanwhile European gold and silver mining had not properly recovered from fourteenth-century production difficulties. Much bullion was being exported to the Levant as well, perforce to correct an adverse trade balance.

After 1460, the bullion situation eased thanks in part to devaluations and the use of copper in currency. There was also recourse to that still rather suspect Italian innovation, credit expansion.[32] It was part of the Culture of Calculation.

Around 1450 there had surfaced in Genoa, against a background of economic malaise, previsions of an African "El Dorado" forever exuding gold. They became a legend which could still inspire Walter Raleigh to explore the Guinea coast 150 years later. A quest for gold was certainly one motive behind the early Portuguese voyages. Others included access to India and Cathay. Also it was "in Asia that the legendary Prester John, the hoped-for saviour of the Holy Places, had been sought for until the early decades of the 14th century".[33] Such hopes could overcome fear of the unknown, not least of a perhaps impossibly "torrid" equatorial zone. Motivational, too, was a desire to outflank Venice as well as the Ottomans.

Technological Advantage

With exploration and expansion, a question that always comes up is how far were they initiated or sustained by technical inventiveness, equipment-wise or procedural. In this case, any answer leads one further to ask how far the Renaissance encouraged this spirit of enterprise within what we know as applied science.

If at this early stage the Renaissance virtuosi were little aware of the explorers and colonizers, the same applied in reverse, even to Christopher Columbus. Still, he did navigate by dead reckoning based on guestimated ship's speed and compass headings. Since *c.* 1290, the compass had sometimes been used in the Mediterranean. It had been known in Iran in 1232. The principle was familiar in China two centuries before that.[34]

The Renaissance also saw much development of nautical cartography and, of course, the printing thereof. The Hellenic–Egyptian geographer and cosmologist, Claudius Ptolemy (lived *c.* AD 150), had previously remained the main fount of received wisdom. But he lost credibility once his presumption that the Indian Ocean was landlocked was proved wrong. A map of its western side drawn in 1558 by Diego Llomen was amazingly accurate, considering the exact determination of longitude was still two centuries away.[35] The Flemish mapmaker, Gerardus Mercator, produced his rectilineal world map in 1569.

The sixteenth century in particular also saw multiple advances in ship design. The kind of vessel most remarked by historians in regard to the reconnaissance phase is the caravel, a very adaptive vessel of between 70 and 200 tons. But by 1600 some ships employed on North Atlantic escort duties were of 800 tons as were the larger Mediterranean cargo ships.

Still, the asset customarily seen as most readily persuading indigenous peoples that the European arrivals were Gods, even though they did have to die, is the firearm. Gunpowder, essentially a mix of saltpeter with sulphur and charcoal, was being produced in China by the ninth century AD. Roger Bacon (*c.* 1214–94), the Franciscan scientist at Paris then Oxford, was aware of it. At the Battle of Crecy in 1346 the English used gunpowder as the propellant in very light cannon able to distract if little else. The Turks used heavy cannon to help gain Constantinople. By 1500, the Chinese lagged well behind both Ottomans and Europeans in firearms development.

Through the sixteenth century there were debates within the Renaissance literati about whether devices so "devilish" ought ever to be used. By 1600, however, pragmatism prevailed.[36] During that century, too, the Europeans had installed quite powerful cast guns on board warships. Since the traditional Mediterranean oared galley was unable to fire broadsides, the Battle of Lepanto (1571) between Christians and Turks was the last major action involving such ships. For a while Spain had kept one fleet comprised of galleys in the Mediterranean and another with sail and broadsides on her Atlantic seaboard.[37]

Gunfire and Germs

Under the 1494 Treaty of Tordesillas between Spain and Portugal, an imaginary line was drawn 370 leagues west of the Azores. Its central provision was that Spain could lay claim to any territories west of that line and Portugal to any to eastward. It favoured the Portuguese more than had the Papal Bull of 1493 which had set the dividing longitude 100 leagues west of Cape Verde.

Nevertheless, the Portuguese were slow to colonize their allotted east side of South America. Conversely, they had established by 1550 no fewer than 50 forts and factories beyond the Cape of Good Hope, the furthest out being Nagasaki. Among the earlier such outposts were Goa (1510) and Malacca (1511). This was merchant adventurism on the grand scale. It did not devastate pre-existing political and social orders.

In the Caribbean, Spanish cattle farmers began to settle in Hispaniola as early as 1493. Most chillingly extraordinary, however, was how Spain's *conquistadores* collapsed the great Aztec and Inca empires. Thus in 1532 at Cajamarca in the Peruvian highlands, 168 Spanish soldiers under Francisco Pizarro defeated without loss to themselves 80,000 Incas, killing several thousand. Atahuallpa, the Aztec demi-god head of state, was made prisoner and later treacherously murdered.

It is tempting to assume that gunpowder always held the key to such disproportionate success. But Jared Diamond has well pointed out that, since the only firearms the Spanish troops had were a dozen harquebus, their weaponry advantage at Cajamarca lay in their swords and lances being pitched against Inca clubs.[38] Yet not even Toledo steel could have overturned unaided a numerical advantage of nearly 500 to one. Another factor must have come more strongly into play to unnerve the Incas. It will surely have been how closely the European descent coincided with the spread of a whole cocktail of novel diseases: smallpox, measles, influenza, typhus, bubonic plague . . . Of no such import strategically but dreadful in its social consequences was the surreptitious migration of syphilis from the New World to the Old. Complex though the latest genetic evidence is, it does seem to confirm a longstanding suspicion that this is what caused the outbreak in Europe in 1495.

Also instructive to note is the attention Western Europe paid to the mid-Atlantic islands: tropical or sub-tropical; inhabited or no. Let us look specifically at the Canaries. As early as 1300 or so, preliminary reconnoiters were undertaken. In 1402 a French expedition established what proved but a short-term presence. Portugal undertook probing actions, 1415–66. But by 1493 the Spanish had captured the entire archipelago from a Guanche (i.e.; native Canarian) population perhaps 75,000 strong.

During this colonization, the Guanche numbers were decimated. The causes were multifarious. Sheer culture shock. Unfamiliar plant, animal and human diseases. Warfare, Exporting as slaves. Deforestation. Feral imported animals, especially rabbits. Guanche neglect of sea-faring. Foreign plants . . . Guanche genes figure still in the genetic profile of the Canaries people. A comment has been, however, that "so slight is the Guanche strain that it probably would not be credited but for nostalgia amongst the present-day citizens for what is unique about their islands and their history".[39]

2 Un Long Dénouement

Hard upon Renaissance expansiveness comes the bitter fissuring of Latin Christendom. The Northern Renaissance had been leading the exploitation of the printed word. Martin Luther, the friar from Saxony who set the Reformation finally in motion with his 95 theses in 1517, secured ascendancy within its diversified leadership by tireless recourse to the printed vernacular. From 1518 to 1525, he published more works in German than did the next 17 most prolific evangelizers put together.

In his lifetime, he published five times as many as all the Catholic protagonists did.[40]

No doubt northern winters encouraged reading and reflection; and maybe sombre skies and pastel landscapes called forth puritan ethics and aesthetics. Southern Europe went rather for oral communication and colourful display. Perhaps, too, the evil spirits of northern mythology afforded an evocative setting for the strong Protestant sense of sin.

Almost everywhere, however, Protestantism made inroads in cities and towns. Whether inroads led to takeover sometimes turned on the local strength of the quasi-feudal nobility. That Cologne stayed staunchly Catholic was down to city management.[41]

Early on, a loose-knit movement voicing a chiliastic "divine law" egalitarianism emerged in the Tyrol, the upper Rhine and the dissected plateau South Germany largely consists of. In the Peasants' War of 1524–6, it was crushed by imperial forces spurred on by none other than Martin Luther, thousands being killed.

One could have imagined that a worsening climate post-1550 would have stimulated sectarian sentiment all round. In fact, heightened zeal preceded the worsening. Desiderius Erasmus, the archetypal moderate, died in 1536. That same year, John Calvin completed his formulation of the austerely deterministic Protestantism he propagated from Geneva. By then, too, the Lutherans had formally endorsed separation from Rome. Moves on the Counter-Reformation side included the Jesuits being founded in 1540–1 and the Inquisition revived in 1542.

That great commercial centres like Cologne, Florence, Genoa, Marseilles, Seville and Genoa continued within Rome's obedience shows that Catholic doctrine was not read as categorically anti-capitalist. Generally, however, the mainstream Protestant ethic "was more welcoming to the amassing of capital and lending it with interest".[42]

1571 and 1588

A geopolitical complication was France. Her continuing disposition towards an independent Catholicism was recharged by the emergence of the Huguenots, an indigenous Calvinism very generally ascendant in the southern half of the country through the late sixteenth century. Externally, the threat to the French monarchy contingently came from a Hapsburg-led coalition, Catholic and maybe, in part, Protestant. The grand alliance Philip II of Spain organized to confront the Turks at sea could be read as having this ulterior motive. The French reaction was to strengthen ties with Turkey.

The naval showdown occurred off the Greek coast in October 1571. The Turkish fleet was shattered at Lepanto, thanks to the superior Christian, especially Venetian , shipborne gunnery. Received wisdom has been that this fleet action (60,000 casualties incurred) was to be of little import strategically.[43] Yet a Turkish navy triumphant could have presented Venice with what she dreaded most, a blockade of the Adriatic. The initiative might thus have been regained, on land as well as at sea, against Christendom as a whole.

Come 1584, Philip II tentatively decided to invade England, partly because of Elizabeth's firming support of the Dutch Calvinist rebellion against Madrid. None the less, triangular diplomacy continued between Madrid, Paris and London, albeit with no side disposed to trust the other two. Then the execution of Mary Queen of Scots in 1587 made an invasion attempt inevitable despite the unwillingness of Paris to lend military support.[44] As oft-recalled, about half the Armada dispatched was lost to English naval enterprise and awful weather. This episode marked the onset of close to a century-and-a-half of Spanish retraction around Western Europe.[45]

Retreat from the Reasonable

Much more violence was engendered by intra-Christian tension than this summation can encompass. Important for now is seeing everything within the context of a general flight from reasonableness and reason, indeed from the positivism the Renaissance extolled.

Among the uglier manifestations was witch-burning. It besmirched Christendom for a good two centuries from the death in 1431 of Joan of Arc, Catholics and Protestants sharing culpability pretty much *pari passu*. As Hugh Trevor-Roper stressed, this craze cannot be seen (*pace* nineteenth-century liberal historians) as mere "delusion detached from the social and intellectual structures of the time".[46]

Astrology should also be mentioned. A revival had started during the twelfth century Renaissance. In 1348, the medical faculty of the *avant garde* University of Paris averred that the Black Death had been triggered by Mars interacting with Jupiter after a joint conjunction with Saturn. Then come the Renaissance proper, the astrologers flourished exceedingly, sustained specifically by the anti-clerical aspect of the humanist ethos and by the collateral revival of Plato.

Not that astrology ever lacked enemies. Savonarola opposed it fiercely as epitomizing the corruptive pagan elitism of the times. In due course, too, the Protestant leaderships turned against the astrologers, the suave Philip Melanchthon surprisingly excepted. Catholic attitudes hardened soon as well. The crunch came with the Papal Bull, *Coeli et Terrae*, issued by the combative Sixtus V (1585–90).

None the less, astrology was bound to retain a certain credibility until European opinion came broadly to accept the Copernican anti-geocentric revolution. This took time. Nicolas Copernicus (1473–1543) published *De Revolutionibus Orbium Coelestium* in the year of his death. Melanchthon's *Doctrines of Physics*, published to confront it in 1553, was printed ten times before the Copernican text was twice. Not before the turn of the century can one say Copernicanism had a strong constituency. As late as 1633, after all, Pope Urban VIII ruefully referred his erstwhile friend Galileo to the Inquisition for flouting a 1616 proscription against treating the new cosmology as other than mere hypothesis not to be affirmed or denied. At the outset, Luther and Calvin had condemned the Copernican challenge to a literal interpretation of "the Book".

Influential, too, may have been wider apprehension about how social mores and motiviations were being undermined by various aspects of the new learning. In 1611, the English "metaphysical" poet, John Donne, gave vent to such concern. His *Anatomy of the World* warned how the Copernican "new philosophy calls all in doubt". Cosmology with "all coherence gone" could induce a collapse of the social order as "every man alone thinks he has got to be a Phoenix".[47]

Retreat to Moderation

In due course, as always, the mood changed. Through the late sixteenth century, the foci of religious strife had been France and the Netherlands. But in the former, the

Edict of Nantes of 1598 granted the Huguenots basic religious freedom plus special status in sundry respects, including their managing the state-subsidized militias in the 200 towns they effectively controlled.

In 1609 in the Netherlands, the seven northern provinces (the core area of the Dutch-speaking Calvinists) signed a 12-year truce with the Spanish Army. These "United Provinces" went on to secure general acceptance of their statehood within the Treaty of Westphalia of 1648. The success of their bid for independence had rested considerably on three factors. They had linguistic and religious commonality. The wide rivers, indented coasts and islands of Holland and — above all — Zeeland enabled "Sea Beggar" irregulars to frustrate continually Spain's highly professional army; and their privateers operated to effect from Germany. Then at a critical juncture in the eighties, Philip II saw fit to divert forces to back the pro-Hapsburg Catholic League in France.[48]

The main aim of the Westphalia settlement was to terminate conclusively the Thirty Years' War, a conflict largely waged across Germany and the cause of enormous misery. It had all started with a revolt by a Protestant nobleman against a Catholic king in Bohemia. Aided by its mountain rim and flat central plain, the region had well preserved its Czech identity. Nevertheless, not long after the reign (920–9) of its first great king, St Wenceslaus, it had accepted the overlordship of the Holy Roman Emperor. For a while from 1355, indeed, Charles IV of Bohemia was the Emperor. After 1438, however, the Emperors were always Hapsburgs. By that date, too, the Bohemian identity was being militantly expressed by the Hussite militant Protestants, a complex and powerful tendency in Bohemia though also Moravia.

The Thirty Years' War was basically as between (a) the Holy Roman Emperor in Vienna plus Catholic member states, together with the Spanish Hapsburgs; and (b) Protestant member states forcefully backed from 1630 by a Sweden initially under King Gustavus Adolphus (killed in action, 1632). Soon France, too, involved itself deeply on the latter side, still fearful of being sandwiched between the Hapsburgs in Vienna and those in Madrid. In fact, France intervened to wreck the 1635 Prague peace agreement, a general accord which otherwise would probably have stuck. It did so for fear of too much influence remaining with the Emperors.

The involvement of Stockholm and Paris overstretched Vienna. The 1648 peace therefore left the imperial role as little more than ceremonial. The new guiding principle was "to each Prince, his own religion". In other words, the ruler of a constituent statelet of the Empire shall determine its religious identity.

England's shortish Civil War (1642–49) ended with the execution of Charles I. A veteran thereof, asked afterwards if it had been about religion, replied "not to begin with but it came to that in the end". But the excesses of a decade of puritan republicanism notwithstanding, the near concurrence of this war's conclusion and the Peace of Westphalia heralded a general pull back into reasonableness and toleration plus the advance of positive rationalism. The said execution bestirred Europe far less than once it might have. From 1650 a decline in astrology was precipitate and seemed veritably terminal. The decline in witchburning soon was terminal. Meanwhile, scientific academies were springing up around Europe. In England, Charles II incorporated an "invisible college" of scientists as the Royal Society in 1662. The Paris Academy of Sciences held its first meeting in 1666. Peter the Great established an Academy of Sciences in 1725. But the decline of prejudice did not presage global peace.

The Seventeenth Century Crisis

The apostles of military mobility look back to this era as the zenith of the cavalry arm as developed by Gustavus Adolphus, Prince Rupert, Oliver Cromwell and so on. Clearly the advent of the wheel-lock pistol had much facilitated horsemen delivering point-blank a volley of fire immediately before recourse to lance or sabre.[49] All the same, big developments were on-going, too, in fortifications and field artillery. The cavalry came to especial prominence during the Thirty Years' War and the English Civil War because in both the political geography of conflict, the time frames and so on left no scope for the creation anew of elaborate castellations.[50]

A radical change in Europe's economic outlook had come during the sixteenth century with the price inflation consequent upon the import of silver from Spanish America. This flow peaked *c.* 1590 then fell back by half by 1650. Commodity prices perhaps quintupled by 1590, then fluctuated around this level.[51] Contributory to this relative levelling out was the collapse of silver mining in Germany through the Thirty Years' War.

Since before 1939, academic debate has continued about the impact and aftermath of the great inflation. One opinion from France's well respected *Annales* school of moderately Leftist historians has been that, within Western Europe proper as one might say, the price rises stimulated capitalistic development but that across more conservative Central Europe it bolstered the pre-existing feudal agrarian set-up.[52] Either way, there will have been a restraining effect on economic growth.

All of which should be seen within the context of a wider tendency historians have long debated, the "general crisis of the 17th century": meaning, in particular, the global incidence of revolutionary unrest, especially between 1640 and 1660. Various explanations have been advanced, for the European experience in particular. Irregular weather and climate. Pervasive silver shortage. Mercantilism. Heavy government expenditure on defence and patronage. Unrepresentative political institutions. Changing social structures. A widespread belief that all Europe was in the grip of chronic melancholy.[53] A literati sense of cosmic doom: "I was born in this setting of time" . . .

What then of climatic instability? The turn of the century was to see (ahead of the coal-driven Industrial Revolution) the nadir of the Little Ice Age. Four centuries of general cooling had lowered surface air temperatures perhaps two or three degrees Celsius.[54]

Thus far, the historical mainstream has mostly given the Little Ice Age short shrift in its search for "general crisis" causation. In 1965, Eric Hobsbawm, a British historian writing from a neo-Marxist perspective, was dismissive of such extraneous explanations. Yet he showed himself how very general through mid-century was a slowing down or worse of international trade growth, along with lower crop yields and higher mortality.[55] In 1960, Roland Mousnier had deemed it improbable that climate variation weighed in.[56] After a fuller review, he highlighted the "large-scale agricultural calamities" which beset the "Century of Revolts" around the world.

Not least was China affected.[57] Taking advantage of a spate of uprisings from 1627, the Manchu invading from the north-east occupied Beijing, the Ming capital in 1644. Drought was the big problem across the north of the country, that second quarter of the century.[58] Linked to a strong Siberian High in winter?

Europe turned noticeably colder as the century progressed, a tendency consistent with perceptions of a Maunder Sunspot Minimum, 1645–1715. The years between 1693 and 1710 saw the deepest cold, thanks to dust loading of the stratosphere by volcanic eruptions. The bad harvests visited upon northern Europe had geopolitical consequences. For instance, an economy thus weakened left Scotland more ready to accept, however dourly, the political union with England enacted in 1707.

A cluster of cold Baltic winters culminated in the "famine year" of 1696–7 during which, in Finland, a third of the population died. The upshot geopolitically was the organization against a weakened Sweden of a containment coalition reaching from Norway and Denmark round to Russia. The Great Northern War duly broke out in 1700.

Stockholm finally lost the initiative with the defeat of a field force at Poltava in June 1709, after the same bitter winter as induced an embattled France to put out peace feelers during the War of the Spanish Succession. The 1721 settlement entailed the loss of Sweden's possessions abroad, all bar Finland which was surrendered to Russia following briefly resumed hostilities, 1741–2. Yet with Sweden as with Scotland, what many would read as diminished political status ushered in intellectual renaissance. One thinks most readily of Carl Linnaeus and Anders Celsius with the former; and David Hume, James Hutton and Adam Smith with the latter.

The New Statism

At one time, French historians were disposed to celebrate the seventeenth century, the age of Louis XIV (reigned 1645–1715), as an era of harmony at home and influence abroad. However, it was not, and could never have been, that blissful. Christians may have been much less keen to go to war with other Christians to resolve theological contention. But religious orientation would long remain basic to defining national identities.

Therefore in France, severe pressure soon resumed on the Huguenots, the liberal Edict of Nantes of 1598 being subject to shallow interpretations. Huguenot revolts duly broke out as early as the 1620s, one rebel force holding out in La Rochelle 14 months. Eventually, in 1685, the Edict was revoked. In other words, Huguenots lost their entrenched entitlement to religious freedom and communal autonomy. After which, very many (a high proportion talented) fled to Protestant lands.

Dutch Endeavour

The seventeenth century was the golden age, *gouden eeuw,* of Dutch art and of commercial enterprise abroad, coming in the wake of independence effectively being gained in 1609. Throughout the flag followed trade, albeit with due caution. Very gradually, meaning by 1900, the Netherlands gained general control of nearly all the East Indies archipelago.

Following indigenous precedent, the Dutch East India company had soon made the nodal and lava-rich island of Java the hub of its imperium. Perforce this effected the huge disparity in indigenous population growth between Java and elsewhere, the divergence which compromises still Indonesia's prospects of stable democracy. However, the more obvious concern early on was the "phenomenal growth" all over of the smuggling and piracy already endemic. This was partly because Dutch naval power was finally broken by the War of the Spanish Succession. It was also on account of the "oppressive commercial monopoly", the East India company sought to maintain.[59]

On the home front, the Dutch Netherlands remained in a strong position economically through 1700, thanks to a thriving entrepôt trade and associated financial services. However, these facilities lost out as other national economies matured. By 1730, that of the Netherlands was lagging.[60]

Warfare, Economic and Military

The prevailing economic philosophy post-1650 was Mercantilism, its most celebrated practitioner being Jean Baptiste Colbert (1619–83), minister to Louis XIV. Its underlying presumption was that the struggle for all-round strength was essentially "zero sum", one nation's gain being another's loss. Strength would fundamentally derive from home-based manufactures selling within markets, colonial and domestic, protected by tariffs or state-backed monopolies. Bullion amassed through surpluses on external trade as a whole would (a) preserve a solid currency base and (b) help the public purse meet imperative expenses, military or infrastructural. It was hardly a regimen likely long to commend itself to the pacemakers of the Industrial Revolution. Nor, indeed, to Britain's American colonists.

A striking feature of the febrile eighteenth century was the rising cost of war due to technological, economic and administrative advance. In the twenty-five years spent at war between 1688 and 1713, Britain's war budgets totalled £143 million. In the Seven Years' War, 1756–63 (dubbed by Winston Churchill "the First World War"), they came to £161million. From 1793 to 1815, they totalled £1660 million.[61]

In the eight or so multinational wars across this total span, England/Britain and France were regularly on opposite sides. Since Britain's taxation and banking systems were overall better founded, it was able, from the War of the Spanish Succession onwards, to subsidize continental allies. The British government gave Frederick the Great two million pounds, 1756–63 and then £65 million to various allies in the course of the struggle against Napoleon.[62]

Prussian Resolve

The rise of Prussia represented the defiance of Geography through stern resolve. The north European plain afforded little natural protection apart from cold winters. Nor was Prussia out of reach of such pre-existing centres of power as Stockholm, Moscow or St Petersburg, Vienna and Paris. Its origins lay in Brandenburg. This medieval statelet inherited the littoral Duchy of Prussia in 1618 but since there was no contiguity, Brandenburg proper remained landlocked until the Peace of Westphalia, 1648. Under the Peace of Utrecht (1713), the Great Elector of Brandenburg was redesignated the King of Prussia. Since 1415, Brandenburg had been ruled by the Hohenzollern princely family. Through to 1918 they were to be the ruling family in Berlin.

A Hohenzollern aptitude for creative leadership found full expression with Frederick the Great (reigned 1740–86), the misogynist son of a brutish father. He led inspirationally an army welded together by harsh discipline. Otherwise he promoted the values of the Voltairean Enlightenment and mercantilistic modernization. Even so,

"Prussia was not so much a state which possessed an army, as an army which possessed a state".

Frederick's generalship finally wrested Silesia from Austria in 1763. His diplomacy gained further security through expansion with the first partition of Poland in 1772. At a critical juncture, relief had come with Russia's exit from the Seven Years' War in 1762, hard upon the death of Tzarina Elizabeth — a Germanophobe with a super-charged hatred of Frederick. If a geographical singularity was at work, it may have been cultural: a gilded memory, folk and statist, of the Junker tradition.

Right up to 1815, youthful Prussia would be fighting always against elimination as a sovereign entity. That would not be said (pre-1792 at least) of France, Britain, Austria or Russia. Their adversaries never had the staying power to effect this against all the impediments, Geography included. In any case, the spirit of the Enlightenment was not that iconoclastic.

Salt Water Imperialism

Outside Europe, the newly-created maritime empires could be fought over. In the Seven Years' War, William Pitt the Elder, Britain's Prime Minister to 1761, came early to appreciate that the fate of the French in Canada (as then delineated) could be decided in Germany. Meanwhile, the Royal Navy proved highly efficacious at blockading France's Atlantic ports while screening Toulon as well. Command of the sea also underpinned the paramountcy Britain's East India Company achieved on the subcontinent through Robert Clive's sensational victory at the Battle of Plassey in Bengal in 1757.

The problem for France had been that not until 1750 (with the appointment of Wenzel Kaunitz as Austrian Ambassador to Paris) did it become clear beyond reason-able doubt that Vienna had effected what we know as the "diplomatic revolution": a strategic realignment which recognized that the big threat to its position now came not from Paris but Berlin. Until then the French had to keep their army up to scratch, neglecting their navy the while.

After 1763, they corrected this inter-service imbalance urgently, severe budgetary strain notwithstanding. So when in 1778 France joined in the American War of Independence, the Royal Navy no longer felt able to close blockade those Atlantic ports. So when British troops found themselves trapped in Yorktown in 1782, the Navy was prevented by a French naval presence from affording relief. The soldiers' consequent surrender effectively ended hostilities. Within several years, however, Anglo-American trade had recovered. London therefore remained "the great capital of the commercial civilization of the West".[63]

The Industrial Revolution

The last four decades of the eighteenth century, an Industrial Revolution got well under way, with Britain the leader. In 1769 James Watt patented a steam engine, as did Richard Arkwright a spinning machine. Around 1785 Henry Cort introduced multiple changes in the smelting of iron, especially the use of coal instead of charcoal. In Paris in 1799, Philip Lebon demonstrated gas lighting. In 1802 Richard Trevethick made a marginally viable steam car; and a tolerably practicable steamship, the *Charlotte Dundas*, was tried out on the Clyde.

The crux was the more efficient concentration for diversified use of inanimate energy. So why was this so long a-coming? The megalithic Stonehenge temple in Britain was a masterpiece of civil engineering. Yet the energy inputs to its construction were almost entirely manpower. Why four more millennia for combustion engines to appear?

The second of the Laws of Thermodynamics enunciated in 1850 by the German scientist, Rudolph Clausius, says that natural redistributions of energy will always be towards an even spread (local and, in the ultimate, cosmic) in the randomized form we know as "heat". This is what makes it hard to effect the concentrations needed for useful work. True, all animals do so the whole time within themselves. Even so, the mechanical exploitation thus of fossil or renewable energy requires very exact design and construction. Progress along these lines was for long insufficient, beautifully crafted though innovations like the waterwheel were.

At last, from the sixteenth century through the eighteenth, big advances were made in refining components, instruments and other ancillaries.[64] An able-bodied man should manage 0.5 megawatt hours of mechanical work a year. A medieval waterwheel run continuously probably yielded 12. World production of inanimate energy in 1860 was an estimated 1.1 billion megawatt hours.[65] Not that one should exaggerate the extent of this leap forward early on. Watt's steam-engine had an energy conversion efficiency of five per cent. A modern counterpart might reach 40 per cent.

Next, one can ask why the Industrial Revolution was so concentrated in Britain its first half century. Part of an answer has to be that through the seventeenth century the Mediterranean lands had been falling behind across much of the economic spectrum.[66] In Italy, still, political fragmentation all but precluded a technocratic boom.

A specific factor then assuming salience was the distribution of coalfields. All those in Europe are found north of the 42nd parallel. So many come near the surface because the carboniferous strata were subject to gentle Tertiary folding on the fringes of the Alpine mountain building process. This exposed them to much erosion but left the surviving seams accessible with relative ease.

Two countries — Britain and Belgium — were especially blessed by Nature's legacy. Not so France. As late as 1870, her coal output was only equivalent to Belgium's, ten million tons a year. Germany produced 20 to 25 million but Britain 120.[67] The latter was further advantaged by her insularity on a broad continental shelf. Distancing from continental extremism, religious and political, was one aspect. Another was the shipment opportunities afforded by tidal estuaries, with shallow inshore waters conducive to wide tidal ranges.

While Britain's population was not half the French (which was 19 million in 1720), its average income *per capita* was rather higher. In particular, France had lost out with the Huguenot exodus. In certain managerial aspects, however, France will have stayed ahead of Britain. Apparently, French economic growth matched the British percentagewise, 1700–1789.[68]

Enlightenment, Romanticism, Bloodshed

Our current mind sets are considerably legacies of the rationalistic/deistic Enlightenment which burgeoned within the *literati* (in France but also Sweden, Scotland, England, the Netherlands, America . . .) from 1675. This movement was fervently committed to reason though also to toleration, literacy, science, factual knowledge, the abolition of slavery and citizenship. It peaked, one can say, with the *Encyclopédie* Denis Diderot and others brought out in 1772.

The Enlightenment extolled moderation in war. However, the material advances it helped stimulate "made for bigger wars".[69] Thus the attention the French Enlightenment paid to mathematics (following on from Descartes and Pascal) directly facilitated the development of artillery science. Napoleon was to gather about him, figuratively speaking, a remarkable aggregation of mathematical talent: Lazare Carnot, Jean Baptiste Fourier, Joseph Lagrange, Siméon Poisson, Pierre Laplace . . . As the Emperor himself liked to say, "By artillery war is made".

From 1760 or so, Romanticism arose as a current of opinion strongly counter to Enlightenment reasoning. Advanced technology it affected to disdain for its largeness and sameness and for grinding down Nature. Science stood condemned for deeming every problem reducible to a batch of discrete questions. The Romanticist pitch was immoderately epitomized by the laudanum-wracked poet, Samuel Coleridge: "I believe the souls of 500 Sir Isaac Newtons would go to the making of a Shakespeare or a Milton."

Not that this dichotomy connoted a straight fight between progressives and conservatives — Left and Right as they would soon be styled. Let us compare two Leftists. The Anglo-American radical, Thomas Paine, wrote with fearsome energy about everything from American independence to the Creator's beneficence as revealed by cosmology. His *The Age of Reason* (published 1791–2) encapsulates the liberal outlook of his day. His stance overall contrasts with that assumed by Jean Marat, the talented French medic who turned revolutionary only to be murdered in the Reign of Terror, 1793. His "anti-Newtonianism" was especially directed against the *Académie Royale des Sciences*, its having refused him membership.

To scorn reason and deductive enquiry — claiming to set store instead by wholeness, naturalism and sensibility — was dangerous. Take the amoral egotist dubbed by Bertrand Russell "the father of the Romantic movement", Jean-Jacques Rousseau (1712–78) of Geneva. Writing from outside the confines of Parisian salons, Rousseau extolled a "general will" to which every member of a polity should surrender himself . He could tell us little more, save that this abstraction is "always constant, unalterable and pure". Guidance that nebulous left a wide flank open to totalitarian ruthlessness of the kind typified by Mao Tse-tung, Stalin, Hitler and — most immediately — Robespierre. Any would-be dictator could insist he it is who represents the "general will".

Revolution might therefore start with Paine-style liberal radicalism but then slide into Rousseau-style totalitarianism, much as the English Revolution had started in 1642 with insistence that the elective legislature more closely control the royalist executive but moved on to regicide and the establishment of puritan military governance by 1654. The French Revolution started with an effusion of tracts arguing about everything though especially constitutional reform. But in 1793 it finally arrived at manic intolerance, bloodily made manifest. Features the respective pre-revolutionary situations had broadly in common included underlying changes in the social order; an alienated intelligentsia; a *dégagé* and corruptive ruling elite; recurrent crises in public finance; and pervasive tension.

By 1792, action–reaction is strongly at work both within France and as between her and *émigrés* and *anciens régimes* close by. That April, the Constituent Assembly declared war on Austria — home of Marie Antoinette. In response, invading armies made menacing ingress but were repulsed within the year through *levée en masse*. In April 1793, Jacobin extremists launched a Reign of Terror via the guillotine, this under a Committee of Public Safety headed by Maximilien Robespierre. In July 1794, the introduction of still more repressive measures (despite the invasion threat's receding) led to a backlash, the reaction of Thermidor. Soon Robespierre was guillotined. Thereby the scene was set for the rise via the Directory regime of that aggressive young artillery officer, Napoleon Bonaparte. He would generate the charisma needed to become Emperor by bold military adventurism, not all of it successful by standard operational and strategic criteria.

In Paris, always the hub of the revolution, several thousand had died during the Reign of Terror. One consequence of it had been a splintering of the Girondin, a moderate republican tendency with its roots around the Gironde estuary — the old Huguenot heartland and a region with historic links with England. During unrest in conservative Brittany, and especially the Vendée, 1793–6, the death toll ran one or two orders of magnitude higher than in the capital. Very generally, the agrarian sector suffered more than the towns from revolutionary impact.

1815 Horizons

About the final downfall of Napoleon may one just say this. In the aftermath of Waterloo, it still seemed not impossible that troops loyal to him could prevent Wellington and Blücher of Prussia entering Paris. Then reinforced by newly trained recruits, these forces just might have checked further armies advancing from Austria and Russia Definitive judgements on these scores would probably require more specific knowledge in the realms of military psychology and administration than we shall ever now possess. In the event, France's reconstituted civil authority was unwilling to back *la gloire* again.[70]

Had Wellington lost, Napoleon's position would clearly have been stronger but perhaps not decisively so. More for sure would have been a marked lessening of Britain's ability to exert a moderating influence on the Concert of Europe, the consultative process the leading victor states turned to in pursuance of peace keeping. The Concert would have slid under the control of an archly conservative Holy Alliance — Tzarist, Hapsburg, Bourbon . . . And this would have been at a time when Europe and the wider world was in turmoil — ambitious, frustrated, apprehensive . . . For the wars had arrested, accelerated, diverted and distorted the course of History. Enlightenment visions of steady joined-up progress had been set back decades. Had those governing the continent simply ignored or suppressed the turmoil, the scene would have been set for a more massive and ugly explosion of discontent a generation down the line.

The nineteenth century ran from 1815 to 1914. If one dare characterize the "spirit of the age", one can say that until 1850 there was anxiety and confusion as just implied. Then all of a sudden this gives way to an age of optimism. Then hardly less abruptly, pessimism sets in soon after 1870.

The Hungry Forties

Hubert Lamb nominated 1850 as the most appropriate end date for the Little Ice Age in Europe.[71] British grain prices had risen steeply 1836–41, driven by bad harvests due to low temperatures brought on by vulcanism around the world. In Ireland, the wet and warmish summer of 1846 induced an explosive spread of potato blight already endemic. It turned critical the malaise resulting from poor harvests, 1839–44.[72]

Huge protests by the Anti-Corn Law League and the Chartists in Britain were followed in 1848 by uprisings in several continental cities. That year saw, too, Marx and Engels launch their *Communist Manifesto*. Ireland was now gripped by a potato famine which would decrease its population (through mortality and emigration) by a quarter by 1851.

Free Trade

Westminster's repeal of the Corn Laws, the tariffs on imported corn, owed all to the timely statesmanship of Prime Minister Robert Peel. It was to be the keystone of a general British move to trade liberalization, 1820 to 1860. Still the most advanced nation industrially, Britain herself stood to gain most from this. But other European countries closely followed suit. Since 1818, indeed, Prussia had been promoting customs unions among the German peoples so as to pave the way for political unification. However, United States strategy remained to secure with high tariffs balanced economic development (see below).

Pax Britannia?

As interpreted by such ardent advocates as Richard Cobden (1804–65) and John Bright (1811–89) of the Manchester School of British liberalism, "free trade" assumed the aura of a political religion. Supposedly all one needed to ensure boundless prosperity and lasting peace were open world markets coupled with a rejection of imperialism and any military adventures.[73]

No recourse to liberal precepts was needed to realize slavery was a blatant contradiction. Gradually it was abolished throughout Christendom at least: Argentina in 1813; the British Empire, 1833; Russia, 1861; the United States, 1865; Brazil, 1871 . . . Regarding Russia, the allusion here is to serfdom, a practice codified by Peter the Great as closely akin to enslavement. It is extremely unlikely to have been abolished as soon and peaceably as it was but for the Crimean War. Ironically, not only Bright and Cobden but — equivocally and contingently[74] — Gladstone, too, had opposed that recourse to arms.

The *bête noire* for many British liberals was Henry Palmerston: Foreign Secretary, 1930–41 and 1846–51; Prime Minister, 1855–8; and today symbolic of the Pax Britannica. But he proactively backed the emergence of constitutional liberalism in Belgium, Iberia and Italy. He worked gladly to such ends within the Concert of Europe tradition, a disposition he and Gladstone shared. He sought to avert the destabilizing consequences of Russia's waxing too strong; the Ottomans waning too precipitately; and, during the 1857 mutiny, India descending into chaos. Yet he was too impulsive; more secretive than was helpful; and too inclined to see small aggravations out of proportion. Nor did a lack of interest in reform at home lie easily alongside his wish to see it abroad.

These mid-century decades of circumscribed British hegemony were also a time of world economic boom and technical development. The first crossing of the Atlantic by unaided steam power was in 1838. Emigration to the USA from Europe was entirely by steamship by 1878.[75] Forty miles of railway were open in the United States in 1820. Come the civil war in 1861 there were 30,600. Accordingly "many sanguinary battles were fought, either to safeguard them or to gain their control; thus determining to a large extent not only the location of the main battles . . . but their eventual outcome".[76] The advent of railways meant that, on balance, the application of steam power to propulsion had narrowed the advantage of sea over land for heavy movement in war or peace.

Acute tensions in the 1840s between Britain and the USA over where the US–Canadian border should run from the Great Lakes onwards subsided abruptly with the 1846 agreement on the 49th parallel. Come the American Civil War, London was no longer attempting governmentally to intervene in the USA's internal affairs. One consideration was that the dependence of the Lancashire cotton industry on cotton imports from the South was a source of embarrassment to Manchester Liberals. With the Mexico border having been secured in 1848, the United States was free from 1865 to realize her Manifest Destiny, "sea to shining sea".

Between 1806 and 1820, a notion had waxed strong in American literature that the High Plains were a "Great American Desert" inimical to agriculture. Between 1855 and 1870, this waned.[77] In the interim, the opinion emerged that the region would be suitable as a great reservation to displace the "nomadic" Amerindians to.[78] Later this thinking gave way to the even less well-founded idea that rain "follows the plow" or, at any rate, the planting of trees.

Weakening Expansiveness

By 1880, European countries were turning again to tariff protection. In Britain and France, in particular, the predilection was for this taking the form of "imperial preference": a reversion to the mercantilist precept of shielding trade with one's colonies. In Britain this approach split the Tories apart in 1903 just as repealing the Corn Laws had in 1846.

In the USA the high Morrill general tariff passed into law in 1861 in flat disregard of the low tariff predilections of the primary-producing South. By 1864, Union tariffs averaged 47 per cent *ad valorem*. Modest reductions were effected post-war. But with the Dingley Law of 1897, the tariff average well exceeded even 1864's.

Several factors contributed to the general protectionist swing. Cycles in international trade (emergent by 1815 and typically lasting close to a decade) were becoming more synchronized worldwide, therefore more firmly defined.[79] Another consideration was that from 1900, when the American dollar finally came to be based entirely on gold rather than gold and silver, the whole world bar China had converted from the bimetallic principle once almost universal outside the British Empire. Using just the Gold Standard meant that international transactions could be effected with more ease and assurance.

On the other hand, the demonetization of silver narrowed a world currency base already lagging behind trade expansion, in the aftermath of the last big gold strikes — California in 1849 and Australia in 1851. The result was that the quite deep cyclical depression of 1873 continued to cast a shadow until the next major gold discoveries — the great reefs of the Witwatersrand in 1886 and the Klondike placer deposits in 1896. Also to reckon with was a tendency for world trade to be cast more widely.

Britain's percentage share of international trade dropped from 19 in 1880–5 to 14 three decades later.[80]

Debate continues about whether the empire Britain had built up by 1900 was a net asset. Strong arguments have been adduced to the effect that it was so in relation to resources, human and otherwise, relevant to war in 1914.[81] However, we must ask ourselves how much an imperial role may divert a nation's attention from other priorities. Some who have addressed this matter *vis-à-vis* insular and imperial Britain have felt its ruling élite was so *blasé* about technological revolution as to take it on board without making the adjustments collaterally required in education, management and social values.[82]

The fact of the matter is that members of the opinion-forming and policy-making elites cannot give their undivided attention to both realms concurrently. Not least does this truism affect the Left. During my time with the editorial team of the *New Statesman*, 1965–8, it struck me often how all of us were far more exercised and up-to-date about problems "East of Suez" than we were about those of Britain's industrial West Midlands, say. Yet this was the time when the country was acquiring a "Sick Man of Europe" image because of the way our post-war economic adjustment was faltering.

It is something to bear in mind. But in the fast shrinking world of today, it cannot be grounds conclusive for not having a proactive external policy. Besides, John Bright could never be faulted for lack of interest in home affairs. Maybe Richard Cobden, like Henry Palmerston, could.

3 A Savage Climax

Of 1914 it has been said that "the little Eurasian peninsula that was Europe . . . contained too much energy and power for the narrowness of its confines. The very process of imperial activity had . . . served the function of safety valve for the overflowing energy of Europe. There was, in 1914, no more room in the world for fresh conquests".[83]

In 1909, Robert Peary became the first to reach the North Pole[84] as, in 1912, Roald Amundsen did the South. What one may loosely term infilling had proceeded apace too. Witness that 1890 decision by the US Federal Census no longer to treat the "frontier" as a special region. In 1886, the first Canadian Pacific train ran from Montreal to Vancouver, thereby fulfilling a precondition of British Columbia's acceding to the confederation. Russian advocates of the Trans-Siberian railway had been drawing on the precedent since 1879.[85] The formation of the Commonwealth of Australia in 1901 set the seal on that continent's being well and truly spanned.

Europe's political impact on Africa evolved from peripheral to comprehensive between 1881 and 1911. In inner Asia, a British military expedition reached Lhasa in 1904. Worthy of mention, too, are Sir Aurel Stein's expeditions deep into the continent between 1900 and 1914: expeditions with geopolitics partly in mind although they brought to scholarly notice the very roots of Oriental civilization.

Communication in all its aspects was being transformed the while. A global network of cable links, initiated in 1850, was effectively complete with the laying, from 1902, of a bifurcated line from British Columbia to Queensland and to New Zealand. A pointer to a still more interactive future was Guglielmo Marconi's wireless transmission from Cornwall to Newfoundland in 1901. The first petrol-driven cars had appeared in Germany in 1885–6. Large-scale production was under way in several countries within 15 years. Henry Ford's Model-T made its debut in 1904.

Yet this pre-1914 era seems to have had less of a sense of ecological overload and resource depletion than obtained a century before or does today. For instance, little heed was taken of the warning from the celebrated Swedish chemist Svante Arrhenius (1859–1927) of the need to constrain induced greenhouse warming. Of more import, however, was a warning by the British scientist, Sir William Crookes, in 1898. He said the world could not replenish much longer the nitrogen its plants took from the soil, given the depletion rate of the Atacama nitrates. Then in Germany in 1913, the Haber–Bosch process was perfected. It effected the commercial production of ammonia from nitrogen and hydrogen. This soon became the chief source of nitrogen fertilizer.

The contribution to world shrinkage made by the internal combustion engine facilitated the two most devastating wars in history, 1914–18 and 1936–45. One speaks of "1936" because a disjointed struggle to check fascism really goes multinational with the civil war in Spain. The Second World War largely originated in Europe West of the Urals and is mainly prosecuted there. The First World War was thoroughly Eurocentric in origin and outcome. Over 10 million died during it from war-related causes, not to mention the 20–40 million worldwide who succumbed to influenza shortly afterwards. In 1936–45, the aggregate toll exceeded 60 million. The call to arms has very generally been a harder sell since 1918 and the more so since 1945.

Mediterranean Strategist

Winston Churchill held high office in both conflicts; and in both promoted a strong Mediterranean strategy. Under his ancestor John, Duke of Marlborough, England/Britain had consolidated its Mediterranean interest, notably by capturing Gibraltar in 1704.

In World War One Winston initiated the Allied attempt (February to November, 1915) to force the Turkish Straits to (a) undermine Turkey's central government in Istanbul and (b) create a supply line to Russia. The channels the great glacial overflow had suddenly created (see 2 (2)) were longish yet very narrow. The Dardanelles is 40 miles long but nowhere more than ten miles across. The Bosporous is shorter but similarly narrow.

As the early loss of three British capital ships confirmed, a battle fleet in littoral waters was in peril. Furthermore, the British and French could not hope to project and sustain ashore an army sufficient to overrun Constantinople/Istanbul, somewhere which would have been defended street by street, floor by floor . . . The Turks were in a militant mood. In 1908, their government had renewed repression of the Armenians. In febrile 1915, this escalated to massacre level.

A Sea of Opportunities

Out of World War Two, Churchill's Mediterranean vision emerges more creditably. A decisive reality in this regard was the technological backwardness of Fascist Italy as manifest in inadequate warplanes, tanks and even ships. Though Mussolini might boast of fascist Futurism and of commanding "eight million bayonets", the Italian economy used only 2.3 million tons of steel in 1937. Germany's consumed 14.9 and Britain's 10.9. Italy shared with Germany a desperate need to access the oilfields of Iraq and Iran.

In the autumn of 1940 and again in May 1941, the Prime Minister insisted tank reinforcements be directed to Egypt not Singapore. Meanwhile *c*. 250,000 Italian troops in Cyrenaica were facing 36,000 British-led in Egypt. But that autumnal rein-forcement had included 50 Matilda infantry tanks adequate to withstand Italian weaponry and grind down field works. This undercut the Italian concept for Cyrenaica's defence which had been to turn coastal villages and townlets ahead of Benghazi into strongpoints. The province duly fell to a British offensive that winter, 175,000 of the Italians being taken prisoner.

This was followed, February to May 1941, by the conquest of Italian East Africa, another initiative Churchill had pressed for. It yielded 250,000 more prisoners of war. These and other offensive successes regionally impacted on world and especially American opinion considerably. Meanwhile the defence of Malta received as much priority as possible, given its potential as an offensive base.

Questioned more is Churchill's insistence on sending troops to Greece, a country whose army was conducting a slow but sure offensive into Albania in response to Mussolini's unprovoked aggression from that quarter in October 1940. In February, the Prime Minister ordered the army victorious in Cyrenaica to advance into Libya no further. Instead it should make ready to divert troops to Greece. Athens itself was dubious about accepting in-country British forces sufficient to provoke German involvement but not to contain it.

The most critical judgement to make is how decisive the Greek distraction was in causing Hitler to delay fatefully his invasion of the USSR, Operation Barbarossa. The archives proffer no proof conclusive either way. So one must follow the actual sequence. In February 1941, British troops started to arrive in Greece. On 1 March, Bulgaria agreed to let the Wehrmacht advance to the Greek border. Then on the 27th a coup in Belgrade led straight to the renunciation of a pact with the Axis. On the 1st of April Barbarossa was postponed a month to mid-June. On the 6th a German-led blitzkrieg begins against Yugoslavia and Greece. On 1st May, the day the last British troops withdrew from the Greek mainland, Barbarossa's D-Day was finally fixed for the 22nd of June.

Liddell Hart became persuaded that the Yugoslav coup was what finally led to the Barbarossa postponement. But he did draw attention to how a severe winter had produced a late thaw with flooding and waterlogging to early June.[86] However, it is far from clear why this necessitated a solid month's delay. The Wehrmacht was to advance on the whole front, Baltic to Black Sea, whereas the watery conditions will, while they lasted, have been quite concentrated in the upper Bug/Pripet Marshes area. Less tangible is what impression the British had made on the Yugoslav coup leaders. But we know British agents were working to encourage a coup. The presence of Commonwealth troops in Greece will have been contributive too.

The next test came in Crete where 50,000 Allied troops had concentrated albeit with little armour and a very weak coverage of anti-aircraft guns. All in all, the Germans landed 22,000 troops by air or — towards the finale — by sea. The Luftwaffe presented itself in remarkable strength considering how unsuitable so much of the Balkans terrain was for airfields. As Crete fell, the Allies were able to evacuate 16,000 troops but suffered heavy naval attrition in the process. The Germans took 5,000 casualties, very largely in their one operational air division. Hitler ruled out further major airborne onslaughts, for example against Malta.

After the United States entered the war, a big issue became the Second Front, Operation Overlord. In meetings with Molotov then Roosevelt in May and June 1942 respectively, the Prime Minister articulated British anxiety about too early a launch. Yet neither he nor his staffers were explicit as to what "too early" excluded. In fact, they could not see the Allies establishing assured sea and air control over and around the English Channel before late 1943. Nor would a sufficient panoply of specialist amphibious equipment be ready until 1944.

The agreed Allied landings in North Africa therefore assumed considerable importance as, the British insisted, a proposed campaign in Italy did. The Wehrmacht had 30 divisions in France and the Low Countries in the Spring of 1944; and 23 in Italy plus near to ten in Yugoslavia. The Allies could disembark six divisions in the first seaborne lift to Normandy along with the descent by parachute and glider of three airborne divisions. It would take another week to double the total. So if a sizeable part of the Wehrmacht's strength in Southern Europe could have been diverted beforehand from there to Northern France, in the absence of an Italian campaign, they might have made an Allied descent on the Channel coast much more problematic.

In the event, the invasion of Sicily then the Italian mainland from July 1943 was very firmly an Anglo-American affair. Yet as late as the Casablanca conference that January, the US Joint Chiefs of Staff had aired the possibility of closing down the whole Mediterranean adventure.[87] But this really bespoke their unhappiness at what they divined as British, and especially Churchillian, coolness towards the Second Front, an attitude they saw as derivative from such episodes as Passchendaele and Dunkirk. It is hard to believe such memories had counted for nothing or that they had not been revivified by the "excellence" of German defensive warfare in North Africa.[88] All the same, it was not unreasonable to bear in mind that Overlord might fail. Several studies published in 1994 considered how different German (and especially Hitlerian) command decisions might have effected this consummation.[89] Besides which, a critical contribution was made by the Allied meteorologists within a run of stormy weather unprecedented for June. Their D-Day forecasts represented a brilliantly successful blend of American sanguineness and British caution.[90]

An Ulterior Motive?

The other dimension in this whole argument is whether a proactive Mediterranean strategy could have done anything to curb Soviet aggrandizement in the Balkans. In the summer of 1943, Churchill threw British support for resistance in Yugoslavia firmly behind the Partisans under Josef Tito. He did so because of their ruthless effectiveness in the conflict with the Nazi New Order in Europe, a conflict which continentally still looked evenly balanced. But the resultant materials and personnel, however welcome, did nothing at that stage to deflect Tito from his adherence to Stalinist Communism.

This meant the Trieste peninsula was poised to become the subject of territorial dispute. It also meant that Tito would not condone any transit by the British or American armies through Yugoslav territory. Therefore Churchill had to give up on the hope he had come to entertain of an Allied landing near Trieste followed by a march to Vienna to forestall the Red Army. The Trieste problem apart, the terrain to be negotiated by-passing Yugoslavia was simply too tough going.

It was not clear, in any case, what impact the capture of Vienna, even if feasible, would have had on developments across the Balkans at large. Towards the end of 1944, however, Churchill did make one move aimed at defining the limits of Soviet imperialism. He ordered the dispatch of British (including Indian) troops to confront the Communists in Greece. (See 9 (1)).

Midway and Stalingrad

The Battle of Midway in June 1942 does not compare with Stalingrad in the scale of commitment and suffering. The death toll was a few thousand, not one or two million. But had things gone otherwise, Japan would likely have seized and fortified the Hawaiian archipelago. Her imperium would have encompassed major centres of oil, rubber and tin production. It would still have been vulnerable to the USA's sea power (not least her submarines) but more in the longer term.

Had the Wehrmacht triumphed at Stalingrad (or at the huge tank-led Battle of Kursk that Spring), it might well have consolidated a flanking defence from Voronezh or Saratov to Stalingrad and Astrakhan, a line the remaining Red Army might have found hard to penetrate, not least because of weak infrastructure to its own rearward.

Given either or both of these eventualities or, indeed, a collapse of Overlord, a "1984" triadic stand-off would likely have developed. In a consequent rush for Weapons of Mass Destruction (WMD), North America would probably have been first with nuclear fission, then thermonuclear warheads as well as with intercontinental bombers for their delivery. But the Germans and Japanese could assuredly develop, separately or together, other WMD — meaning, poison gases and bio bombs.

Ultimately the test would have been whether, in the context of such a gridlock, the spirit of constitutional liberalism waned before that of fascism did. Optimism on that score does not come easily. In short, the Eurocentric era would have ended in the nastiest way imaginable.

Notes

1 R. A. Beck, "Climate, Liberalism and Intolerance", *Weather* 48, 1 (February 1993): 63–4.
2 Henry Vignaud, *American Historical Review* XVIII, 3 (April 1913): 505–9.
3 Elena Lourie, "A Society Organised for War, Medieval Spain", *Past and Present* 35 (December 1966): 54–76.
4 Jaime Vicens Vives (Frances López-Morillas, trans.), *An Economic History of Spain* (Princeton: Princeton University Press, 1969), pp. 108–10.
5 Oliver Leaman, *Averroës and his Philosophy* (Oxford: Clarendon Press, 1968), p. 5.
6 John Herman Randall, "Scientific Method in the School of Padua", *Journal of the History of Ideas*, 1 (1940): 177–206.
7 Philip Jones, *The Italian City State. From Commune to Signoria* (Oxford: Clarendon Press, 1997), pp. 255–6.
8 Wallace K. Fergusson, "Humanist Views of the Renaissance", *American Historical Review* XLV, 1 (October 1939): 1–28.
9 Map of "The Diffusion of Printing" in David Rundle (ed.), *The Hutchinson Encyclopedia of the Renaissance* (Oxford: Helicon, 1999), p. 338.
10 Brian Goodey, "Mapping Utopia", *The Geographical Review* LX, 1 (January 1970): 15–30.

11 Charles Homer Haskins, *The Renaissance of the 12th Century* (Cambridge, MA: Harvard University Press, 1927), p. 11.

12 R. I. Moore, "Literacy and the Making of Heresy, *c.* 1000–1150" in Lester K. Little and Barbara H. Rosenwein (eds.), *Debating the Middle Ages* (Oxford: Blackwell, 1998), Chapter 21.

13 E. W. Webster (trans.), *The Works of Aristotle* (11 Vols.) *Meteorologica* (Vol. III) (Oxford: Clarendon Press, 1908), Book 1, Section 14.

14 *Ibid.*, Section 10.

15 *HCC*, pp. 232–3.

16 Astrik L. Gabriel in Francis Lee Utley (ed.), *The Forward Movement of the 14th Century* (Columbus, Ohio State University, 1961), p. 82.

17 *Ibid.*, p. 108.

18 Joseph R. Strayer, *On the Medieval Origins of the Modern State* (Princeton: Princeton University Press, 1980), p. 56.

19 Dauvit Brown, "When did Scotland become Scotland?", *History Today* 46, 10 (October 1996): 16–21.

20 Emmanuel Le Roy Ladurie (Siàn and Ben Reynolds, trans.), *The Mind and Method of the Historian* (Brighton: Harvester Press, 1981), p. 274.

21 HCC, p. 98.

22 B. J. Graham in B. J. Graham and L. J. Proudfoot (ed.), *An Historical Geography of Ireland* (London: Academic Press, 1993, Chapter 2.

23 D. M. Bueno de Mesquita in J. R. Hale, J. R. L. Highfield, B. Smalley (eds.), *Europe in the Later Middle Ages* (London: Faber and Faber, 1965), Chapter X.

24 Denis Cosgrove "The Myths and Stones of Venice: An Historical Geography of a Symbolic Landscape", *Journal of Historical Geography* 8, 2 (April 1982): 145–69.

25 Margaret Mann Phillips, *Erasmus and the Northern Renaissance* (London: Hodder and Stoughton, 1969), pp. 33–4.

26 John Ruskin, *The Stones of Venice* (Orpington: George Allen, 1885), p. 8.

27 Johan Huizinga (James S. Holmes and Hans van Marle, trans.), *Men and Ideas* (London: Eyre and Spottiswoode, 1960), p. 271.

28 J. B. Bury, *The Idea of Progress* (London: Macmillan, 1920), Chapter 2.

29 John H. Pryor, *Geography, Technology and War Studies in the Maritime History of the Mediterranean* (Cambridge: Cambridge University Press, 1992), p. 191.

30 Robert S. Lopez, "Hard Times and Investment in Culture" in Karl H. Dunnenfeldt (ed.), *The Renaissance, Medieval or Modern?* (Boston: D. C. Heath, 1959), pp. 50–61.

31 Steven Runciman, "The Diffusion of Greek Culture", *The Geographical Magazine* XIX, 11 (July 1946): 494–501.

32 Pamela Nightingale, "England and the European Depression of the mid-15th century", *Journal of European Economic History* 26, 3 (Winter 1997): 631–51.

33 J. R. S. Phillips, "Geography in the 15th Century" in *The Medieval Expansion of Europe* (Oxford: Oxford University Press, 1988), Chapter 11.

34 Frederic C. Lane, "The Economic Meaning of the Compass", *American Historical Review* LXVIII, 8 (April 1969): 605–17.

35 J. H. Parry, *The Age of Reconnaissance* (London: Weidenfeld and Nicolson, 1963), Plate 47.

36 J. E. Hale, "Gunpowder and the Renaissance: an essay in the history of ideas" in Charles H. Carter (ed.), *From the Renaissance to the Counter-Reformation* (London: Jonathan Cape, 1966), pp. 113–44.

37 Parry, *The Age of Reconnaissance*, Chapter 7.

38 Jared Diamond, *Guns, Germs and Steel* (London: Jonathan Cape, 1997), p. 74.

39 Alfred W. Crosby, *Ecological Imperialism. The Biological Expansion of Europe, 900–1900* (Cambridge: Cambridge University Press, 1986), Chapter 14.

40 Marc U. Edwards, *Printing, Propaganda and Martin Luther* (Berkeley: University of California Press, 1994), p. 1.

41 R. W. Scribner, "Why was there no Reformation in Cologne?", *Bulletins of the Institute of Historical Research* XLIX (1976): 217–41.

42 David E. Sopher, *The Geography of Religions* (Englewood Cliffs: Prentice-Hall, 1967), p. 41.

43 For example, Fernand Braudel (Sian Reynolds, trans.), *The Mediterranean and Mediterranean World in the Age of Philip II* (London: Collins, 1972), Vol. II, Chapter IV.

44 De Lamar Jensen, "Franco-Spanish Diplomacy and the Armada" in Charles H. Carter (ed.), *From the Renaissance to the Counter-Reformation* (London: Jonathan Cape, 1966), pp. 205–22.

45 Paul Kennedy, *The Rise and Fall of the Great Powers* (London: Unwin Hyman, 1988), Map 4.

46 H. R. Trevor Roper, *The European Witch Craze of the 16th and 17th Centuries* (Harmondsworth: Penguin, 1969), p. 112.

47 For fuller consideration of the impact of all the new learning, see *EC*, pp. 45–54.

48 Pieter Geyl, *A History of the Low Countries, Episodes and Problems* (London: Macmillan, 1964), Chapter 1.

49 Alan Shepperd, "The Thirty Years' War" in James Lawford (ed.), *The Cavalry* (London: Roxby, 1976), pp. 73–83.

50 Geoffrey Parker, "The Military Revolution, 1560–1660, a myth", *Journal of Modern History* 48, 2 (June 1976):195–214.

51 Charles P. Kindleberger, *A Financial History of Western Europe* (Oxford: Oxford University Press, 1993), Fig. 2.1.

52 Stanislas Hoszowski, "Central Europe and the 16th- to 17th-century Price Revolution", in Peter Burke (ed.), *Economy and Society in Early Modern Europe. Essays from Annales* (London: Routledge, 1972), pp. 85–103.

53 Angus Gowland, "The Problem of Early Modern Melancholy", *Past and Present* 191 (May 2006): 77–120.

54 *Oceanus* 29, 4 (Winter 1986/7): 38.

55 E. J. Hobsbawn in Trevor Aston (ed.), *Crisis in Europe, 1560–1660* (London: Routledge and Kegan Paul, 1965), Chapter 2.

56 *Past and Present* 18 (November 1960): 25.

57 Roland Mousnier (Brian Pearce, trans.), *Peasant Uprisings in 17th-century France, Russia and China* (New York: Harper and Row, 1970), pp. xvii, xx, 312 and 331–2.

58 Zhaodong Feng *et al.*, "Temporal and Spatial Variations of Climate in China during the last 10,000 years", *Holocene* 3, 2 (1993): 174–80.

59 C. R. Boxer, *The Dutch Seaborne Empire, 1600–1800* (London: Hutchinson, 1965), pp. 104–5.

60 C. H. Wilson, "The Economic Decline of the Netherlands", *Economic History Review* IX, 2 (May 1939):111–27.

61 Paul Kennedy, *The Rise and Fall of the Great Powers* (London: Unwin Hyman, 1988), Table 2.

62 *Ibid.*, p. 98.

63 D. A. Farnie, "The Commercial Empire of the Atlantic, 1607–1763", *Economic History Review* XV, 2 (December 1962): 203–17.

64 A. R. Ubbelohde, *Man and Energy* (Harmondsworth: Pelican, 1963), pp. 25–7.

65 Carlo Cipolla, *The Economic History of World Population* (Harmondsworth: Penguin, 1964), Table 4A.

66 For example, Carlo Cipolla, "The Decline of Italy. The Case of a Fully Matured Economy", *Economic History Review* V, 1 (1952): 178–87.

67 N. J. G. Pounds, *An Historical and Political Geography of Europe* (London: George G. Harrap, 1947), Fig. 6.7.

68 Florin Aftalion (Martin Thom, trans.), *The French Revolution. An Economic Interpretation* (Cambridge: Cambridge University Press, 1990), p. 191.

69 J. U. Nef, *War and Human Progress* (London: Routledge and Kegan Paul, 1950), p. 328.

70 David Chandler, *The Campaigns of Napoleon* (London: Weidenfeld and Nicolson, 1966): Chapter 93.

71 H. H. Lamb, *The Changing Climate* (London: Methuen, 1966), pp. 65, 74–5.

72 C. O. Grada, "Irish Agricultural History: some recent research", *Agricultural History Review* 38, 2 (1990): 165–73.

73 Ian Bradley, *The Optimists* (London: Faber and Faber, 1980), Chapter 5.

74 William Partridge, *Gladstone* (London: Routledge, 2003), pp. 76–83.

75 Raymond L. Cohn, "The Transition from Sail to Steam in Immigration to the United States", *Journal of Economic History* 65, 2 (June 2005): 469–95.

76 Ernest F. Carter, *Railways in Wartime* (London: Frederick Muller, 1964), p. 24.

77 Ralph H. Brown, *Historical Geography of the United States* (New York: Harcourt Brace, 1948), pp. 370–1.

78 Stephanie Lemenager, *Manifest and Other Destinies* (Lincoln: University of Nebraska, 2004), pp. 50–3.

79 William Ashworth, *The International Economy, 1850–1950* (London: Longmans, 1952), pp. 133 and 183–4.

80 *Ibid.*, p. 166.

81 Avner Offer, "The British Empire, 1810–1914: a waste of money?", *Economic History Review* 2 (1993): 215–38.

82 Martin J. Wiener, *English Culture and the Decline of the Industrial Spirit, 1850–1980* (Cambridge: Cambridge University Press, 1981), Chapter 2.

83 René Albrecht-Carrie, *The Meaning of the First World War* (Englewood Cliffs: Prentice-Hall, 1965), pp. 42–3.

84 Peary's claim was well corroborated by the 2005 Anglo-Canadian expedition. See George Wells, *St Edward's Chronicle* (Autumn 2007): 57–61.

85 J. L. Black, "The Canadian Pacific Railway as a Model for the Trans-Siberian Railway", *Sibirica* 4, 2 (October 2004): 196–200.

86 B. H. Liddell Hart, *History of the Second World War* (London: Cassell, 1970), pp. 132–5.

87 *Ibid.*, p. 438.

88 Michael Howard, *The Mediterranean Strategy in the Second World War* (London: Weidenfeld and Nicolson, 1968), pp. 45–6.

89 For example, Williamson Murray, "Could We Have Lost D-Day?", *Quarterly Journal of Military History* 6, 3 (Spring 1994): 6–21.

90 Neville Brown, "The Weather for Overlord", *Journal of Meteorology* 19, 189 (May/June 1994): 141–9.

5 | Cold War Origins

It has been proposed that, during the descent into World War Two, Geography worked more to Hitler's advantage than was then appreciated. What should now be considered is the part it played in the slide into Cold War. The nub is how Stalin defined *les limites naturelles* of superpower hegemony. Manifestly, his basic criterion was the Kremlin's ability to retain overriding control over every regime thus embraced. In the interwar period, he had conspicuously failed to sustain Soviet hegemony over China and later Spain, the former being too big and the latter too distant.

Since when, new factors had entered the equation. One was the destruction and death inflicted on the USSR by total war. For some time, census studies have put war-related fatalities at close to 30 million, a fifth of the 1941 population.[1] Furthermore, in 1946 the malaise engendered was aggravated by acute drought extending from Moldavia through the Black Earth region to the Volga. Admittedly, the grain deliveries officially recorded fell only an eighth compared with 1945. But assuming this record be accurate, it is explicable in terms of peasants retaining little for local consumption. Famine and migration were therefore widespread.[2]

Anyone still reluctant to locate ecological stress and armed strife within the same syndrome should read Mikhail Gorbachev on his boyhood in Stavropol in the north Caucasus. The family cabin had an earthen floor. In winter a calf or two slept on that floor while humans slumbered on the big Russian stove. In 1933, half the village starved to death in the struggle against collectivization. Then came the Stalinist purges, his own grandparents being held awhile on concocted charges. Next came the war, his father at the front and a harrowing spell of Nazi occupation. Finally, the 1946 harvest virtually collapsed. That situation had been aggravated by "no care" being "given to the earth" because of "precedence to heavy industry which worked mainly for armaments manufacture".[3]

Another disturbing circumstance was the nuclear bomb. Stalin's show of indifference at Potsdam to the successful Los Alamos test proves nothing since agents had forewarned him. Nor need one read overmuch into his 1946 averration that the "atomic bomb" was a mere terror weapon or, as Mao Tse-tung put it, "a paper tiger". In fact, news of the Hiroshima bomb was greeted in Moscow with dismay. For it showed the USSR had still to achieve superpower parity despite huge sacrifice.[4] In

1946, the Soviet Foreign Minister, Viacheslav Molotov, berated the Tito regime for shooting down two planes belonging to the American nuclear monopolists.

Meanwhile, the Kremlin pressed ahead with its own bomb. At the war's end, informed American estimates put the first Soviet test at between two and 20 years away. In the event, detonation occurred in September 1949. It was of a plutonium bomb. A Briton temporarily turned traitor — the distinguished physicist, Klaus Fuchs — had handed over an invaluable plutonium design in 1945.[5]

Awareness that nuclear weapons were an ominous departure did not mean they figured forthwith in Soviet military doctrine. In 1946, Stalin suggested to Milovan Djilas (soon to emerge prominently as a dissident Yugoslav moderate, anti-Tito and anti-Stalin) that lavish investment in a Volga-Don canal was warranted because the Black Sea fleet could retire to cover Stalingrad in the event of further hostilities.[6] That proposition was operationally archaic and in no way doctrinal. But it pointed to the kind of doctrine Stalin might have been comfortable with.

Yet the USA, too, was slow to make nuclear capability a theme organic to military science. Till 1950, there was little bar an emphasis on striking the "vital first blow",[7] an emphasis which in retrospect seems misplaced in relation to the geography of intercontinental air war and, prospectively, the arithmetic of nuclear stockpiles.

Zhdanov[8]

A broader interpretation of Soviet motivations may hinge on the fortunes of Andrei Zhdanov, First Secretary of the Leningrad Communist Party (1923–44). In mid-1946 he returned to the Kremlin's inner circle after two or three years being sidelined. His allotted task was to intensify the drive for ideological purity and contrived enthusiasm. Thus the Society for the Dissemination of Political and Scientific Knowledge was constrained to adopt a charter committing it "to inculcate national pride, to combat obsequiousness to foreign culture . . . and the falsity of bourgeois democracy".

An infamous beneficiary from this pitch was the biologist Trofim Lysenko. He asserted that environment can induce within living individuals changes which can forthwith be transmitted to their progeny. The great majority of Western specialists dismissed this as charlatan. The post-Stalin Soviet scientific establishment broke Lysenko between 1956 and 1965.

Among other Stalin/Zhdanov/Lysenko absurdities had been the planting, from 1948, of belts of trees hundreds of miles long. The stated aims were to speed up the water cycle while checking soil erosion, aims realisable only most narrowly.

Zhdanov has been seen as the architect of a forward foreign policy. A sounder interpretation might be that he sought consolidation abroad just as at home. He was quite the most active Soviet delegate at the launch, in September 1947, of the Communist Information Bureau, alias the Cominform. Its declared purpose was full data exchanges between the Communist parties of the USSR, Eastern Europe, France and Italy. It represented a consolidation to be assessed in relation to the Communist withdrawal from coalition governments in Belgium, France and Italy; and to the July crisis over the Marshall Plan (see below).

Come the year's end, there were signs that Zhdanov was again out of favour with Stalin, a renewed lapse from grace accentuated by the Belgrade–Moscow split which climaxed early in 1948. The leading role he was accorded in the second Cominform plenary that June was probably intended to oblige him to denounce his erstwhile friends, the Yugoslavs. He died, ostensibly from heart trouble, in August. During 1949, ideological fervour somewhat abated awhile. Meantime hundreds of people were purged from official and party posts in Leningrad.

Poland Focal

Clearly Stalin had sought at the outset a military stake in Europe over and beyond a due share in the occupation of Germany and Austria. Witness the Soviets' studied indifference to the Warsaw rising of Poland's Home Army in August 1944 which began soon after Red Army troops reached the Vistula there. Answerable to the government in exile in London, the Home Army was the country's biggest resistance movement.

Unfortunately, continual goading by Nazi terror had made it prone to phantasize. In January 1943, Gro-Rowecki, the veteran soldier still in command till July, earnestly pressed the London Poles to have the Polish airborne brigade drop in country, a request reiterated in July 1944.[9] An aerial projection that far would have been parlous initially, then unsustainable yet irreversible, and very hard to join up with Home Army deployments.

In July 1943, Polish *angst* had been compounded by the Germans discovering in Katyn forest near Smolensk the bodies of thousands of Polish officers Stalin had murdered in 1939–40. This made dialogue with Moscow a doubly unwelcome prospect as did again the USSR's insistence on the Curzon line being its western boundary. Yet neither Polish rejectionism nor stiffened Wehrmacht resistance nor, indeed, imperfect Home Army command-and-control could excuse the Soviets doing not a thing to assist a rising their own broadcasts had called for. Evidently Stalin was looking to what perforce transpired, the Home Army rising up and being crushed. Two hundred thousand Poles died, fighting midst rubble and often from sewers. **(See 8)**.

Uranium Access

No doubt the Soviets felt a strong presence within Europe could enable them to hold the whole region hostage should they ever be faced with American nuclear suasion. However, a more particular aim will have been to bolster supplies of uranium ore. Bohemian pitchblende had been *the* source of radium and uranium earlier in the century and still counted. But the USSR also began prospecting in its Central Asian republics. So in 1950 the CIA guestimate was that a third of Soviet uranium supplies came from within the USSR and the rest from Eastern Europe, especially East Germany and Czechoslovakia.[10]

Within three years of Stalin's death in 1953, most of the quasi-colonial binational enterprises the other Eastern European governments had been required to form with the Kremlin had been abolished. The main exceptions were those created to operate uranium mines in Romania, Bulgaria and East Germany.

Germany and Czechoslovakia

The limits of Moscow's new imperium were determined partly in the light of the

West's reaction though more in relation to what Stalin himself deemed manageable. This determination was made manifest between 1946 and 1948.

Germany was a test case. Whether Stalin ever wanted to encompass this power house is doubtful. In any case, the possibility receded. A referendum held in March 1946 among Social Democrat (SPD) cardholders in West Germany and West Berlin decisively rejected any fusion with the Communists (SED). Then that autumn elections were held for Berlin's city assembly. On a 92 per cent turn out, half the votes went to the SPD and a mere fifth to the SED, blatant intimidation by the Communists notwithstanding. Meanwhile on 6 September James Byrnes, US Secretary of State, publicly intimated that the US military presence in South Germany was henceforward to keep the Soviets out rather than the Germans in. He also announced the fusion of the American and British zones. This "Bizonia" marked the inception of the Federal Republic to be.

Nevertheless, 1947 was a year of deep crisis, not least in West Germany. The winter, January to March, was regionally the worst since the "hungry forties" of the previous century. It was followed by a summer drought that, in France for instance, caused bread grain yields to fall 30 per cent in what was rated the worst crop year on record.[11] Come October, a basic ration scale of 1550 calories a day was not being reached everywhere in Bizonia. Meantime, politics in France and Italy waxed more bitter.

That summer, General George Marshall, by then Secretary of State, launched the European Recovery Program — alias the Marshall Plan. Suffice for now to note that the pumping in of American and Canadian economic aid helped stave off the "spectre of Communism" *à la* Marx's *Communist Manifesto* of 1848. In Italy, the "Popular Front" hard Left did less well than originally expected in the April 1948 elections. In France, industrial militancy subsided as the trade unions splintered and contracted.

Meanwhile, the cohesion of the "Soviet bloc" was being underpinned by economic and social reconstitution along Marxian lines. This would have made it hard to accede to the Marshall Plan even had Moscow been indulgent. In fact in July 1947, the Czech government pulled out of the first Plan conference hard upon strong Soviet representations.

The Moravian Communist leader, Klement Gottwald, had formed a coalition government in 1946 after what were accepted as free and secret Czechoslovak elections. By early 1947, however, the Communists had switched to a campaign of forceful intimidation, a campaign the Marshall Plan démarche clearly reinforced. It culminated in a veritable coup early in 1948.[12] Czechoslovakia will have been brought firmly inside the bloc for its uranium yet also because it otherwise looked on the map like a stiletto thrust straight into the Kremlin's new rimland.

The South Slavs

Against this *coup de main* can be set a virtually concurrent break between Yugoslavia and the USSR. The course of that dispute was in its turn interactive with the armed struggle in Greece. In December 1944, British troops were deployed against an uprising in Athens as the Communist resistance to Axis occupation redirected itself. This intervention was decidedly unpopular with the Roosevelt administration *inter alia* yet, as Winston Churchill observed, drew no criticism from *Pravda* or *Izvestia*, the two main Soviet newspapers.[13]

Pro tem insurrection was quelled but in 1946 the Communists resumed hostilities, now in the bleak northern hills where cross-border support could come from Albania, Yugoslavia and Bulgaria. The backdrop was that Tito in Belgrade and his Sofia counterpart, the redoubtable Georgi Dimitrov, were looking towards a communist South Slav federation, perhaps with its kingpin a united Macedonia carved out of Yugoslavia, Bulgaria and Greece. In the inter-war era, the IMRO Macedonian insurgents had fought fiercely with the governments in Athens, Belgrade and Sofia, not least (to 1923) that of Bulgaria's reformist Peasant-Party. The sanguine hope now was that a federal entity could accommodate Macedonian aspirations but also face down pressures from Moscow and the West. Dimitrov never broke with Stalin as overtly as did Tito. After all, his country had no Mediterranean coastline. Nevertheless, there were a succession of quarrels between Sofia and Moscow, two capitals sometimes on very fraternal terms historically speaking. Initially a big bone of contention was Stalin's hostility to the Greek Communist insurrection. Dimitrov himself died, ostensibly from natural causes, while on sick leave in Moscow in July 1949, just as the Greek civil war was ending. (See 9 (1)).

Iran

About Iran, Mackinder had been contradictory in 1919. His readers were told Alexander the Great's Persian campaign "was really based on sea power" (p. 117). Yet they were also told Iran was one of the Heartland's "western regions" (p. 334).[14] In August 1941, the Soviets and British occupied the country, their underlying aim being to open an additional supply route to the USSR. Their occupation divided the land on a North-West versus South basis, an arrangement broadly reminiscent of the agreed carving out of spheres of influence in 1907.[15] The Tudeh party (i.e. the Iranian hard Left) waxed strong in the north-west during those wartime years. So in November 1945, the Democrats (as the local Tudeh party styled themselves) rose in force in Azerbaijan, proclaiming its autonomy within the Persian state. Its capital, Tabriz, fell on 16 December after the Soviet army had blocked government attempts to reinforce. However, the insurgents were fatally compromised by Moscow's abruptly announcing in March 1946 that its troops were leaving Iran altogether.

China

One could have assumed that the Chinese Communists' switching their focus to Yenan with the Long March (1934–5) had been to consolidate a rapport with the Soviets. Yet one would have been wrong. In 1936, a Kuomintang (KMT) general with Maoist leanings kidnapped Chiang Kai-Shek in order to extract from him certain guarantees. Stalin scoffed at this "Japanese plot".[16] Subsequent Stalinist policy was more or less consistent with this pitch. In 1938 five flights of Soviet aircraft (with Soviet pilots) were flying alongside the KMT Nationalists.[17]

In the absence of solid Soviet backing, the Maoist prospects would depend the more on how the balance of advantage in-country between them and the Kuomintang would be effected by a conflict with Tokyo which, in July 1937, became full-scale war. Soon Japan seized much of the alluvial plain of Eastern China, a phase which culminated that December with 40,000 murders and countless other atrocities during the infamous "Rape of Nanjing", Chiang's erstwhile capital. In response, the Kuomintang adopted a strategy of "scorched earth" linked to evacuation westwards. One subset of this strategy was breaching the dykes of the Hwang-ho causing it to reach the sea south of the Shantung peninsula instead of to its north. In the consequent inundations, 900,000 Chinese died while two million more were made homeless. Alas, there is no sign this delayed the Japanese more than several weeks.

Yet in other ways the retreat westwards was epic. A key target area was the "red basin" of Szechwan, a feature which fits into the regional tectonics acutely: "No other part of China is so protected and enclosed by Nature".[18] The exit to eastwards is via the Yangtse gorges.

The one-time treaty port, Chungking became the KMT wartime capital. But that accolade was a daunting measure of the adaptation required in a province of 20 million people then ranked as China's most backward. The foot-binding of girls had died out there but 20 years before. A tenth of the crop area was given over to opium. Warlords and secret societies held much sway.

So government had to be firmly *dirigiste*. Opium production was suppressed. An ordnance industry was developed. Trunk roads and airfields were built, the latter providing for United States Army Air Force (USAAF) deployments to proffer air defence by 1943 and bomb Japan by 1944.[19] The latter year Szechuan withstood a heavy Japanese ground offensive. Chungking was closely threatened but never entered.

Not that Chinese resistance nationwide was without its ambiguities. Provincial governments sometimes made accommodations with the occupation forces. Moreover, the Japanese attempted to generate puppet regimes as they lately had in Manchuria. The first, Mangjiang, was established as early as November 1937, it supposedly being a Mongolian confederation. Otherwise such contrivances were more to the South and East.

More generally, there was often across that vast country a "live and let live" disposition with the Japanese troops pulling back into cities and airfields, emerging in strength only when punitive action was undertaken. The Maoists in particular still lacked the heavy weaponry storming walled cities called for; and, indeed, were desperately short of weapons and ammunition overall. Then again, Yenan, as Mao's capital, was throughout the war "a symbol of hope, a beacon for progressive-minded young Chinese and Western sympathizers alike".[20] Yet its vivacity was despite the main war zone (i.e. the limits of Japanese-held cities and airfields) being very close by for much of the time. Had Tokyo so resolved, its aircraft could have bombed Yenan to smithereens, driving all the Communist cadres into the cave villages in the arid loess hills.

Nevertheless, sympathetic Western progressives have long insisted that the Communists assumed a salient role in the resistance. Jack Belden was reporting from China for a decade to 1942 and again from 1946. He later stressed how strong Maoism became, militarily and politically, in Shansi–Hopei–Honan–Shantung, a region well to eastward where alluvial plains and mountains in sharp juxtaposition afforded varied cover and diverse resources. He also averred how the Maoists regularly sabotaged vital railways, their defensive ditches and forts notwithstanding. More tellingly, he recalled the "Three Alls" punitory strategy the Japanese felt constrained to apply in North China from the winter of 1941–2. To wit, "Kill All; Burn All; Loot All".[21] It is said that 60,000 Chinese were killed in the hunt for American airmen who had crash-landed after the Doolittle raid on Tokyo in April 1942.

This Sino-Japanese war was plenty tough enough on all sides despite the qualifications expressed above. An authoritative Japanese source conservatively puts the killed in action at 1,300,000 Chinese (Kuomintang and Communist) and 500,000 Japanese.[22] Other Chinese war-related deaths were a good ten million more. Then in terms of political sociology, the most radical outcome was the dissolution of the social structure of North China, the creation of a vacuum of identity which was ideal for Communist progress. Perhaps no less impacting was that 1941 to 1943 widely saw the droughtiest time for a century.

On 14 August 1945, the USSR and Kuomintang China signed a treaty of alliance. Its provisions included continued Soviet access to Manchurian transport facilities on terms too neo-colonial for any Chinese Communist to live with comfortably. However brave a face the Maoists put on things, it was clear the Kremlin still sought to treat them as upstarts.

Notes

1 Susan J. Linz in Linz (ed.), *The Impact of World War Two on the Soviet Union* (Totowa: Rowan and Allansheld, 1985), Chapter 1.
2 Sheila Fitzpatrick in Linz (ed.), *The Impact of World War Two on the Soviet Union*, Chapter 8.
3 Mikhail Gorbachev, "My route to Green", *New Scientist*, 190, 2550 (13 May 2006): p. 51.
4 David Holloway, *Stalin and the Bomb* (New Haven: Yale University Press, 1994), p. 127.
5 *Ibid.*, p. 196.
6 Milovan Djilas (Michael B. Petrovich, trans.), *Conversations with Stalin* (London: Rupert Hart-Davis, 1962), p. 141.
7 Lawrence Freedman, *The Evolution of Nuclear Strategy* (Basingstoke: Macmillan, 1981), Chapter 1.
8 The themes addressed in this section are explored more fully in D. C. Watt, Frank Spencer and Neville Brown, *A History of the World in the Twentieth Century* (New York, William Morrow, 1968), Part 3, Chapter 3.
9 Jan M. Ciechanowski, *The Warsaw Rising of 1944* (Cambridge: Cambridge University Press, 1974), p. 139.
10 Holloway, *Stalin and the Bomb*, p. 177.
11 Henry Bayard Price, *The Marshall Plan and its Meaning* (Ithaca: Cornell University Press, 1955), pp. 30–1.
12 Hugh Seton-Watson, *Neither War nor Peace* (London: Methuen, 1960), pp. 202–5.
13 Winston Churchill, *The Second World War* (London: Cassell, 1959) (abridged version), Chapter XX1.
14 Mackinder, *Democratic Ideals and Reality* (London: Constable, 1919).
15 N. J. G. Pounds, *An Historical and Political Geography of Europe* (London: George C. Harrap, 1947), Fig. 132.
16 Philip Short, *Mao* (London: Hodder and Stoughton, 1999), p. 349.
17 John Erickson, *The Soviet High Command* (London: Macmillan, 1963), pp. 390 and 400.
18 Albert Kolb (C. A. M. Sym, trans.), *East Asia, Geography of a Cultural Region* (London: Methuen, 1971), p. 272.
19 H. J. Richardson, "Szechwan during the war", *Geographical Journal*, CVI, 1–2 (July–August 1948): 1–25.
20 Short, *Mao*, p. 534.
21 Jack Belden, *China Shakes the World* (Harmondsworth: Penguin, 1973), Chapter III, pp. 11 and 12.
22 *Kodansha Encyclopedia of Japan* (Tokyo: Kodansha, 1983), pp. 199–202.

6 | Emergent Influences

1 War in Three Dimensions, 1908–1980

The air is the domain that initially owed everything to the internal combustion engine. In 1888 came the first motorized airship. Fifteen years later, the Wright brothers famously effected powered flight by heavier-than-air machines, albeit for under a minute at a height of several yards. By 1913, altitudes of 15,000 feet and ranges of several hundred miles were attained. No fewer than 100,000 warplanes had been built by 1918.

Apparently the term "air power" was coined by H. G. Wells in 1908 in his novel *The War in the Air*, a tale about interhemispherical conflict between air fleets comprised of not aeroplanes but airships. This depiction bespoke a confusion Wells never resolved about the status of the warplane as such.[1] Seen just as a doctrinal exercise, however, it does foresee control of the air proffering scope for independent and decisive strategic action. It thus anticipates the seminal Smuts report of 1917. This warned that "the day may not be far off when aerial operations ... on a vast scale become the principal operations of war". Out of these findings was born the Royal Air Force (RAF).

Nevertheless, the first decisive use of the RAF related to an alternative perspective, that which has seen air power as closely complementary to mobile surface action. A future Chief of Air Staff noted how, as the formidable German offensive of March 1918 gathered momentum, "there took place one of the most bitter and intensive rearguard actions ever fought by British aircraft, with results the importance of which it is difficult to overstate and at a terrible cost in casualties". Over two days a breakthrough near Bapaume was averted "by a concentration of every available aircraft on low-flying action".[2] The losses were largely due to a German heavy machine gun designed to engage armour also, *Tank und Flieger*.

F. W. Lanchester was a British engineer remarkable not only as a pioneer in aerodynamics but also as an apostle of air warfare viewed in the round. In 1916, he argued that aerial interdiction of a rearward transportation network could create an impassable zone 100 to 200 miles deep, out of which might come a "bloodless victory".[3] This utopian expectation notwithstanding, he insisted military aviation should be dubbed the "fourth arm": one standing alongside the infantry, cavalry and artillery, not over against the Army or Navy as such. For "command of the air can never be taken to carry a meaning so wide" as "command of the sea".[4]

A signal exercise in deductive logic within the 1916 study was what has come down to us as the Lanchester square law: "the fighting strength of a force may be broadly . . . proportional to the square of its numerical strength" multiplied by the capabilities of its units. If side A is twice as strong as side B, the former can concentrate its fire twice as intensively while enduring return fire half as intensively. Attrition rates percentage wise therefore contrast even at the outset by a factor of four.

The influence of this law on events will be at its most clear-cut when the opposing forces freely access one another — i.e. in the air, on the high seas or on flat, open land-scapes. Though Lanchester was primarily concerned with contests between air fleets, he demonstrated his hypothesis via a subtle interpretation of Nelson's battle plan at Trafalgar. Where, as in most land campaigns, the constraints Geography imposes are tighter and more diverse, the relationship between relative numbers and effective "fighting strengths" tends to be towards linear.[5]

Some have carried a critique further, perceiving a disposition to see damage inflic-tion merely in terms of physical attrition — demoralization and disorganization being discounted.[6] Besides, whichever formula is used, linear or square, attrition gaps are very liable to widen as battle progresses. Still, a lot depends on leadership attributes, not least how well Geography is anticipated and responded to. In any case, the square law cannot be applied definitively if the two sides are at all "asymmetrical" — visibly different in character. As a rule, they are.

Air defence of the homeland must always be the bottom-line task of military avia-tion. A classic instance is the 1940 Battle of Britain. Air Chief Marshal Dowding, head of RAF Fighter Command from 1936, worked from the basic geometrical fact that the effective radius of action of his interceptor fighters — the Spitfires and the Hurricanes — was something under 150 miles. His sectoral defence network was reliant for early warning on ground-based radar stations extending from Dunbar through Dover to Cardigan Bay. That summer, the RAF beat back the Luftwaffe, the latter losing 2300 warplanes (many of them bombers) whereas the former lost 900 fighters.

Yet the British could well have lost air control of south-east England, with dire consequences, had Reichsmarshal Goering recognized the cardinal importance of those exposed radars. So could they had not Dowding so firmly resisted, on 15 May, a proposal to take 75 more fighters out of his order of battle and dispatch them into the collapsing situation in France. So could it if a switch to blitzing London, rather than the RAF's forward airfields, not been made by the Luftwaffe in early September (see below). So might it have, indeed, had not Luftwaffe pilots, with Goering's encour-agement, refrained from shooting up RAF lads descending by parachute: an example of how a sense of chivalry engendered in that pristine aerial domain sometimes found meaningful expression. War with clean hands, not just in a clean shirt.

The Bomber Offensive

However, the two supreme tests of strategic air power to 1945 were the bombing offensives by the Allies against Nazi Germany and by the Americans against Imperial Japan. The nuclear attacks on Hiroshima and Nagasaki apart, the former has attracted more controversy. Even during the war, doubts were expressed by the Archbishop of Canterbury and other British church leaders, along with Leftist opinion in general. Since when, debate has raged between those who have condemned the offensive in principle while also insisting it failed operationally; and those who insist its moral validity is confirmed by its operational success.

One should bear in mind that, in the campaign against Germany, the British air staff and then the American started out committed to tolerably selective targeting — i.e. of war-waging assets not cities per se. As the point was put by Ira Eaker, the first commander of the 8th United States Army Air Force based in Britain, "We must not let History indict us for using the strategic bomber against the German man in the street".

Through 1943, however, the 8th USAAF proved unable to launch daytime precision strikes against nodal targets, well-defined but well defended deep within Europe, without incurring mortifying losses. Moreover, RAF Bomber Command had had to accept by early 1942 that it still lacked the target acquisition technology needed for night operations against strictly defined objectives. Never mind that nocturnal ingress was appropriate to the RAF's gradual build-up from smallish beginnings. Nor that "the USAAF by day; the RAF by night" suited the respective corporate personae of these two air arms.

For the British, a big problem early on was the smoggy Ruhr. This erstwhile aviation meteorologist well recalls a forenoon watch in Scotland during the 1958–9 winter. Fine anticyclonic conditions countrywide were marred only by an acute haze affecting a short stretch of the east coast. Extrapolation indicated that the air in question could well have resided over the Ruhr 36 hours before. In 1942, over half Germany's heavy industry was Ruhr based.

Throughout, the weather was taxing. Forecasting upper winds was especially problematic, patchy data being fed into predictive models still under development. Freezing levels might typically be *c.* 10,000 feet in high summer. Cloud above could soon induce heavy wing icing; and, even in clear air, carburettor icing could develop. The cruising altitude of the RAF's Lancaster bomber was 15,000 feet. Then again, planes returning in the small hours were liable to encounter radiation fog on the runways.

Even so, the chief challenge to aircrew morale inevitably came from enemy interception rates. Among the determining factors, by night or day, were relative mass, depth of intrusion, and electronic advantage. Averages over time exceeding three or four per cent a mission were deemed non-sustainable. Meanwhile, airfield density became a constraint on bomber build-up. In the heavy bomber zone from Oxfordshire to East Anglia and Yorkshire, there was usually a military airfield every several miles.

Come early 1944, however, technical advances were well in train which would shift the balance of advantage decisively towards the Allies. Targeting accuracy at night had made quantum advances. No less decisive was the revamping with a Rolls-Royce engine of a hitherto indifferent American fighter, the Mustang. This afforded quantum gains in speed and range at altitude. Covering the USAAF bombers on their daylight raids, the Mustang cut losses dramatically.

More generally, it at last gave the 8th Air Force the control of German skies by day needed to build a campaign of relatively precise attacks. By definition, any such mission would involve fairly small numbers of aircraft; and in air space still contested, these might be overwhelmed. The RAF had been liable, until 1944, to suffer disproportionate losses in night raids by small numbers of heavy bombers. The losses on the Dambuster raid in May 1943 were eight out of 19. This absolutely critical consideration has been very consistently ignored by critics who would have had the bomber offensive rely on precision attacks throughout.

A conspicuous positive for the Allies in the new situation was the Luftwaffe's desperate shortage of synthetic aviation spirit from May 1944 as its very dispersed production pattern became more bomber accessible. Another gain was to be the crippling from that autumn of Germany's railway and canal systems. Moreover, the programme for "secret weapons" had been put back some months overall, most particularly by the RAF raid in May 1943 on the development complex at Peenemunde on the Baltic. The weapons fatefully delayed included the V-1 cruise missile; the Messerschmitt 262 jet fighter; and the "oyster" pressure mine laid on the seabed in shallow waters. Besides, a redirection of the Luftwaffe heavily towards home defence ensured for the Allies air superiority in other theatres, West and East.

About half Nazi Germany's production of ball and roller bearings was coming from Schweinfurt, a Bavarian town of 60,000 people. So in October 1943 the USAAF had subjected it to heavy raids. However the results obtained had not been proportional to the terrible losses incurred. Besides, about 15 per cent of Germany's requirement was coming from Sweden, a fraction largely comprised of the scarcer kinds of bearings. What with this and a dispersion of production, Germany was able to avoid a serious bearings crisis.[7] Overall, her industrial growth went steeply into reverse from mid-1944.

Another swing argument unremarked in the literature but arguably important is the impact on German senior management of the growing Allied domination of the skies. Might this help explain, for instance, why the Third Reich strategy for aircraft production became so hapless a muddle? And did it contribute, perhaps weightily, to the Führer himself becoming ever more *dégagé*? From mid-1943, he was bunkered down in East Prussia visiting Germany proper seldom and, most significantly, never going to bombed cities. Meanwhile, the impulses he "sent through the machine of administration grew weaker" within that fragmenting maze.[8] During his frightful final days back in his Berlin bunker, he gave himself over to drawing visions of "Germania", the new Berlin destined to arise once victory was his.

Everybody now allows that the Führer's life was one long vengeance quest. Retribution for the horrendous paternal beatings he (like Stalin) had endured. Revenge against those (Jewish or whoever) who declined to admit him to art school on the strength of his amazingly bland paintings . . . In 1940, he allowed himself to be goaded by an RAF attack on Berlin into a diversion of the Luftwaffe, from 7 September, to continually attacking London rather than Fighter Command's forward stations. Yet during the exceptionally fine weather since 30 August, the latter had been brought almost to breaking point.[9]

In January 1944, to the horror of Willy Messerschmitt and the rest, he ordered that the Me 262 jet fighter be fitted out for bombing instead of interception, a decision not formally rescinded till March 1945. Next, he was rejecting military entreaties that the V-1 offensive be directed not against London (on the off-chance of killing Churchill) but against our overcrowded invasion ports. Maybe the continual Allied bombing helped tip his decisions irrationally in favour of revenge lust rather than operational advantage. Witness, too, how Hermann Goering was reduced awhile to a blubbering wreck by the 'firestorm' raid on Hamburg in August 1943.[10]

The sadness is that the combined bombing offensive was responsible for well over half a million deaths plus a welter of cultural erosion. Moreover, those of us who firmly endorse this campaign as being essential to victory can still deplore how the momentum it built up led to the new modes of electronic precision being employed to effect not just selective strikes but also, as late as February 1945 and mainly by the RAF, the over-killing through force concentration of central Dresden. In August 1951, it was my sombre fortune to cross by train Bomber Command's "aiming point" — a circle two miles across. The East German authorities had removed by then all the shattered buildings and their rubble heaps. Roads and rail tracks had been remade. Otherwise the middle of Dresden was a lunar landscape, a Sea of Tranquillity. After Dresden, Sir Arthur Harris, head of Bomber Command, ordered attacks on some ten smallish cities by then of far more historical than operational importance.[11] Also unfortunate, in my view, was the neglect almost throughout of leaflet dropping. Still, one has to recognize how unpopular this stratagem was with aircrews.

The Airland Battle

More generally the air power problem facing Western democracies through two World Wars and afterwards was its role in theatre conflicts sometimes in very distant places. Quite a critical criterion in mobile mechanized warfare is the number of warplanes assigned per army division in the field. In the overland blitzkriegs against the Low Countries and France in 1940 and then the USSR in 1941, the Luftwaffe had but 20 to 25 warplanes for every division in the Wehrmacht's order of battle. Yet in central Europe in 1985, the corresponding ratio was 65 for NATO and 50 for the Warsaw Pact.

Nor is this by any means the whole story. Tactical aircraft have had more scope than, say, Main Battle Tanks (MBTs) for incorporating technical progress. Finance reflects this. In 1943, the cost of a Sherman MBT would be, at 1975 prices, $160,000. An MBT in use today might have cost then $550,000.[12] Obversely, the flyaway cost of an F-14 Tomcat, a strike fighter in USN service from 1975, was $17,000,000, over 50 times the cost (at constant prices) of a P-47 Thunderbolt escort fighter in 1943.[13]

Account should also be taken of the introduction of thousands of helicopters into armies, air arms and navies. So should it of surface-to-air weapons figuring much more in anti-aircraft defence than in 1940–1. So should it of how warplanes had tended to become more multirole.

A salient truth about regular warfare over tolerably open landscapes between 1914 and 1985 was that ultimate victory was very consistently gained by the side enjoying air superiority. However, the aerial function identified as primary by Lanchester had repeatedly failed to live up to expectations. The allusion is to "interdiction" — i.e. crippling enemy forward forces by strikes to their rearward. Relative failure has resulted even where choke points have been presented by topography and/or the transport nodes the enemy was reliant on. Take, for example, peninsular Italy, 1943–5 and the Korean peninsula, 1950–3.

In the not dissimilar situation of armies in retreat, troop management could count for much. During the Arab *débâcle* of June 1967, Saad Shazli (Egypt's chief of staff in 1973) withdrew his division across the Sinai. More famously, after the Battle of Alamein, Erwin Rommel led the Axis armies back from Egypt across Cyrenaica and beyond mainly along ill-defined desert tracks. Later Sir Freddie de Guingand, the

British 8th Army's Chief of Staff, commented apropos the interdiction of those forces that "we had visions of the retreat being turned into a rout . . . in the event the results were most disappointing".

The man in the cockpit of a plane roaring down the runway into blue skies beyond long epitomised the independent air power many subscribed to. Yet in 1967 the then Controller of Aircraft at Britain's Ministry of Technology observed that the ability to operate at all satisfactorily in hostile air space had first exercised air staffs "in the late 1940s, once they had realised the potential impact of the guided missile in all its applications". Indeed, the proponents of missilry "could have predicted the demise of the manned combat planes in the tactical role" at latest by 1964–7, the maritime scene excepted.[14]

Yet in the period 1940–80, no general trend emerged in the attrition of warplanes in contested skies. Geographic and other variables precluded it.[15] Of most fundamental concern therefore was these machines being immobilized on the ground. Take, for example, Major General J. F. C. Fuller, a leading apostle of armoured mobility. Eighty years ago, he characterized the warplane as "a kind of animated rocket fired . . . from an aerodrome . . . , the future of air power is to be sought not in the air but on the ground . . ."[16] Instances in this study have included the onset of Barbarossa; the Middle East wars of 1956 and 1967; and, in staff appreciations, the erstwhile Central Front across Germany.

Dispersal can be among the partial answers unless the topography be utterly constrictive, which is not often the case. When the Falklands War of 1982 started, the Argentinians had 30 Pucara light strike aircraft dispersed around 20 to 50 improvised airstrips on the islands. But 19 were to be destroyed on the ground as were another six in the air. On the other hand, when the Poles and then the Finns had recourse to this stratagem in war in 1939 they did so to some advantage. Still, it has usually to be reserved for lightish warplanes of limited speed and range, not least in view of the long runways normally required otherwise, especially for landing. Nine thousand feet became the norm for NATO strike fighters in Germany, a dependency far more susceptible than anything Fuller had imagined. However, the Gripen, which the Royal Swedish Air Force has had in service since 1992, was designed to be less demanding — a prime manifestation of Stockholm's resolve to exploit militarily an ambience of hard rock, strong and straight highways, forest and winter darkness.

Vertical Take Off and Landing by monoplane (notably Britain's Harrier strike fighter) or helicopter can, of course, dispense with runways completely but when doing so incurs penalties in range and other aspects of performance. Besides, all aircraft depend on substantial infrastructural support.

All this, plus the general advance of electronics, largely explains why the missile (surface or air launched) was by 1980 steadily gaining ground against the customary strike fighter; and why, of course, the Remotely Piloted Vehicle was doing likewise for reconnaissance and target acquisition. That the corollary was closer army–air integration was recognized in the Airland Battle doctrine endorsed by the United States Army and the USAF in 1982–3.[17] Included in the new thinking was explicit emphasis on Follow On Forces Attack (FOFA) — i.e. an aspect of interdiction.

The Land Battle per se

Evidently the quest for decisively mobile warfare had its modern origins in the most defining armed conflict of the twentieth century — the Western Front across France, 1914–18. It had generated a singularly macabre landscape along with a distinctive culture — one sometimes sublime yet always obscene, a searing infliction on the younger generation.

Nobody really foresaw things would pan out thus. The only person ever credited with doing so was the Polish Jewish banker, I. S. Bloch. Yet stimulating as his 1899 sortie into war studies was, its prediction of how a Franco-German conflict would evolve actually proved less than insightful.[18] He did expect Germany to seek to reduce France before her other big adversary, Russia, could fully mobilize. He did not anticipate a transgression of Belgian neutrality. Instead, he believed the French line of forts opposite Germany would impede an advance sufficiently. Such troops as forced their way through the gaps would be engaged by French mobile reserves. Insufficient German units could survive well enough to allow of an investment of Paris.

In the event, of course, the Germans advanced through Belgium only to be finally checked, essentially-speaking, by an outpouring of the Paris garrison — *les taxis de la Marne*. Two lines of trenches were then extended adversially from the North Sea to the Swiss border.

Soon these lines were densely manned due to very full mobilization by the various belligerents. An obvious yardstick of density was the width of front held by a division. For it was an all arms formation of 10,000 or so men which, very much in the tradition of the Roman legion, was considered independently capable of sustained combat in whatever situation. In the Ypres salient, a very hard fought part of the Western front, there was a British division every one or two miles by 1917. On the UN Main Line of Resistance in Korea, 1951–3 (another very static front) there was to be one division every eight miles or so.

On the Eastern Front in World War One, overall densities were a lot less than in France. Therefore the campaigning throughout could be more fluid. At the risk of sounding flippant, one can find an analogy in youngsters playing a scratch game of football. The more who turn out, the slower the goals may come.

From the outset, those concerned sought a way out of the Western Front gridlock. Poison gas was introduced, first by the Germans in 1915. The result was simply that trench warfare became even more macabre. In 1916, the British introduced the tracked fighting vehicle — "the tank". Come August 1918, some 400 were deployed for the Amiens breakthrough. Over half this fleet was incapacitated (by hostile action or simple breakdown) within several days but its influence on events had been critical.

In time for their great offensive that March, the Germans had evolved a solution conceptually simple but almost decisive, one derivative from the long-standing Prussian tradition of giving much autonomy to local commanders in order to cope with the exigencies of mobile war on the North European Plain.[19] Attack not in thin gray lines but in many small groups, the leader of each being left to advance opportunistically and adaptively. A very similar philosophy was to be enunciated brilliantly in a reflective retrospect Rommel published in 1937. He particularly stressed utilising terrain and weather; being tireless in reconnaissance; confusing the enemy; and, not least, "Even in the attack, the spade is as important as the sword".[20]

Complementary to J. F. C. Fuller in the inter-war debate in Britain and beyond about armoured warfare was B. H. Liddell Hart. The latter underwent on the Somme something of an eerily psychological crisis — acute, lonely and mysterious.[21] He duly emerged from the trenches resolved that such horrors must never revisit mankind.

Writing at length in 1927, he averred that the armoured mobility a tank might proffer was already attainable even over rough ground, provided it was not too hilly, wooded or watery. He added the rider that Germany, keen to develop the tank having been so worsted by it, might seek to do so via friendly access to the secluded plains of "Russia" (i.e. the USSR), the other great power in purgatory.[22] Come the turn of the decade this was happening.

By 1960, he was firmly persuaded that, since Napoleonic times at least, "the Defence has been gaining . . . over the Offence".[23] He believed that to overcome equivalently modern adversaries defensively deployed across representative terrain, a numerical superiority of five to one would normally be needed along the main axes of advance, a stipulation liable to mean a two or three to one superiority across the front as a whole.[24] He ascribed this envisaged state of affairs to the way improvements in firepower, mobility and communications had continually reduced the number of troops needed to hold a mile of front in battle. He dismissed the stunning supremacy of the Offensive in 1939–41 as an aberration due to the neglect of certain elementary principles by defending commanders.

That a typical division was becoming more potent was well enough evidenced. Witness the growth in logistic demands. In 1941 a panzer division on active operations might use 350 tons of supplies a day. In the Korean War, the corresponding figure exceeded 1000 tons.[25] Surely, however, higher attrition rates were liable to disadvantage progressively the weaker side, usually the one resting on the defensive.

Besides, a lot will depend on the specifics of firepower, mobility and inter-communication. Towards the end of the Middle Ages, dramatic developments in firepower and substantial ones in overland mobility much assisted the reduction of baronial castles. In 1914–18, bullets favoured the Defence but shells more the Offence. If any generalization can be made about the situation post-1945, it has to be that modern firepower and communications could much enhance surprise attack against airfields but also a range of other facilities. Moreover, surprise would be inherent. If one had known an onslaught was coming Thursday daybreak, its onset would still have engendered the most agonized surprise.

Nor can one accept Liddell Hart's neglect of Geography in both its linear and its qualitative aspects. Almost certainly he read too much general import into the Allies' tough experience in 1944 of the Normandy bocage with its sunken roads, small fields, tight hedgerows and stout buildings. Then again, unduly wide divisional sectors would always oblige divisional commanders to depend on mobile defence, a mode much reliant on counterattack. To all of which one could add that, whatever their pitch *vis-à-vis* the all-round defence of their homeland, our Muscovite protagonists favoured (as in 1914 and before 1941) the grand offensive or counteroffensive on the North European plain.[26] They might not have been too impressed by a NATO overly reliant on the attributes of non-nuclear Defence.

Insurgency Experience

A dimension of war of enduring concern is insurgency.[27] Quite the most celebrated example early last century was the Arab revolt against the Ottomans, a campaign always associated with Lawrence of Arabia. Lawrence was an outstanding diplomat, morally and physically courageous and highly empathetic. Maybe, however, he was never himself the great tactical innovator he wished to be seen as.[28] None the less, two brilliant paragraphs in his *The Seven Pillars of Wisdom* encapsulate the doctrine of that particular campaign, operating as it was out of what today's military would call "ungoverned space", ungoverned and, without aviation, unobservable.

Lawrence depicted it as "a war of detachment . . . contain the enemy by the silent threat of a vast unknown space, not disclosing ourselves till we attack . . . not seek either his strength or his weakness but his most accessible material". Seek superiority, especially of relevant weaponry, "at the critical point and moment of attack". Emphasize the cutting of "empty" stretches of railway. Seek perfect "intelligence".[29]

Still, the way of the insurgent after 1918 was to be expounded most influentially by Mao Tse-tung. No matter that much of what he said, too, was in reaction to evolving local circumstances. In 1929, he warned that the fact that, especially in South China, the Red Army very considerably comprised *elements déclassés* was not an argument for promoting "large-scale actions by roving insurgents". Better to work from the premise that a revolutionary army would necessarily be operating from within a selected base area and so be inherently vulnerable. However, this circumstance could be offset by turning a big "encirclement and suppression" campaign waged by the enemy against one's own side into a number of small separate campaigns of encirclement and suppression waged by the latter against the enemy, " . . . blockade within blockade, the offensive within the defensive, superiority within inferiority". Military, political, economic and geographical conditions together "constitute only the possibility of victory or defeat; they do not in themselves decide the issue".

In the famous statement published in 1938, he saw advantage in a strategy of *Protracted War*. In another that same year, *On the New Stage*, he averred that in "a big semi-colonial country such as China", one could, during protracted war, "encircle the cities and isolate them", the prime aim being to deny them to the enemy. How much positive importance the revolution should attach to urban areas is nowhere spelt out. But he does seem oddly relaxed about the ghastly likelihood that often a city will "be held only after it has changed hands a number of times". His simile, "The popular masses are like water and the army is like a fish", implied that the revolution was nourished most essentially by China's vast rural hinterland. Being dependent on it for nourishment in every sense made it vital for the guerrillas always to behave in exemplary fashion.

Yet given that he was already ensconced in Yenan with its hinterland lightly peopled to west and north, Mao's emphasis was liable to change. Addressing the Sixth Plenum of the Central Committee in March 1938, he fervently insisted that the Yenan base area had "been built up by the barrel of a gun. Anything can grow out of the barrel of a gun".[30]

A contrasting situation was that facing George Grivas (1898–1974), the Greek general who masterminded the EOKA insurgency against the British in Cyprus, 1955–9. A hard-line conservative, he was concerned not with social revolution but with *enosis*, the union of Cyprus with Greece. The contest was essentially one for world opinion; and, of course, the insular battlefield was three orders of magnitude smaller than China. The *Preparatory General Plan* Grivas promulgated from 1953 duly revealed differences of emphasis from Maoism though also similarities. The British "must be continuously harried and beset until they are obliged by international diplomacy exercised" through the UN to resolve the Cyprus problem. The harassment will be by small sabotage and strike groups operating from safe havens as well as by popular passive resistance. The number of organized groups will be limited because "the terrain should appear empty". Agents of the colonial authority were liable to be executed. The smuggling of arms "is in hand".[31]

Opportunity Knocks Seldom?

The EOKA campaign began hard upon the considerable success of the Vietminh in the first Indo-China war, this not long after the huge Maoist victory on the Chinese mainland. Accordingly, a disposition was burgeoning to regard insurgency as the irresistible wave of the future. In reality, of course, such gains as it then registered no more proved that than did events between 1939 and 1941 show the blitzkrieg was forever unstoppable.

Britain's far-flung experience, 1950–80, affords in its totality a helpful perspective. By 1955 or thereabouts, the back had been broken of the Chinese Communist insurgency which had begun in Malaya in 1948. About the same time, the Mau Mau uprising within the Kikuyu tribe in Kenya was crushed. When British troops pulled out of South Arabia in 1967, the successor regime in Aden was not the Cairo-based Front for the Liberation of the Occupied South Yemen (FLOSY). Instead it was the more authentically indigenous National Liberation Front (NLF), an alternative London disliked a sight less and ultimately positively encouraged. Nor had 'nibblekrieg' aggression by Ahmed Sukarno in Indonesia destroyed Malaysia. Nor did a concurrent bombing campaign in Hong Kong come to anything. Nor was Sinn Fein on the way to uniting Ireland by force. And by 1967 at the latest, organic *enosis* was sliding off the agenda in newly independent Cyprus. (See 9 (1)). And Britain's retention for military reasons of two Sovereign Base Areas was non-contentious. From 1975, a rebellion in the Dhofar province of Oman faded fast, consequent upon Beijing's withdrawal of support but also action by British special forces — the Special Air Service.

In short, every counter-insurgency campaign Britain engaged in during those three critical decades was considerably successful at least. Some would argue, of course, that this was because London could usually enlist indigenous support by offering the end of empire. Yet this argument can be two-sided. Insurgents are always happy to claim all the credit for imperial withdrawal, effected or impending.

Latin America had by 1870 almost completely emerged from outright Iberian colonialism or backlashes to restore it. But a sense of United States dollar diplomacy was to become pervasive. Even so, through the early 1960s, attempts to generate Marxian insurgency in rural areas went badly. You could say this was a function of pronounced social backwardness. You could also say that what was then relevant, too, in many areas was low population density, considerably lower for instance than what Fidel Castro had triumphantly worked with in Cuba. Bolivia, where in 1967 the charismatic though ruthless Ernesto "Che" Guevara was killed, was a conspicuous example of both these factors in play. Each had worked otherwise for Lawrence, the most particular differences being that he did not have aerial surveillance and strike to contend with.

With Guevara gone, the continent's highest profile insurgent was the Brazilian Communist, Carlos Marighela, alias "Carlos the Jackal" (d. 1969). He put his stamp on all the different groups with his tract, *A Handbook of the Urban Guerrilla* (1968). In this, he eclectically formulated a strategy for urban unrest to spearhead the revolution. Among its raft of options were kidnappings, raising false alarms, and turning peaceful demonstrations into violent confrontations through the use of snipers and bombers.[32]

Going down this path could trigger a Right-Wing backlash, one perhaps too careless of liberal democracy to prepare the ground for it.

Latin America was exceptionally urbanized by developing world standards. As early as 1965, the urban population was reckoned to have overtaken the rural. Some aspects of the continental geography may have favoured this. So might have pre-colonial city experience. But neither of these factors applied in the case of Uruguay. Nevertheless 45 per cent of the population lived in the capital, Montevideo, and 30 per cent in other urban areas. Here and indeed elsewhere, a determining influence may have been the classical Mediterranean civic tradition transmitted via the Iberian homelands. In Uruguay's case, Britain was also a big influence through the formative nineteenth century.[33]

Though so strongly urban, Uruguay had waxed prosperous the first half of the last century with its sheep and cattle. It could also boast a thriving modern culture. But come the sixties, it faced weakening export markets; and across the decade 1962–72 inflation (so often a Latin American problem) was recorded as 6460 per cent. But kidnappings and assassinations by the Tupamaro National Liberation Front did not prevail; and in 1973 the military intervened more directly in government. There, as everywhere, Marighela had posthumously succeeded only in as far as Latin American statehood got a bad name awhile for Rightist extremism.

What then of Marxist-Leninist theoreticians classically saying that a "proletarian" revolution is bound to succeed eventually even if there are a few false dawns? Closer to reality may be the proposition that an insurrectionary opportunity lost may never recur. Almost all the large-scale insurgencies which flourished after World War Two — e.g. in Greece, China, Indo-China, Malaya, the Philippines, Poland, the Ukraine and belatedly Algeria — did so in territories fought over during that conflict. The traumas of occupation and liberation had been radicalizing and polarizing. Guerrilla cadres were promoted, armed and assisted by outside sponsors. Then the departure of the Axis powers left a transient administrative vacuum and sometimes deeper social divisions. Often, too, weaponry had been dumped.

However, it is the middle-rank peasant farmer rather than the *sans culotte* labourer who was likely to spearhead any authentic revolt in the countryside. The former had too much at stake to ignore adverse tendencies. He was also more exposed to political influences from the towns and cities.[34] Yet this social stratum also stood to benefit greatly from the Green Revolution in agriculture proliferating through the developing world by 1967. (See section 2). It was to be a powerful antidote to insurgency. Mao's family had been yeoman farmers, kulaks if you will.

Lastly, guerrilla leaders have often been hamstrung by the divergent geopolitical interests of a sponsoring state. Thus in April 1970, the bazaar area of Amman in Jordan leading down to the Roman amphitheatre was parcellated by road blocks colourfully controlled by the Palestinian insurgent groups, the police and the army in various permutations. The large Soviet embassy was clearly masterminding a steady flow of military supplies to these urban guerrillas.

Come September, King Hussein resolved to act decisively against a situation bizarre in those terms. He sent in a task force headed by Bedouin tanks. The Palestinians fought hard and unitedly but failed because Moscow had supplied no anti-tank weapons — mines or rocket launchers. That would have conceded to them too much autonomy.[35] Still, neither there nor anywhere else was complacency warranted *vis-à-vis* the longer term. It is less still now.

Naval Power

Competitive naval building between Britain and Germany, c. 1897 to 1914 and then between Japan and the USA, 1917–21, showed how readily the High Seas lend themselves to arms races, qualitative and/or quantitative. In response, about ten multinational agreements were reached on naval limitation between 1908 and 1938. That these exercised some constraining influence seems clear. But their texts did tend to be too conservative. A gibe often made at admirals and generals is that they are always preparing to fight again the war just ended. One could as well say of arms controllers that they are always trying to prevent it.

Most significant of the said agreements was the Washington Naval Treaty of 1922 signed by the USA, Britain, Japan, France and Italy. Primarily it covered capital ships (i.e. battleships and battle cruisers) and aircraft carriers. Cruisers were not addressed comprehensively until the London Naval Conference of 1930 when Britain, the USA and Japan accepted limitations. Until then, the focus in the Washington Treaty on larger vessels had encouraged the proliferation of cruisers, a tendency furthered by rather dated thinking about scouting and the protection of trunk ocean highways.[36] Moreover, submarines were not covered.

Then again, the Washington distinction between capital ships and carriers failed to anticipate how rapidly the latter would emerge as *the* capital ships of the future. The naval Battle of Midway in June 1942 has been identified (in Chapter 4) as a global turning point, this by dint of the Americans being decisively victorious. Everything had hinged on the timely location of the opposing carrier forces by the respective reconnaissance flying boats plus, of course, the Americans having broken Japanese naval codes.

Battleships contributed nothing material to the outcome at Midway. So coming after the successful strike by a British carrier against the Italian fleet in harbour at Taranto in November 1940 and then, of course, the Japanese carrier task force attack on Pearl Harbor some twelve months later, Midway confirmed the ascendancy on the High Seas of strike aircraft as opposed to big guns.

However, a difficult question still posed continually was the comparative advantage of air power exercised in the maritime environment being sea- as opposed to land-based. The mortifying loss of ships to land-based air attack off Greece, Crete and Malaya left the British with memories that resonated as the recapture of the Falklands got under way in 1982. The Ministry of Defence in London was acutely conscious at the outset of how vulnerable our ships were to Argentinian aircraft operating from bases ashore which we might clandestinely watch closely but felt inhibited politically about actually attacking.

Moreover, when like could be compared with like, through World War Two and for some time after, carrier aviation was inferior performance-wise to its counterparts ashore. The F-4 Phantom, which entered service with the United States Air Force (USAF) and United States Navy (USN) from 1960, was the first American example of strike fighter equivalence.

Low priority funding will sometimes have held naval aviation back. More fundamental, however, have been the sundry constraints imposed by flight deck operation, including the need for strong undercarriages and arrester gear. A favoured remedy has been to build bigger hulls. Take the French nuclear-driven aircraft carrier, *Charles de Gaulle*, which entered service in 2000. At 40,000 tons, it has a displacement one-third larger than have the vessels it is replacing. But its flight deck area is 50 per cent larger, allowing of the readier handling of larger aircraft.[37]

More historically speaking, the American debate was about the 70,000-ton *Forrestals*, constructed between 1955 and 1961, versus the 30,000 ton *Essex* class, built in World War Two. A *Forrestal* could carry three times the aviation spirit while the aircraft it earmarked for attack missions were a good 50 per cent more numerous and generally larger. Weather statistics appeared to show that in, say, the Norwegian Sea or the Straits of Taiwan, a *Forrestal* could operate its aircraft 345 days a year but an *Essex* on only 220.[38]

On the other hand, neither ship could recover more than one aircraft at a time. And each might too easily be turned by enemy action into an obligation rather than an asset. One recalls the fire which broke out accidentally in the *Forrestal* herself in the Vietnam theatre, a fire which claimed 129 lives plus 57 aircraft destroyed or badly damaged. Indeed, even the flight deck of the 90,000 ton *Enterprise* (the USN's only nuclear propelled carrier) looked disconcertingly congested with but half its full complement of planes parked there as economically as possible.[39] Yet another difficulty was that neither the *Forrestal* nor the *Enterprise* could then negotiate Suez or Panama. Besides which, a task force built around one or two more attack carriers might need, on active service against a technically sophisticated adversary, to occupy a circle a good 300 miles across, a requirement that could make deployment difficult in narrow seas. Moreover, force dispositions optimized for warding off aircraft or longish-range missiles might be less adapted to thwarting submarines or Fast Patrol Boats (FPBs). One thinks particularly of the 6th Fleet in the Mediterranean, an important instrument of regional deterrence or modulated diplomacy.

FPBs of a few hundred tons displacement, each able to launch several missiles individually equivalent perhaps to eight-inch shells, figured in the planning of various navies by 1975. The Egyptian sinking of the Israeli destroyer, *Eilat*, in October 1967 was the first use in earnest of this weapons system. The strong performance of Israel's new FPB force during the 1973 war confirmed its potency. Since even in fair weather FPBs rarely operate more than 150 miles from base, they can be seen doctrinally as an expression of littoral power. This they have in common with minefields laid inshore as well as coastal artillery. Operation in the 'brown water' along the adversary's coastline assumed much importance in USN discourse several years back. Derivatively the notion is Mahanian. though Mackinder thought similarly about the Narrow Seas around Britain.

However, public awareness throughout the West is very particularly of the oceanic threat submarines have posed. Admiral Sergei Gorshkov (who retired in 1985 after three decades as C-in-C of the Soviet navy) reckoned that in World War One the manpower the Allies dedicated to retaining adequate control of the North Atlantic was 20 times that deployed by Germany via her U-Boat fleet. For tonnages respectively lost, the ratio was close to 100. Nor were things much different in World War Two.[40] For this reason alone, the West could never have sustained another Battle of the Atlantic against a large and modern Soviet submarine fleet. The choice would therefore have been between tactical nuclear recourse involving geographic escalation or else outright surrender. So would it in the event of a blockade of key ports by means of the multisensor sea mines available by 1980.

The Allies had looked fairly close to defeat on the North Atlantic early in 1943; and had looked closer still in the Spring of 1917. Yet both times round, a turnabout ensued which was quite diametric. However, the circumstances were contrasting. In 1915 and 1916, the Germans waged a restricted submarine war against commerce. What precisely the restrictions were varied. However, the general thrust was towards as unrestrained a campaign as relations with neutrals allowed.[41] Then in February 1917, all restrictions were removed. As expected, one result was the entry into the war, that April in fact, of the USA. But another was a quarter of the ships leaving British ports that month failing to complete a round voyage. The progressive introduction from May of a convoy system made an invaluable contribution to breaking this gridlock. Without it, Britain and therefore France could well have been blockaded into submission before American power could weigh in.

In March 1943, Allied shipping losses to the U-boats surged to 600,000 tons but by May had fallen back to 220,000. Admiral Doenitz, their C-in-C, duly recalled all his submarines from the North Atlantic for extended rest and recuperation. As things transpired, this meant the Allies had won the Battle of the Atlantic. A deciding factor had been a deployment of Catalina ultra-long-range flying boats and adapted Liberator bombers in order to close a mid-ocean surveillance gap. Meanwhile in the Pacific, an American submarine offensive was building up which was virtually to cripple Japan's mercantile marine by the end of 1944.

By 1945, big advances were on-going in active sonar — acoustic sensing through sea water. This registers ranges and bearings by the reflection off solid surfaces of the sound waves the sonar was emitting on specific frequencies. In good conditions, submarines up to 15 miles away could now be fixed, several times the 1918 norm. Passive sensing (i.e. of sounds emitted by the target) later made considerable strides with the advent of computerized discrimination. They sometimes read bearings across ranges well in excess of what active sensing could manage. But their performance was less consistent. Nor could they gauge range except perhaps roughly, this by collating data from two separate platforms.

The sensors just considered could variously be mounted in ships, submarines or helicopters. In addition, considerable work was done on the development of powerful long-range shore-based sonars. But since they operated on Very Low Frequencies (VLF), their ability to register small images was the more problematic. Besides, all sonars were and are compromised to whatever extent by gradual or abrupt changes in salinity, temperature, oxygen content or marine life.

Furthermore, by 1975–6 the USA, the USSR and Britain had in service well over 100 attack submarines that represented quantum gains in depth of transit, submerged speed and agility by dint of having nuclear propulsion and the novel hull designs associated with it. These boats may operate well below the thermocline, the discontinuity (acoustic and otherwise) between the well mixed surface water and the steadier flow below. Moreover, some boats were understood to achieve 35 to 40 knots submerged — i.e. close to the speed of a homing torpedo and considerably faster than any frigate or destroyer could be built for without entering the uncertain realm of revolutionary hull design. This is because of how the generation of surface bow waves absorbs more energy with rising speed, this at a much more than proportional rate. The total drag exerted on a fleet destroyer under way increases 20-fold between 16 and 27 knots.

So, subject to the specifics of any warlike crisis, submarine nuclear propulsion proffered for the nations accessing it a formidable riposte to the advances in Anti-Submarine Warfare. This applied even though such a boat might necessarily have a surface displacement several times that of a diesel-driven counterpart.

Yet overall maritime nuclear propulsion (naval or mercantile) had not been taken up by 1980 to the extent anticipated by some in the pristine heyday of "atomic" enthusiasm 25 years previously. The *Enterprise* had cost twice as much to build as a *Forrestal*; and, when on operations, needed full replenishment every twelve days regardless. Moreover, a complex web of proscriptions limited nuclear-driven access to foreign ports.[42]

Discussion of base facilities remained inadequate, with the term "base" being casually applied to anything from a coral staging post to a vast complex like Singapore, Pearl Harbor or Norfolk, Virginia. And yet Mahan still seemed justified in cautioning that modern logistics made warships even more base dependent. For political reasons, the USN stressed how independent was its 6th Fleet in the Mediterranean of earmarked shore facilities. Nevertheless, it did acquire a modicum, notably at Piraeus from 1972. Of less enduring pertinence, in the wake of Taranto and Pearl Harbor, was Corbett's averration that "a properly defended anchorage is not more easy to injure than it ever was".[43] How well can one judge, in any case, what is "properly defended" against all contingencies?As Mahan further noted, the availability of good natural harbours does have significance, mercantile and naval.[44] The essentially table-land African continent is deficient in this respect. The main exception regionally is the Maghreb which is geologically an extension of the Tertiary folding of Southern Europe.

Sometimes the said deficiency can be corrected by a heavy construction programme. Such was undertaken in the big Polyarnoy base facility near Murmansk as the new Soviet navy was built up after 1950. There was something of an ideological dimension in this since granite excavation, concrete and steel were seen as the visual expressions of Communism dourly correcting "the errors of Nature". Analogous is the military's heavy application of concrete across North Korea from 1953.[45]

2 Social Prediction

It is salutary to trace how the long-standing discourse about the pressure exerted on our planetary environment by humankind was revamped from the early sixties. A major obstacle to a synoptic understanding was that the new "environmentaliam" was never properly related to "strategic studies". Yet the two interests burgeoned about the same time. Moreover, their *modus operandi* were very similar. Each was factual; numerate, predictive, and, within set limits, rational. Both had their well-defined schools of "hawks" and "doves", the former being less inhibited about the exercise of overriding power — economic or military.

Within the planetary interest, the dialectic between market-driven hawks and restraint-minded doves was very visible and direct. A highly contentious exemplar of dour dovishness was *The Limits to Growth*, a study by the Systems Dynamics group at MIT. In its World Model Standard Run, it predicated no major change in the physical, economic or social relationships which historically have shaped Man's exploitation of this planet. On that basis, accelerating resource depletion would mean per capita output of food and industrial products starting to fall early in the twenty-first century. So by 2050, world population would be in precipitate decline.[46] Nor was this scenario improved by doubling the resource base postulated for 1970. Instead the population crash starts earlier because of a dramatic rise in pollution.[47] The use of an aggregated index for pollution came in for criticism.

Herman Kahn aided by colleagues at his Hudson Institute went on the opposite tack several years later. In a "surprise-free" projection from 1976 to 2176, they foresaw world population rising from four billion to 15, while average income per head rose from $900 to $20,000 a year at constant prices, this hopefully with a marked narrowing of the income differential (currently a 100 to 1) between the wealthiest one per cent and the poorest 20 per cent.[48] However, they freely acknowledged that, at this early stage, "the evidence is often more suggestive and heuristic than sufficient".[49] This was a work born out of creative tension between mathematical projections and free-ranging speculation. Had he not died too young, Kahn might well have bridged to good effect the gap between strategic studies and societal development. But neither this study nor the MIT one anywhere mentions climate change.

Throughout the first two post-war decades, the Malthusian pressure of population on resources, especially grainstuff, was a recurring theme. In 1965 and 1966, food output per head in Africa; South and East Asia; and Latin America was little, if any, higher than the average for the late 1930s. But then, in 1967, aggregate crop yields in these lands rose a good 6 per cent.

This leap forward owed something to higher prices and better weather as well as to secular efficiency gains. In addition, however, severalfold increases in output per acre were locally being registered in various countries (notably, in this first instance, India and Pakistan) by the use — in conjunction with intensive irrigation and lavish applications of fertilizer and pesticides — of new "dwarf" varieties of wheat, rice and maize. This Green Revolution was not without hazards. Nevertheless, it was very much a "positive" overall.

Even so, the spectre Thomas Malthus had so graphically identified had not been laid. In 1968, Dean Rusk, then US Secretary of State, still felt constrained to warn us that land hunger might be a direct and repeated source of strife by the 1980s should the population explosion slide out of control in the interim. The fierce border clashes ("the football war") associated with the infiltration into Honduras in 1969 of 250,000 people from overcrowded El Salvador would be a clear-cut instance. Likewise Ibo emigration from their congested heartland in the Eastern Region of Nigeria was the backdrop to the Nigerian civil war of 1967–70, a conflict that claimed (largely through famine) over a million lives. Overcrowding of the tribal reserves was one determinant of the South African situation. Palestinian insistence on a "right of return" was then a central feature of the Arab–Israel conflict. Periodic strife within Indonesia owed much to demographic imbalance. The twin islands of Java and Madura embraced but 7 per cent of the land area yet bore, as of 1962, two-thirds of its population. (See 4 (2)). In some situations (e.g. Nigeria and Israel–Palestine), matters were complicated further by a disposition within the respective parties to secure demographic superiority in anticipation of further conflict, political or military.

The decade 1965–75 saw more focus on urbanization as a dimension of the demographic crisis.[50] The urban population of the newly developing world was taking off from 25 per cent of its total in 1970 to a predicted 36 per cent by 1990, a trend accelerated by the capitalization of agriculture. All too expressive thereof was the filth, congestion, ugliness, unemployment and anomie of the shanty towns characteristic of, in Barbara Ward's graphic phrase, "the cities that came too soon". In Francophone West Africa, the shanty districts were colloquially named after the surplus *bidons* or steel drums much used in their construction. For the *bidonvilles* of Dakar read the *bustees* of Calcutta or the *barricades* of Lima or the *favellas* of Rio. Insurgency in Latin America duly switched towards the cities.

There was also unrest in the cities of the West as the 18 to 25 age group used freedoms lately gained to extend liberation further. The interactive influences bearing on them included Vietnam, university expansion, the contraceptive pill, television news, Earth-rise on the Moon, and the erosion of Wild Nature. The notion of a worldwide urban syndrome gained ground. As a ranking British social scientist put it, the troubles being encountered "seem to be particularly severe . . . in Japan, India and the USA; very likely also in China . . . The worst difficulties occur in areas that have populations of one million or more . . . ".[51]

Awareness of climate change came very late on. A major reason initially was inertia within the meteorological/climatological profession. In 1977, the then Director of the UK's Meteorological Office (himself a cloud physicist of distinction) insisted that much more research was needed to predict "the marginal affects of Man's intervention in climate", effects which "may well be masked by natural variation". He also anticipated that the threat to stratospheric ozone from ChloroFluoroCarbons (CFCs) or other chemicals would prove to "have been greatly exaggerated".[52] To quite an extent, of course, professional inertia was reinforced — if you like, rationalized — by the coolish interlude that set in around 1940 worldwide. (See 15).

A shift in professional and popular attitudes towards recognition of "greenhouse" as a salient concern came with the American summer drought of 1988. From the Great Plains to the Atlantic seaboard, the aridity and heat were prolonged, all this hard upon what had widely been the driest Spring for fifty years. Yet the international community of strategists were little bestirred, despite having a surplus of mental energy with the Cold War winding down. An IISS directing staff perspective on future strategic directions was published in 1992. With verve and insightfulness, it certainly well argued a need to break out of the nuclear straitjacket. Even so, it mentioned climate change not at all.[53] Nor was this at all untypical.

Still, some reassurance and even inspiration can be drawn from the fact that, in the long years in which the greenhouse effect was being so widely ignored or discounted, it was the subject of solemn warnings by the two men often spoken of as fathering the American and Soviet hydrogen bombs respectively. In 1958, Edward Teller and a colleague noted that, if hydrocarbon fuel combustion drove up the level of atmospheric carbon dioxide (CO_2), this would act like "the glass in a greenhouse" with melting ice caps the consequence.[54] Ten years later, Andrei Sakharov was warning that CO_2 "from the burning of coal is altering the heat reflecting qualities of the atmosphere. Sooner or later, this will reach a dangerous level".[55]

Notes

1   See the author's *The Future Global Challenge* (New York: Crane Russak, 1976), pp. 49–50.
2   J. C. Slessor, *Air Power and Armies* (London: Oxford University Press, 1936), pp. 105–6.
3   F. W. Lanchester, *Aircraft in Warfare, the Dawn of the Fourth Arm* (London: Constable, 1967), pp. 187–8.
4   *Ibid.*, p. 145.
5   D. L. I. Kirkpatrick, "Do Lanchester's equations adequately model real battles?", *Journal of the Royal United Services Institute for Defence Studies* 130, 2 (June 1985): 25–7.
6   J. N. Merritt and P. M. Spreu, "Negative Marginal Returns in Weapons Acquisition" in

Richard G. Head and Ervin J. Rokke (eds.), *American Defense Policy* (Baltimore: Johns Hopkins Press, 1973), pp. 486–95.

7 Alan S. Milward, *The German Economy at War"* (London: Athlone Press, 1965), pp. 121–3.

8 *Ibid.,* p. 131.

9 R. J. Ogden, "A Battle of Britain — A Meteorological Retrospect", *The Meteorological Magazine* 119, 1420 (November 1990): 260–6.

10 Adolf Galland, *The First and the Last* (London: Methuen, 1955), pp. 222–3.

11 A. C. Grayling, *Among the Dead Cities* (London: Bloomsbury, 2006), p. 73.

12 *Defense and Foreign Affairs Digest,* 11, 1978, p. 26.

13 W. D. White, *US Tactical Air Power* (Washington DC: Brookings Institution, 1974), Fig. 4.2.

14 Christopher Hartley, "The Future of Manned Aircraft" in *The Implications of Military Technology in the 1970s*, Adelphi Paper 46 (London: International Institute for Strategic Studies, 1968), pp. 28–37.

15 Neville Brown, *The Future of Air Power* (London: Croom Helm, 1986), pp. 26–9.

16 J. F. C. Fuller, *On Future Warfare* (London: Sifton Praed, 1928), p. 202.

17 John W. Woodmansee, "Blitzkrieg and the Airland Battle", *The Military Review* LXIV, 8 (August 1984): 21–39.

18 I. S. Bloch, *Is War Now Impossible?* (London: Grant Richards, 1899), pp. 95–7.

19 Weichong Ong, "Blitzkrieg: Revolution or Evolution", *Journal of the Royal United Services Institute for Defence Studies* 152, 6 (December 2007): 82–7.

20 Erwin Rommel, *Infantry Attacks* (London: Greenhill Books, 2006), p. 61.

21 Alex Danchev, *Alchemist of War* (London: Weidenfeld and Nicolson, 1998), pp. 58–9.

22 B. H. Liddell Hart, *The Remaking of Modern Armies* (London: John Murray, 1927), Chapter IV.

23 B. H. Liddell Hart, "The Ratio of Troops to Space", *Journal of the Royal United Services Institute* CV, 618 (May 1960): 201–12.

24 B. H. Liddell Hart, *Deterrent or Defence* (London: Stevens, 1920), Chapter 16.

25 Martin Van Creveld, "Logistics since 1945: From Complexity to Paralysis?" in Brian Holden Reid and Michael Dewar (eds.), *Military Strategy in a Changing World* (London: Brassey's 1991), Chapter 16.

26 A. A. Sidorenko, *The Offensive* (Moscow: 1970), translated and published by the USAF, Washington DC.

27 Julian Lewis, "Double-I, Double-N", *RUSI Journal* 153, 1 (February 2008): 36–40.

28 Tom Hill, "Reassessing Lawrence. Architect of a Guerrilla Campaign?", *Journal of the Royal United Services Institute* 151, 6 (December 2006): 73–7.

29 T. E. Lawrence, *The Seven Pillars of Wisdom* (London: Jonathan Cape, 1935), p. 194.

30 Stuart R. Schram, *The Political Thought of Mao Tsi-tung* (Harmondsworth: Pelican, 1969), Chapter V.

31 Charles Foley (ed.), *The Memoirs of General Grivas* (London: Longmans Green, 1964), Appendix 1.

32 Robert Moss, "Urban Guerrilla Warfare", *Adelphi Paper* (London: The International Institute for Strategic Studies, 1972), Appendix.

33 Peter Winn, "Britain's Informal Empire in Uruguay in the 19th Century", *Past and Present* 73 (November 1976): 100–26.

34 Eric W. Wolf, *Peasant Wars of the Twentieth Century* (London: Faber and Faber, 1971), p. 292.

35 See the author's "The Jordanian Civil War", *The Military Review* L1, 9 (September 1971): 38–48.

36 "Cruisers and Washington" in Antony Preston, *Cruisers, 1880–1980* (London: Arms and Armour, 1980), pp. 67–80.

37 Jeremy Tyler, "What Lessons can the United Kingdom learn from the French Aircraft Carrier, *Charles de Gaulle?*", *The Naval Review* 95, 3 (August 2007): 214–19.

38 See the author's *Strategic Mobility* (London: Chatto and Windus for the Institute for Strategic Studies, 1963), Chapter V(2).

39 Arthur Hezlet, *Aircraft and Sea Power* (London: Peter Davies, 1970), Plate 12.

40 S. G. Gorshkov, *The Sea Power of the State* (Annapolis: United States Naval Institute, 1977), pp. 100–1.

41 Arthur Hezlet, *The Submarine and Sea Power* (London: Peter Davies, 1967), p. 65.

42 Michael Pugh, "Nuclear Warship visiting: storms in ports", *The World Today* XLV, 10 (October 1989): 180–3.

43 Julian Corbett, *Some Principles of Maritime Strategy* (London: Longmans Green, 1919), pp. 208–9.

44 A. T. Mahan, *The Influence of Sea Power on History, 1660–1783* (London: Sampson Low and Marston, 1890), p. 83.

45 Benjamin F. Schemmer, "North Korea buries its aircraft, guns, submarines and radars inside granite", *Armed Forces Journal International* 121, 13 (August 1984): 94–7.

46 Donella H. Meadows *et al.*, *The Limits to Growth* (London: Earth Island, 1972), Fig. 35.

47 *Ibid.*, Fig. 36.

48 Herman Kahn, *The Next 200 Years*, (London: Associated Business Programmes, 1977), p. 55.

49 *Ibid.*, ix.

50 See the author's *The Future Global Challenge* (New York and London: Crane Russak for Royal United Services Institute, 1977), Chapter 6, The Urban Implosion.

51 Ursula K. Hicks, *The Large City: A World Problem* (London: Macmillan, 1974), p. 3.

52 B. J. Mason, "Man's influence on Weather and Climate", *Journal of the Royal Society of Arts* CXXV, 5247 (February 1977): 150–65.

53 John Chipman, "The Future of Strategic Studies: beyond even Grand Strategy", *Survival* 34,1 (Spring 1992): 109–31.

54 Edward Teller and Albert L. Latter, *Our Nuclear Future . . . Facts, Dangers and Opportunities* (London: Secker and Warburg, 1958), p. 167.

55 *New York Times* translation of Andrei D. Sakharov, *Progress, Co-existence and Intellectual Freedom* (New York: New York Times Book Service, 1968), p. 49.

The Eastern Question

7 | The Muscovite Heartland

Twice Orgerd, Grand Duke of Lithuania (1341–77), was repulsed in front of Moscow, having advanced along the great morainic (i.e. post-glacial) ridge Napoleon was to make his axis of withdrawal, 1812–3 (see below). Each was cognizant of the centrality of Moscow between the headwaters of rivers respectively flowing into the Baltic, the Black Sea, the Caspian and Lake Ladoga. The early rulers of Moscow had used their situation, skillfully and often deviously, not to exploit their strength but compensate for their lack thereof. This they had done in interaction first with the Rus then with the Mongols.

Moscow had been sacked a second time by the Mongols in 1382, an event intended to make neighboring princes the more submissive awhile. But the Mongols themselves were by then distracted by internal dissension and by the threat Tamurlane presented. In 1392, they turned to Moscow for military support, agreeing in exchange to its annexing the Grand Principality of Nizhniy Novgorod, along with the strategic Volga–Oka confluence. This made Moscow dominant within East Russia (as opposed to West Russia or Lithuania). Soon after which, the south Russian khanate (alias the Golden Horde) was undermined by the depredations of Tamurlane, notably the destruction of Saray and Astrakhan (1395–6). This virtually precluded the Horde's retaining effective control of East Russia.

Thus was the scene set for the unification and expansion of Great Russia or Muscovy with Moscow indisputably its capital. A milestone in this progression (which was studded by orgies of official terror) was the final surrender in 1478 of Novgorod.

This ancient city had achieved by 1400 a population of perhaps 400,000, despite lacking Moscow's nodality in relation to the new geopolitics, its ambient soils being too poor for agriculture. It relied on food imports from better lands which Moscow controlled access to.

The military decline of the Golden Horde was ultimately to be made irreversible by gunpowder. However, this consummation was delayed by recourse to the Crimea as an offensive base area. The Crimean Tartars were receiving some Russian tribute until 1700.

Well before that, an authentic Russian state had been established with the Church still playing a lead role in kneading a national identity. This it did in divers ways: the icons; the monasteries; the Old Church Slavonic language for liturgy and formal dialectic; the Chronicles, not always accurate or complete but alluringly distinctive; the exotic wooden churches of the North; the renaissance in religious painting, 1380–1500 . . . Throughout the area covered by the Chronicles (i.e. from Poland to Novgorod), it is possible to make an elaborate compilation of the climatic erraticism evident there as elsewhere in Europe in the fourteenth and fifteenth centuries.[1] The foundation of the strong Tzarist tradition of natural science?

In 1328, Moscow's ruler assumed the title of Grand Duke and was assigned by the Tartars to collect tribute from territories nearby. Ivan III, Grand Duke 1462–1505, not only subdued Novgorod but expanded the duchy generally, notably at the expense of Lithuania. He also rebuilt the Moscow Kremlin, employing Italian engineers. Ivan IV, alias "Ivan the Terrible" (reigned 1533–84), was the first Grand Duke to be formally acknowledged as Tzar.

A map of Moscow crafted *c.* 1600 shows a rather commodious walled city extending evenly from the Kremlin and Red Square.[2] Tzar Ivan built an administration answering simply to himself. He used the threat of land confiscation to oblige the *boyars* (late medieval large landowners, often at court) to contribute to a military build-up. His capture, in 1552 and 1554 respectively, of Kazan and Astrakhan, two relict Tartar capitals on the Volga, opened the way to the East. His progress westwards was checked principally by Sweden. This widely read Tzar introduced printing to Russia.

Alas, a manically cruel streak was evident even in boyhood. Witness his hideous treatment of animals and birds. When he grew older, this trait surfaced most during surges of rage, these to be followed by swings to mortifying repentance as when he killed his own son with his bare hands. Peter the Great and Stalin could similarly be unspeakably savage, indeed on a vaster scale. In the governance of all three "there were features not far short of genius, in their statecraft, in their determination and in the iron control they exercised over their people. Yet their genius was flawed, if not by actual madness, at least by a lack of mental balance".[3] This verdict is graciously restrained.

Peter the Great

In 1700, some 40,000 Russian troops were defeated by 12,000 Swedish in a snow-storm at Narva. More decisive, however, was the defeat of 28,000 Swedish by 40,000 Russians at Poltava (north-east of the Dnieper bend) in June 1709. As usual, several factors played in. The Russians most energetically built earthworks to consolidate a defence line already strong by dint of thick woodland, marsh and stream. All throughout, Peter I ("the Great", grandson of Ivan the Terrible) epitomized courage and zest, a leader from the front. His antagonist, Charles XII of Sweden, was unable to command what was bound to be a complex attack as closely as he normally would have, this because of a foot wound incurred ten days before. Besides, this overly confident young man had not realized how forcefully the Russian army had been reformed since Narva. The soft core of the old army had been 20,000 Moscow-based *streltsy* musketeers — an unprofessional and indulgent coterie. In the course of eliminating them, Peter had executed 1200 and systematically mutilated many more.

Moreover, the Swedish fighting at Poltava had been (a) diverted from a march on Moscow, and (b) generally worn down by a ruthlessly adept scorched earth campaign waged by Peter through the bitter winter of 1708–9. Admittedly the Great Northern

War (1700–1721) between (a) Russia plus up to six other northern polities and (b) Sweden was to drag on a while yet. But Poltava had been a turning point.[4] The peace settlement would confirm the loss of all Sweden's Baltic possessions bar Finland, a country finally taken over by Russia in 1742.

Nearly seven feet tall and mighty strong with it, Peter the Great (reigned 1672–1725) had exhibited since childhood a lust for cruelty and obscenity that suffused his whole personality. But he combined this with an insatiable yen for physical labour, craftsmanship and matters mechanical. Being reared from the age of ten at a hunting lodge near Moscow nurtured these enthusiasms. Witness his beloved sailing boat, apparently a gift from Elizabeth I of England to Ivan IV. Life at the lodge also left him with a savage and abiding contempt for the *boyar* and *streltsy*-ridden Moscow establishment. In so far as both men grew up in an ambience of marginalized privilege, there is some correspondence between Peter the Great and Osama bin Laden.

Technocratic Transfer

Peter drew more determinedly than any predecessor on the West's skills in governance and science. Early on, he and 250 retainers did a three-year Great Embassy of Europe. Though its declared main purpose was to drum up an anti-Turkish alliance, there were broader aims. It arrived home just before New Year's Day 1699 on the old calendar, 1699 being the year the eschatologically-minded anticipated the showdown with the anti-Christ. The big *streltsy* purge began forthwith.

Cosmopolitan arts and crafts were much encouraged.[5] Elite education was developed along Western lines; and the Academy of Sciences was founded in 1725. Central government was rationalized drastically. Russia's territorial domain was not vastly extended but a considerable Baltic coastline was acquired to flank the building of Peter's new capital, St Petersburg, from 1712. Work also started on a Black Sea naval base at Taganrog on the Sea of Azov. In due course, too, Russia was divided into eight and later ten *gubernii*, each absorbing several of the customary provinces. Historians still debate whether they ever had enough suitably professional staff.

Between 1695 and 1725, fifty-two new ironworks were opened, a quarter of them in the Urals. The latter were to account for 40 per cent of all Russian production which divided pretty equally between State and private. In 1725 Russia led the world in iron output and, indeed, supplied most of Britain's iron imports.[6]

For Peter, the canons of good government were speed, compulsion and — by no means least — taxation. His "guiding stars were reason and enlightenment as he understood them, not popular preference or approval".[7] Through the rest of the Tzarist era, only one Russian ruler approaches him in visionary stature.

Catherine the Great

Catherine II, alias "Catherine the Great" (reigned 1762 to 1796), was a German princess married to the very inadequate Peter III. She succeeded him after his violent death while in the custody of military loyal to herself. She was industrious, intellectually curious and optimistic. She carried through at local level pertinent administrative and juridical reforms. The erraticism of Russia's eighteenth-century climate led to "68 very hungry years" putting a strong premium on better local management.[8]

Catherine looked forward more positively than had any previous Russian ruler to the abolition of serfdom, a designation still applicable to over 50 per cent of the population in a broad zone encompassing St Petersburg, Smolensk and the Oka–Volga confluence; and to over 20 per cent almost everywhere else.[9]

Yemelyan Pugachov's year-long Cossack and peasant rebellion of 1775 (extending from the Caspian to the Urals) worked in favour of change. The Cossack frontiersmen on the thinly peopled southern steppes had been identified since the fifteenth century as a people to reckon with. The colonies they formed around the Dneiper, Don and Ural rivers were martial in a rather democratic way. They enjoyed their peak leverage in the seventeenth century — i.e. before that gunpowder revolution had impacted fully.

Old Believers redux?

Peter I and Catherine II were each involved in Church–State relations, respectively reigning near the onset of and into the long decline of the Old Believers, a groundswell of protestation against Church reform along what were considered too Western lines. They had strongly emerged in opposition to the reformism engendered by a didactic Patriarch Nikon and Tzar Aleksey in the 1650s. Allowing for the mood swings and ambivalences characteristic of such situations, it seems fair to say that Old Believer attitudes at times took over about half the priesthood and laity.

Peter had been brought up along customary Orthodox lines, showing an aptitude for church singing. He wanted no truck with "divine right of kingship" but did want his subjects to cast off "fatalistic resignation, devotion to outward forms of worship and ingrained superstition",[10] these all being — in his view — barriers to technocratic modernization. Sectarian Old Believers could be liable to double taxation, dress regulation and sundry other pressures. Also, Peter abolished the patriarchate, substituting for it a synod he himself appointed. Better education of at least the senior clergy was set in train. Church administration was streamlined. The monasteries, some of them alternative loci of power on the boundless Russian landscape, came under tighter supervision.

Catherine II continued this general trend, most notably by the promotion of liberals within the Church as elsewhere. But the apprehension she so readily evinced when they took freedom of expression at face value defined the limits of her enlightened despotism. Her successor bar one, Alexander I (reigned 1801–25), started out imbued with Enlightenment ideals but by 1812 was being taken over by German mysticism, Lutheran pietism and Quakerism. He sought to resolve his confusion by turning zealously autocratic, especially in foreign policy.

Paris and St. Petersburg

Peter the Great reputedly said "Europe is necessary to us for a few decades, then we can turn our backs on her." In terms of geopolitics, that is not quite how things worked out. Catherine the Great encouraged the First Coalition to form up against Revolutionary France, though she herself focused instead on Poland's elimination through partition, a process ostensibly complete by 1795.

After Catherine's demise, Russia was to fight France within the Second Coalition. She did so with considerable success in north Italy. Then under the irascibly capricious Tzar Paul, she aligned briefly with Napoleon (1800–1). Next she was to join in the Third Coalition against France, 1805–7. Then in the wake of Bonapartist triumphs on land plus a terrible 1806 harvest, St Petersburg entered into alliance with Paris at Tilsit. This involved her joining Napoleon's Continental System for economic warfare against Britain.

In 1810, however, Tzar Alexander (son of the murdered Paul) took Russia out of the Continental System, one consideration being the discomfiture its constrictions were causing the landed classes. Prominent among several others was French encouragement of resurgent Polish nationalism. Mistrust deepened on both sides, with good reason on both sides. Classic action–reaction.

1812

The invasion of Russia by Napoleon's multinational army (i.e. only half French) of 600,000 was the upshot. Its progress to tragedy has been related many times, most philosophically by Leo Tolstoy. Suffice now to reflect on how such a force was reduced to 20,000 "scarecrows" by the time it left Russian territory.

Napoleon had identified well enough some of the difficulties facing him, above all a compelling need to fight decisive engagements yielding large numbers of Russian prisoners. For a start, there were only three roads, all of them bad, across the Pripet Marshes, a feature 200 kilometres broad. Elsewhere the road quality was very consistently poor. Then as regards foraging, the Russians could again use a "scorched earth" strategy to make the more taxing the low agricultural potential and the 1811 poor harvest. The continental Russian climate was strongly seasonal: hot summers, wet autumns and cold winters. But Napoleon was misled on the question of climate aberration (see below). He himself misread the Russian mood. And he failed to compound aright the problems he faced.

Quite the biggest action was Borodino, a setpiece in which the two sides fielded some 130,000 men apiece and had remarkably similar orders of battle — i.e. mixes of arms. Some well qualified to judge have said Napoleon twice missed a great opportunity by failing to press home a tactical advantage.[11] As it was, the French had over 30,000 *hors de combat* and the Russians 40,000 in what became an artillery-driven slog.

Borodino had been a battle site because, as he neared the Dvina, Napoleon finally opted to strike at Moscow not St Petersburg. The latter could hardly have burned down on occupation with the Tolstoyian "inevitability" Moscow did. All round, it could have been a sounder operational base, open to support by sea (save in icy midwinter) as well as being better able to exploit fissures in the Russian body politic. Assuming one thinks Napoleon was in any sense right to invade at all, it would have been the best objective. Wintering in Smolensk was another alternative, pondered then and since.[12]

Many commentators, starting with Napoleon himself, have stressed exceptional winter severity as contributing to catastrophe. Yet the first adverse weather was an abnormally hot high summer which killed off tens of thousands of the horses and made many men sick too. All the same, the winter was unexpectedly early and deep. It was so within a secular spell of renewed cooling that the emperor's meteorological adviser, Pierre Laplace, had overlooked. Laplace was a brilliant mathematician and, indeed, astronomer but also someone with altogether too uncritical a faith in the possibilities of scientific prediction. He duly averred that Russian winters "really" began in January whereas, in 1812, savage cold first onset early in November. The French retreat from Moscow had started on 19 October. Cruelly ironic is how a temporary thaw on 20 November almost caused a desperately depleted Grande Armée to be trapped the wrong side of the river Berezina. Surreally symbolic is how closely its retreat did follow the line of that southernmost terminal moraine from the last big glaciation, the Würm/Wisconsin.[13]

Napoleonic Aftermath

The ultimate collapse of Napoleon left Russia better placed, so it briefly seemed, to influence Western European events. Tzar Alexander much wanted a functioning Holy Alliance (Russia, Prussia, Austria . . .) to stamp out radical reform throughout Europe and, ideally, Latin America. Faced with British discouragement, however, that dismal vision dissolved within a few years.

In its place, apprehension waxed that the decay of the Ottoman Empire would proffer temptations to Great Power irredentism, Muscovite most particularly. The smashing of the Turkish fleet at Navarino in 1827 by British naval power (aided by French and Russian) was seen in London as facilitating an independent Greece, hopefully sufficiently well founded to withstand either threats or blandishments. This issue effectively marked the beginning of the Eastern Question as a recurring dilemma in Britain's external affairs the next 80 years.

It was the ultimate cause of the Crimean War (1854–6), the proximate one being the French persuading the Turks to allow Rome custody over the Church of the Holy Nativity in Bethlehem, customarily an Orthodox preserve. In 1853, the Russians defeated the Turkish fleet at Sinope and marched into Moldavia and Wallachia. France and Britain duly dispatched to the Crimea an expeditionary force 60,000 strong. Though ill-managed (certainly on the British side), it proved able to outpoint what should have been in that locale a much stronger Tzarist army. This was hardly a brilliant demonstration of Russian strategic mobility in the forenoon of the railway age.

For the duration, the Allies had waged economic warfare via naval blockade, especially in the Baltic. At the same time, they threatened the key Russian naval base at Kronstadt. The overall results are hard to gauge. While the ordinary people of the Baltic provinces suffered badly, big landowners wherever still tended to profit from a war boom.[14] Early in the conflict, the British abandoned privateering as well as the prerogative of searching neutral ships likely trading with the enemy, a concession intended to underpin political support in Europe and the USA. At all events, under the 1856 Treaty of Paris, the Russians made big concessions. They relinquished (a) their right to re-establish a naval presence in the Black Sea and (b) the guardianship of Christian subjects of the Sultan. More basically, Moscow accepted that the Ottoman crisis (plus the status of the Straits) were the concern of what became known as the Concert of Europe: a softer, more contingent rendering of the Holy Alliance.

A realization domestically that Russia had thus been worsted on its very doorstep fed into peasant rage about the impositions of serfdom. Often this rage found immature expression: millenarian beliefs in the imminence of paradisal translation; suspicion that the Tzar had ordained the end of serfdom but the landlords had never let on, . . . But an urban intelligentsia was now waxing more powerful and was seeking reform in every direction. Within several years, serfdom was being abolished.

Somebody who chose a career in the Russian army partly to seek redress for the Crimean humiliation was Nikolai Przhevalsky (1839–88). More than anyone else, Russian or British, he epitomized Rudyard Kipling's "Great Game": the Anglo-Russian rivalry in which exploration (natural science or archaeology) served to pave the way for Inner Asia dominance. His destinations included Turkestan, Tibet and the Amur. His first priority was always military reconnaissance. He was deeply imbued with a Social Darwinist scorn for indigenes, not least the Chinese.[15]

Between 1856 and the next Tzarist crisis of extroversion, which came in 1877–8, pan-Slavism waxed strong. It can be defined as the quest for an exclusive sphere of influence in the Balkans. But the emergent Balkan nationalities were not enamoured, not even Bulgaria which would owe almost everything to Russia, in national regeneration and liberation from Turkey.

The resumption of Russo-Turkish hostilities in 1877 brought a Russian army to the landward outskirts of Constantinople and, in action–reaction, a Royal Naval task force to the Golden Horn. The Great Game was more or less at its climax. Russia's annexation, in 1884, of the city of Merv in Turkmenistan caused apprehension in Calcutta and London. But Merv lies 600 miles west-north-west of the Khyber Pass; and the beeline crosses what was then an extremely primitive Afghanistan. Lord Salisbury, the Foreign Secretary, quipped "Mervousness does not stand the test of large-scale maps". Once again a threat to Kronstadt was presented.[16]

Russian fiscal management could not have coped with a commitment as demanding as a probe towards India. Ever since Catherine II's time, imperial designs had been financed by foreign loans and a depreciating paper currency. Moreover, depreciation was repeatedly made uncomfortably erratic by bouts of warfare, the Russo-Turkish war of 1877–8 most certainly no exception. Not until a switch to the gold standard had been completed in 1899 was the currency able to proffer as much crisis stability as the more general economic backwardness would allow.[17] The British, of all people, should have realized this.

If there was a challenge to the Raj in India or, indeed, the Near East, it perhaps lay more in the aplomb with which St. Petersburg struck deals with the mainstream Islamic leadership regionally as the imperial frontier extended across Central Asia. This propensity served to strengthen religious leaders within their respective communities. One longer term result was that a million Moslems served in the Tzarist army in the First World War.[18] Another effect may have been increased Russian influence in Persia where much of the Shia religious leadership became persuaded awhile that the more conservative and rather mystical Muscovite Christianity might be a lesser evil than the West's values, however defined.

Not that Russian progress into Central Asia was purely or even predominantly a question of soft power. The Kazakh steppes, in particular, were progressively brought under the control of forts intended to check Kazakh internecine conflict (as well as forays by the Dzhungar dwelling to eastwards) but also, from 1865, to constrict transhumance, thereby obliging the Kazakhs to become more sedentary. Thenceforth, too, widely intrusive Russian settlement was encouraged. The number of immigrants imposed on the Kazakhs averaged 60,000 a year, 1890 to 1915. They usually settled where rainfall was in excess of 20 cm a year.[19]

For just over a century from 1765, a polylateral struggle was waged in the Caucasus. The local participants were Orthodox Georgians, Armenian Christians and the mainly Shia Azerbaijanis. Each fought within themselves as well as against those without. Steep mountains, forests and snowfields afforded many defensive positions. The external powers involved were Turkey, Persia and Russia and, potentially (in 1854–6 and 1877–8), Britain. By 1878, all Georgia and most of Armenia and of Azerbaijan were in Russian hands. Meanwhile, oil was becoming a factor in grand strategy. From 1883, a railway and from 1900 a pipeline linked Batum on the Black Sea to the "eternal fires" of Baku by the Caspian. Extensive Russian colonization had also been taking place there.

The Russo-Japanese War

The Treaty of Berlin, 1878, committed the Concert of Europe more firmly to stabilization in the Balkans and Asia Minor. Therefore Russia turned her attention more to the Far East. However, this was still a region too remote to lend itself to policies being controlled with finesse from St Petersburg. The upshot was a strategy which was overly bold and poorly co-ordinated.

Two primary concerns were the acquisition of an ice-free port and the forging of the Trans-Siberian rail link. The seaport of Vladivostok, founded in 1861, was not ice-free in winter. So in 1898 the Chinese were persuaded to lease the Liaotung peninsula (Port Arthur and Dairen included) to Russia for 25 years. Meanwhile, in 1891, work had started on the Trans-Siberian railway to Vladivostok. Mostly it would follow the discontinuous but tolerably straight belt of Black Earth soils fringing the southern limit of the Siberian *taiga* (see below). But the last section eastward would pass through very tough terrain while closely circumventing Manchuria. Still, by 1902, the whole route was completed single track, save that everything and everybody had to be ferried across Lake Baikal.

China had ceded Korea and Taiwan to Japan after defeat in war, 1894–5. So a defensive compact was made with Russia in 1896. It included the provision of a corridor for the Russians to construct a very straight rail link to Vladivostok via Harbin. This Chinese Eastern Railway, notionally under a Chinese President, effectively bisected Manchuria. Then came the Liaotung deal just mentioned. A rider to it provided for a corridor for what became known as the South Manchurian Railway from Harbin to that peninsula. The country around Dairen, in particular, was then strongly developed by the Russians. Manchuria duly became a cockpit for geopolitical rivalry between St Petersburg and Tokyo. Geopolitical rhetoric soon ensured that Korea figured too.

Given a diplomatic impasse, Japan made undeclared war against Russia from February 1904. It was fought, on both sides, with courage and a correctitude based on mutual regard. The Japanese captured Port Arthur in December 1904 and Mukden in March 1905. The latter had cost them 40,000 casualties and the former 60,000. The death toll at Port Arthur included two sons of Nogi Maresuke, the Japanese field commander.

Yet it was the naval Battle of Tsushima that finally determined things. In May 1905, a 40-ship task force from the Russian Baltic fleet arrived off Japan, having originally been dispatched by the Tzar to Port Arthur. The weary armada engaged Japan's navy in the historic straits between Kyushu and Korea. After being badly worsted by the Japanese big ships, this Russian force was all but obliterated overnight by torpedo boats, the total Japanese losses being but three of the latter. The development by 1886 of operational torpedoes had already engendered great anxiety in established naval circles since they were recognized to be extremely lethal against all warships yet eligible for delivery by small craft. Russia had failed to anticipate that Japan's new-found navy would effect what still stands as the supreme application of this stratagem.

The strategic railway programme notwithstanding, the Russian land campaign had been compromised by logistic overstretch. In the case of the Baltic task force, such problems were aggravated by logistic diplomacy. This especially applied in regard to the half million tons of quality coal needed *en route*, this preferably being received in sheltered harbours. International law was unclear about how far non-belligerents could facilitate this without compromising their neutrality.

For some time past, the Russian navy had depended on Britain as a singularly good supplier of hard coal. However, London was reluctant to collaborate too directly because of (a) the 1902 treaty with Japan and (b) the casualties incurred in her North Sea fishing fleet through its being fired on by a jittery Baltic task force. The Russians had their first coaling off the Danish coast and their second (following hard bargaining with Madrid) at Vigo in Spain. Afterwards France (as an ally of Russia) edgily made harbours available in West Africa, Madagascar and Camranh Bay in Indo-China. Lüderitz in German South-West Africa also became a stopover point.

The Hamburg–America line provided the colliers throughout. However, Kaiser Wilhelm vacillated all too characteristically about how far to endorse this business deal. Needless to say, when this ill-balanced task force of mainly "rusty, undermanned, antiquated hulks" eventually arrived in Japanese waters it was very unready for war.[20]

Pre-Revolutionary Crisis

American mediation helped end hostilities in September 1905. The terms were not too mortifying for St Petersburg if only because Tokyo was too exhausted to be more insistent. Even so, the Liaotung lease was transferred to the latter as was southern Sakhalin and much of the South Manchurian railway. So a plutonic upheaval in the Russian body politic had been brought closer. Already unrest was almost endemic, having been aggravated by widespread famine, especially in 1891–2 but also in 1897, 1898 and 1901.

A decidedly negative manifestation had been the anti-Semitism that climaxed in 1891–2 with the eviction of thousands of Jewish artisans from Moscow; and the clearance of Jews from a wide swathe of land inside the western border. A contingently more positive one had been the growth of Anarchist sentiment led by Mikhail Bakunin (1814–76) and then Prince Peter Kropotkin (1842–1921), a gracious Russian aristocrat renowned as an explorer of Siberia and advocate of non-violent social transformation. Romantic visions of the *mir*, the peasant tradition of collective self-management at village level, was a defining feature of Russian Anarchism.

Leo Tolstoy (1828–1910) was another aristocrat who came to advocate, at a high literary level, non-violent social transformation guided by rural values. Yet while he was crafting *Anna Karenina* (1873–6) there was an upsurge of political killings led by the Land and People, a St Petersburg faction largely comprising disaffected young aristocrats. It assassinated Tzar Alexander II in 1881, then was itself destroyed in the backlash. The vacuum thus left in the radical cosmos was substantially filled by Marxism.

The 1905 Revolution

In December 1904, a strike crippled the Baku oil fields. Then in January 1905, a huge but basically loyalist crowd gathered outside the Winter Palace in St Petersburg, living standards its main concern. Troops opened fire killing, that "Bloody Sunday", at least 130 demonstrators. Come the summer, rural rebellion was spreading again — most violently in the peripheral and national minority areas. In October, a general strike paralyzed the whole country. By then, too, scores of local Soviets were being formed, notably in the railway sheds. That autumn the Tzar accepted a manifesto providing *inter alia* for freedom of expression and extending the franchise to the peasantry.

Barely a year later, however, he was to backtrack on a commitment to have the Duma (the national assembly) approve all legislative bills. Much the same attitude can be seen in yet another failure so to reform the armed services as to avoid further humiliating defeats. Granted, in 1906–7, between 50 and 80 per cent of all troop commanders were replaced. However, this purge "seems to have been directed as much against those who had been indecisive in dealing with civilian revolution and mutiny . . . as against those with questionable ability to lead troops in battle".[21] And yet "Russian rearmament profoundly affected western Eurasia after 1909 and significantly contributed to the road to war in Europe in 1914".[22]

Not that all seemed lost for Tzarist Russia come December 1905. That month, a General Strike call had little effect. Instead, a succession of violent uprisings were savagely put down, including one by Moscow Bolsheviks. There the tables were turned by troops loyal to Tzar Nicholas (as still the great majority more or less were) entering by the one rail line not on strike.[23] Once things had quietened, agricultural progress made the running. Within several years, productivity gains had redounded considerably to the benefit of the peasantry. There was also promise medium-term in the industrial sector with its large firms and statist links.

World War One

Historians readily concede that the Imperial Russian Army "made a tremendous contribution to the Allied cause".[24] At the outset, its offensive actions distracted the Germans in their drive towards Paris, though whether they actually lessened its strength depends on what one sees as the limiting factor. Was it the number of troops available? Or was it (*à la* Schiefflen Plan) the capacity of the regional transport network? At all events, one of the two Russian offensive thrusts, that into Galicia, was a success in itself. The other led to disaster at Tannenberg midst the Masurian morainic lakes of eastern Prussia. After ten months of hostilities, Russian casualties (prisoners of war included) are said to have topped 3,800,000.

Yet indirect support of other allies continued quite strongly through 1916. By then, however, enormous problems were building up in terms of weapons and munitions shortages; gross mismanagement of the draft; unrest across the home front; weakening central government . . . The immediate outcome was the leftist revolution of February 1917. Leftist but also grass-roots patriotic. One popular aim was the constraining or removal of the "German dynasty", the Romanovs.[25]

Initially the uprising was confined to Petrograd — to use St Petersburg's wartime name. Yet soon land seizures were proliferating fast and far. Next, the stakes were raised by the Germans' dispatching Vladimir Lenin (1870–1924) from exile in

Switzerland to Petrograd's Finland Station to arrive in April 1917. Even so, into the summer the Bolshevist position looked decidedly shaky as Kerensky sought to curb them as extremists and defeatists. Lenin withdrew to Finland awhile. Leon Trotsky (1879–1940) was in prison.

Then patriotic sentiment played in favour of the Hard Left. An attempt by Kerensky to reactivate the Galician front, this in fulfillment of an Alliance commitment, crumbled. Worse, his politically naïve Commander-in-Chief, Lavr Kornilov, was persuaded by people acting with some Allied support to essay a counter-revolutionary coup in Petrograd acting in defiance of Kerensky himself. The move was foiled by popular resolve with the railway workers tearing up key tracks. The result was a decisive accession of strength for the Bolsheviks in Petrograd and well beyond. Hence the October Revolution. Alexander Kerensky fled for good. Kornilov was killed in action the next year.

The Siberia Legend

One might conclude this perspective on Tzarist Russia's emergence and demise with a review of how the mental mapping of Siberia has evolved to date, bearing in mind that this vast territory closely matches the "heartland" as Halford Mackinder pristinely perceived it. Peter the Great's averration that his tzardom was "an imperium" led on to 1724 and his dispatching Vitus Bering (1681–1741) of Denmark to delineate the land bridge Peter presumed linked to America – what was coming to be known as "Siberia". Bering showed a bridge did not quite exist. His other accomplishments included identifying Kamchatka as a strategic feature.

On first sight, the resemblance between Siberia and the Canadian Shield is marked. All the same, the shield's geology is the more consistently ancient. The hydrocarbon-bearing rocks of West Siberia are relatively young. And yet there is a pronounced topographical uniformity across the West Siberian plain: "the largest extent of flat land in the world", it extends 1200 miles across in both directions and is never above 700 feet.[26]

The *taiga* forest, so prevalent across this region, was to become the prime *motif* of a school of exquisitely subtle landscape painting. It thrived in the early nineteenth century, in line with the general evolution of Russian attitudes towards Siberia and just several decades behind, medianally, the great English landscape school.

For centuries, Russians had settled across that long tongue of steppic grassland extending roughly along 55°N through west Siberia between the *taiga* and the salt steppe or desert. Yet before 1917 "the enormous forested tracts of Siberia were scarcely touched. There forest fires did more damage than did deliberate felling by Man, especially along the line of the Trans-Siberian where the present-day dominance of secondary birch forest bears witness to the results of sparks from steam locomotives".[27] Not until a Forestry Directorate was established in western Siberia in 1884 was a start made with lumbering.

The extractive industry Peter the Great had sought to encourage East of the Urals was fur gathering. Initially "the luscious pelts of the Siberian *taiga* were as precious as gold and silver"; further North one went, the more quality improved. However, by the early nineteenth century, this trade was in precipitate decline due to changing fashions but also over-exploitation. Duly, a reaction set in against trite expectations of Siberian abundance. As one St Petersburg bureaucrat put it, "Nevskii Prospect is worth at least five times as much as the whole of Siberia". Regarding official perspectives, however, it is needful to recognize that, by the eighteenth century, Siberia had become notorious for bad administration, too far from St Petersburg to be much else.

Visions of what Siberia could be about were soon diversifying. Landscape painting was one manifestation of this. Another was development of the penitentiary role. Many of the Decembrist rebels of 1825 were there long-term; and not a few became persuaded the region could be rapidly developed by immigrants imbued with more progressive norms (e.g. no serfdom) than obtained in Russia proper. Some drew inspiration from the American model. Others turned to Russian historical romanticism.

Particularly influential in this regard as in others was Alexander Herzen, Leftist revolutionary and (post-1848) "Russian soul" nationalist. He was exiled to the countryside (1834–40) but in European Russia. Never did he set foot in Siberia. Nevertheless, he thoroughly convinced himself that its elemental qualities left scope for fresh starts along egalitarian lines. Further, he was among those who believed midcentury that Russian–American coupling in this frontier context might somehow free the world from old Europe. However, like Mikhail Bakunin, he apprehended that a developed Siberia would claim its independence. That consummation some in Moscow sought to defer indefinitely, by holding back on regional development.

However, defeat in the Russo-Japanese war actually stimulated the development of Siberia. In 1700, some 300,000 Russians are believed to have been living there. By 1905, perhaps 4,400,000 new immigrants had arrived. Then by 1913, three million more had settled, the peak year being 1908. While noting these figures, Fridtjof Nansen (1861–1930) — the great Norwegian explorer, scholar, statesman and Nobel Laureate — celebrated Siberia with characteristic generosity. In so doing, alas, he most certainly overstated the agricultural potential of its largely acidic soils.[28]

The Northern Sea Route along the Arctic coast was also accented by the 1905 defeat though still more so by the advent of Bolshevism, given Stalin's professed conviction that "the soul of Russia lies in the ice and solitude of the Far North". Ironically, however, Terence Armstrong recognized in 1952 the opportunity presented by the diminution of sea ice since 1920, a consequence of regional warming and — supposedly — increased storminess. Yet he also stressed that, however important the Northern Sea Route was in opening up the Far North (much of the Gulag Archipelago included), it had little wider impact. For instance, only one part in 40 by volume of American lend-lease aid in World War Two came westwards down this route.[29]

In more recent times, the poet, Evgenii Evtushenko, extolled his Siberian homeland, a pitch no doubt related to his fervent anti-Maoist stance. Also, in 1974, a Solzhenitzyn now in exile abroad urged the USSR to ease up on its preoccupation with the outside world and focus on "our hope and our reservoir", alias the "Russian North East". Its "boundless expanses, senselessly left stagnant and icily barren for four centuries", need the sacrifice, zeal and love of "a free people with a free understanding of our national mission" to "awaken them, heal them, beautify them with feats of engineering".[30]

However, the portion of Siberia which reasonably qualifies as the "Russian North East" is actually worse, much worse, than a *taiga* plain. It mainly consists of some of the rawest mountains anywhere, usable in places for metal mining but for little else. The muddled romanticism and obsolescent "conquest of Nature" undertones that informed this disquisition were the *alter ego* of a savage critique Solzhenitzyn was enunciating of what the West was using its democracy for, the "continuous seepage of liquid manure, the self-indulgent and squalid popular mass culture" that was threatening to undermine deeper values in Russia as elsewhere. This reaction made this great author more sympathetic to Putin than one might otherwise have expected. Meanwhile the population of the Russian Far East has fallen from 8.5 to 7.0 million the last 15 years. Energetic Chinese immigrants have sadly accentuated the malaise and resentment. They have been under notice to leave.

Notes

1 I. E. Buchinsky (T. C. Marwick, trans.), *O Klimate Proshlogo Russkoy Ravniny* (Leningrad: Gidrometeroizdat, 1957), pp. 74–80.
2 Robin Milner-Gulland with Nikolai Dejevsky, *Cultural Atlas of Russia and the Soviet Union* (Oxford: Equinox, 1989), p. 57.
3 Vivian Green, *The Madness of Kings* (New York: St. Martin's Press, 1993), p. 138.
4 John Adair, "Poltava" in Cyril Falls (ed.), *Great Battles of History* (London: Weidenfeld and Nicolson, 1964), pp. 31–48
5 Lindsey A. J. Hughes, "The Seventeenth Century Renaissance in Russia", *History Today* 30, 2 (February 1980): 41.5.
6 Lionel Kochan, *The Making of Modern Russia* (Harmondsworth: Penguin, 1963, p. 107.
7 Nicholas V. Riasanovsky, *Russian Identities* (Oxford: Oxford University Press, 2005), p. 82.
8 Ye. P. Borisenkov in Raymond S. Bradley and Philip D. Jonas (ed.), *Climate since AD 1500* (London: Routledge, 1992), Chapter 9.
9 Milner Gulland and Dejevsky, *Cultural Atlas of Russia the Soviet Union,* p. 111.
10 B. H. Sumner, *Peter the Great and the Emergence of Russia* (London: English Universities Press, 1950), p. 139.
11 Brigadier Peter Young, "Borodino" in Falls (ed.), *Great Battles of History*, pp. 128–39.
12 David Chandler, *The Campaigns of Napoleon* (London: Weidenfeld and Nicolson, 1967), pp. 858–9.
13 *HCC*, p. 297.
14 Clive Anderson, "Economic Warfare in the Crimean War", *Economic History Review* 2nd Series XIV, 1 (1961):34–47.
15 Kyrill Kunakhovich, "Nikolai Mikhailovich Przhevalsky and the Politics of Russian Imperialism", *International Dunhuang Project News* 27 (Spring 2006): 3–5.
16 I. R. Hill, *War at Sea in the Ironclad Age* (London: Cassell, 2000), p. 189.
17 Olga Crisp, "Russian Financial Policy and the Gold Standard at the End of the 19th Century", *Economic History Review* VI, 2 (December 1953): 156–75.
18 Robert D. Crews, *For Prophet and Czar* (Cambridge, MA: Harvard University Press, 2006), pp. 350–60.
19 I. Stebelsky, "The Frontier in Central Asia" in J. H. Bater and R. A. French (eds.), *Studies in Russian Historical Geography*, 2 Vols. (New York: Academic Press, 1983, Vol. 1, Chapter 7.
20 Lamar J. R. Cecil, "Coal for the Fleet that had to Die", *American Historical Review* 69, 4 (July 1964): 990–1005.
21 John Bushnell, "The Czarist Officer Corps, 1881–1914. Customs, Duties, Inefficiency", *American Historical Review* 86, 4 (October 1981): 752–80.
22 Jonathan A. Grant, *Rulers, Guns and Money* (Cambridge, MA: Harvard University Press, 2007), p. 243.
23 Beryl Williams, "Russia, 1905", *History Today* 55, 05 (May 2005): 44–51.
24 Malcolm Mackintosh, *Juggernaut* (London: Secker and Warburg, 1967), p. 17.
25 Kochan, *The Making of Modern Russia*, p. 241.
26 W. H. Parker, *An Historical Geography of Russia* (London: University Press, 1968), p. 19.

27 R. A. French, "The Russians and the Forest" in Bater and French (eds.), *Studies in Russian Historical Geography*, Vol. 1, Chapter 2.

28 Fridtjof Nansen, *Through Siberia, the Land of the Future* (London: William Heinemann, 1914), Chapter XIII.

29 Terence Armstrong, *The Northern Sea Route* (Cambridge: Cambridge University Press, 1952), pp. 39–40 and 49.

30 Mark Bassin, "Inventing Siberia: Visions of the Russian East in the early 19th Century", *American Historical Review* 96, 3 (June 1991): 763–94.

8 | The Soviet Experiment

One question to ponder is how Stalin came to succeed Lenin. There could be several answers. One must be the latter's awareness of the former's sinister skill as an authoritarian organizer. Shortly before his own death, Lenin did fulminate against Stalin but suggested no alternative.

All else apart, a more empathetic leader might have found the human losses consequent upon world war hard to bear. To the battle casualties in that conflict and its aftermath, add accentuated mortality behind the fronts — notably the global influenza epidemic of 1918–19, a product of military mobilization and civilian debilitation. Reckon in, too, "birth deficits" due to male call up. Overall the world's population in 1920 was 60 million less than it otherwise could have been; nearly half that shortfall was in Russia.[1]

No doubt, too, containing the counter-revolution waged from 1918 onwards (with some support from several Allied armies) called for a bit of iron fist. That August, the Bolsheviks and their associates effectively held the Russian heartland from Archangel to Astrakhan and from Smolensk to the Urals. Though ill-adapted as yet to open mobile warfare, they were already adept at urban fighting. The charismatic revolutionary, Leon Trotsky (1879–1940), had been made Commissar for War several months after directing the Petrograd uprising in November 1917.

In the event, this desperate struggle was all but won by the Bolsheviks with the collapse of socialist and White conservative opposition in the Caucasus and southern Russia in the winter of 1920–1. A large Japanese force would stay in Siberia till 1922 and northern Sakhalin till 1925 but without a clear military aim or much support at home.[2] That deployment had owed something to Tokyo's concern to do more to justify its elevation to Great Power status at this, the start of the League of Nations era.

Early in 1920, the Allies had proposed, as the basis for a Polish–Soviet border determination, the Curzon Line: a cartographic divide which defined as well as any might where Polish demographic dominance gave way to Russian or Ukrainian. However, the Polish nationhood emergent under Joseph Pilsudski was looking for a resolution more like the 1772 frontier which came close to Smolensk. So in April, it having become improbable that attacking the Reds would facilitate the return to power of the Whites, Pilsudski went on the offensive, seizing Kiev awhile. As ex-commander of a Polish Legion the Germans had sponsored, Marshal Pilsudski had brought his troops to a high level of proficiency whereas their Soviet adversaries were still unready, especially in data transmission and logistics, for a war of strategic manoeuvre. The outcome was a crushing Polish victory at the Battle of Warsaw in August 1920 and hence the Riga peace treaty in October. This set the frontier well east of the Curzon Line.

The Warsaw *débâcle* engendered years of recrimination in army and party. Leon Trotsky had done remarkable work in 1918–19 welding the Red Army together. Nevertheless, he had expressed at the time his conviction that it still remained unready to conquer Poland *en route* to bringing Communism to a collapsed Germany, a putative prospect not a few comrades enthused about in the heady midsummer of 1920. [3]

A difficulty throughout had been that Russia (with its limited industrial proletariat) fitted badly the Marxian revolutionary progression. Not only had Marx himself been famously Russophobic around the time of the Crimean War. Not until the 1870s did he or Engels interest themselves much in Russian revolutionary thought; and any such interest was always laced with scepticism. Pertinent is Engels' 1885 comment that "people who boast that they have *made* a revolution always see the day after . . . that the revolution *made* does not in the least resemble the one they would have liked to make".[4]

However, "exporting the revolution" from Russia was a precept Trotsky had often expounded in more general terms. So the Battle of Warsaw was bound eventually to count against him in the bitter rivalry with Stalin evident since 1918. Moreover, that year Lenin had introduced "War Communism", a reworking of the Soviet experiment which was much more centralized and authoritarian politically. In September, the Cheka political police initiated a "Red Terror" — a systematic elimination of political opposition through "mass arrests, interrogation under torture, hostage taking of family and children, and summary mass executions".[5] The organizational aspects of such developments will have been very much Stalin's *metier*.

Collectivized Agriculture

In 1928, the revisionist New Economic Policy of 1922 gave way to the first Five Year Plan. From the outset, a salient theme was the collectivization of agriculture. Pre-1914, agriculture even in western Russia was demonstrably backward compared with that in, say, East Prussia: the former being far too subject to demographic overcrowding, soil exhaustion, under-capitalization and absentee landlords. In 1905–6, peasant disturbances had affected over 75 per cent of districts in eastern Ukraine and middle Volga. That compass coincided quite closely with rural population densities being exceptionally high — 40 to 75 per square kilometre. Agriculture, in fact, was "at its most primitive in the central Black Earth region where the wooden plough and harrow were in universal use".[6] The *loess* was sustaining too many people.

The communitarian tradition in the Russian countryside might have served as the basis for co-operative agreements between individual peasant owners. The route Stalin actually followed, that of enforced collectivization, engendered a fierce backlash which must have extended well beyond the rich peasants — the *kulaks* — so often focussed on exclusively by commentators in the West. From 1929 to 1932, there was veritably civil war in much of the countryside with peasants killing livestock, burning crops and destroying homesteads rather than be collectivized. For instance, by 1933 the number of sheep and goats kept was officially put at 42 million as opposed to 146 in 1928. By then, too, the number of human deaths through famine (especially in the Ukraine) may have reached 14 million; and millions of country people were in *gulags*. Rather over half the 25 million peasant households had thus far been collectivized.[7]

The Great Terror

The campaign against the *kulaks* led on to the Great Terror purges of 1934–8. According to Khrushchev's "secret speech" in 1956 denouncing Stalin, these started in reaction to the gunning down by a Communist dissident of the head of the Leningrad party organization. Obviously, however, they were ultimately manifestations of Stalin's warped personality. Contradictory though this was in many ways, its driving force was a boundless urge to exercise power brutally. For many, this reality was obscured by his abiding predilection for a simple life style, one devoid at a mundane level of ostentatious display.[8]

Very largely, the assemblages of individuals responsible for the mass killings during the Great Terror were those who had performed likewise in the civil war and then the collectivization drive. They were associated with Efim Georgievich Evdokimov (1891–1939). All the key figures were themselves arrested (and most, if not all, shot) early in 1940. Authoritative estimates are that one million people were executed in the Terror and eight million sent to *gulags*. In all probability, not a million of those inmates survived.[9]

In the wake of the 1934 assassination, the purge of Communist party members was especially severe in Leningrad where Andrei Zhdanov was party boss. In the purge of the military, the Far Eastern Army was something of a special case. Thus it was the only such body of troops commanded by a Marshal, "the tough, competent and practically experienced" Vassily Blukher. Initially it was purged less severely. In 1938–9, however, leniency was set aside. Blukher's subsequent "fall and death marks the end of the last tenuous hope of action against Stalin".[10] There had been some intimations of dissident scheming in Siberia.

Overall, insinuations of collusion with predatory foreign states (and especially Germany) waxed prominent in the rationale presented for purging the military in particular. In early 1938, Marshal Mikhail Tukhachevski was framed as a supposed collaborator with the Third Reich. He was chief of the Red Army and a prophet of the full-scale mechanized counter-offensive. His demise was the prelude to a bloodbath in which up to 35,000 army officers were dismissed, imprisoned or, in very many cases, summarily shot. They included 15 out of the 20 marshals and army commanders; and even 200 out of the 406 brigade commanders.

The Winter War

How much all this had compromised the operational efficiency of the Red Army was demonstrated by the "winter war" against Finland from October 1939. Under the spheres of influence provisions of the Nazi–Soviet pact that August, Moscow was free to insist on territorial concessions from Finland as from others in order to gain strategic depth. After Helsinki had failed to concede enough, Moscow turned to war — the ostensible aim now being to set up a "Finnish Democratic Government" comprised, of course, of a Communist rump.

The Red Army deployed 1,200,000 troops in theatre which gave it an order of magnitude advantage in manpower. True, this would never have been easy to exploit across the critically important Karelian isthmus: a mere 40 miles wide at the border and with good defensive terrain bolstered by the simple but well sited fortifications of the Mannerheim Line. Northwards from Lake Ladoga, the scope for open warfare was greater in principle whether or not it could be exploited in practice. Throughout, the Soviets were overwhelmingly superior, on paper, in air and naval power.

However, the Finns fought with much more flair and resolve. Their ski-borne fighting patrols, in particular, became legendary. Moreover, these intrinsic strengths were accented by their being far better equipped to withstand seasonal mean temperatures 8 to 10 degrees Celsius below the then norms (see below). Only utter exhaustion, human and material, induced the Finns to make peace in March 1940.

The Great Patriotic War

Though more recently historians have concluded otherwise,[11] in the immediate wake of the quite heavy fighting between the Soviet army and the Japanese in 1939, the general inclination afterwards was to say the former had been worsted. So taken together with what was known of the Great Terror and then the close coverage of the Winter War, there was by 1941 a widespread disposition to believe the USSR would not long withstand a full-scale German onslaught. A week before this transpired, in the form of Operation Barbarossa, Britain's Joint Intelligence Committee gave her two months. There were plenty of informed people in Berlin who thought likewise. Hence their preparedness to embark on that catastrophe.[12]

It might be reasonable to presume that the Soviet armed services had duly absorbed lessons from the near disastrous Winter War. Yet such was but partially the case. Much prevarication meant that "when the invasion came the Red Army was caught between two types of organization and two types of armament. The old had not been completely abandoned; the new was still being debated".[13] A case in point was the Stalin Line, a rather intermittent cordon of "deep defensive" and "antitank" zones, construction of which began in 1933. It ran from the Gulf of Finland to the Black Sea just inside the then international border; and was at its strongest south from the Pripet Marshes.[14] Little transfer of its assets westward to the new border zone even started until the Spring of 1941. In the short time remaining, it proceeded "in a haphazard and leisurely manner".[15] In the event, this politically-driven preoccupation with forward defence led to 800 Soviet aircraft being destroyed on the ground on Sunday, 22 June 1941 — the first day of Operation Barbarossa.

Stalin's personal obsession with offensive action found expression in plans for an all-out counter-attack sooner rather than later. Initially formulated in the autumn of 1940, these had by May 1941 assumed the form of a huge thrust involving 153 divisions from south of the Pripet Marshes through to Silesia. Inevitably, preparations were hamstrung by limits on railway capacity; poor command and control provision at field level; and other shortcomings apparent during the Winter War but not yet corrected. Throughout, the military were dogged by Stalin's contrary concern not to invite German pre-emption by evident preparation.[16] The Soviets would have done far better to prepare in depth a succession of defence lines backed by mobile forces for sectoral counter-attack.

The awful progress of the German invasion has often been well recounted. Suffice now to identify some salient realities appertaining to it. By that first wintertime the Wehrmacht was investing Leningrad, confronting Moscow, and standing astride the Don valley. The ascendancy they had gained in savage manoeuvre warfare had culminated in October in a big pincer movement closing at Vyazma (half-way between Smolensk and Moscow). This resulted in about a million Soviet losses (dead, gravely wounded or prisoners). In Moscow there were panic riots with much looting that month.

Yet the Wehrmacht's gains were not well underpinned logistically. The Soviet railway system proved much harder to adapt than anticipated. Making due use of the often muddied or dusty Russian roads would have required a level of mechanization Germany was quite unable to achieve or sustain. And one particular option that was thereby ruled out was an overwhelming assault on Moscow. The 70 divisions in that central sector could not have been reinforced and resupplied sufficiently.[17]

That winter the capital was the focal point of a major Soviet counter-offensive undertaken largely by troops fresh from the Far East. Stalin had gambled on the soundness of intelligence about Japan's currently peaceable intentions towards the USSR gathered by the formidable spy ring Richard Sorge had established in Tokyo by 1936. Meanwhile, in the South, the first major amphibious assault by Russia at least since Viking times recovered (at heavy cost) the Kerch peninsula, thereby easing awhile the pressure on a beleaguered Sevastopol. And now Leningrad was destined to withstand siege, terribly but triumphantly.

Again, the colder interlude in Europe's climate from 1939–40 is a factor to conjure. Come early 1941, the German meteorological service (arguably "the best and the brightest" at that time) erred, remarkably like Laplace, in endorsing the view that June was not too late in the season to launch an invasion of Russia. Once again, an interlude of colder years was not perceived. In 1941, heavy October rains were to make most roads impassable. Then hard frosts from early November disorganized further the railways so essential to the Wehrmacht's progress. Meantime, the German soldiers were desperately ill-accoutred for winter compared with the Soviets. Admittedly, that Moscow counter-offensive was only moderately successful. Even so, it strengthened a burgeoning sense of underlying Soviet resolve. As Ilya Ehrenberg, the famous Moscow columnist, averred at the time: "Our victory is clothed in a camouflage dress. You won't notice it at first. Its lips are sealed and its face is powdered with snow."

Nevertheless, from May to August 1942, the Wehrmacht made anew huge territorial gains, these against a Red Army reeling, demoralized and confused like before. Again, too, the Soviet resistance stiffened, this time before Stalingrad, a city located where the Volga and the Don almost merge. The task of defending it (or, to be more accurate, its ruins) was rendered much harder by its standing on the right bank of the Volga.

Swiftly "Stalin's city" assumed an iconic significance matched by few other urban centres in the history of war: Constantinople in 1453; Valetta in 1565; Verdun in 1916 . . . ? At one stage, the remaining Soviet presence within it was a 30–kilometre line of riverside enclaves, each 200 to 1500 metres deep and either weakly connected with or disconnected from one another. Yet on November 19/20, a brilliantly prepared Soviet pincer movement broke into Romanian positions to northwards then to southwards;

and within ten days had linked up, isolating the 330,000 Germans still in and around Stalingrad. Then on 16/17 December, the Volga finally froze hard.

This freezing allowed hundreds of Red Army vehicles a day to drive across to the right bank. The pincer made those trapped German soldiers reliant on an infeasible air resupply. Hermann Goering, head of the Luftwaffe *inter alia* was almost the only top Nazi of substance, according to their macabre lights. Even so, he could be markedly error prone. Witness Dunkirk and the Battle of Britain. Witness now his assurance to Hitler that the troops inside the surrounded enclave — the Kessel — could be kept supplied by air. His target of 300 tons a day was never once attained. One German staff opinion was that instead of the one airfield available inside the Kessel, six would have been needed.[18]

In February 1943, what was left of the 6th Army finally surrendered. Thenceforward the initiative passed ever more assuredly to the Soviets. It did so emphatically after the Battle of Kursk in July and August, mobile attrition on the grandest scale. The Wehrmacht was expecting that its 3000 tanks and self-propelled guns would inflict crippling losses by dint of equipment quality and command skills. Instead, they were fought to a standstill and emerged the more exhausted.

One legend of the Great Patriotic War is the transfer "to Siberia", ahead of the Wehrmacht's 1941–2 advances, of much defence-related production capacity. The quantity and quality of the output of materials thereby sustained represented for Soviet Communism a triumph of popular willpower, Stalin or no Stalin. Take the production of warplanes in 1942, the most difficult year for Moscow. For the USSR the rounded figure is 24 thousand, for Britain 18 and for the USA 25. The German total is 12; the Italian 7 and the Japanese 6.

The Roots of Resilience

In fact, the displacement was not much to Siberia. Of 67 new towns reported in a 1950 Soviet study as being founded during the Great Patriotic War, 31 were in the defined Urals region and all bar four west of the Ob and its tributaries.[19] So the industrial centre of gravity of the Soviet Union was even then decidedly to the west side of Mackinder's world heartland.

Not that one can deny that the Soviets were imbued with an awareness of strategic depth which (as per Dostoevsky and Solzhenitsyn) can metaphorically relate to large wells of sensibility — the Russian soul. Also, depth as a strictly spatial notion has affinities with the Englishman's sense of insular impregnability. In 1940 we realized, whenever we dared ponder, that we could well be defeated by a combination of submarine and aerial blockade without German troops descending on anywhere. Even so, we still drew succour from the knowledge that no hostile forces had landed in strength for nigh on 900 years.

Still more curiously though, in the Russian military in 1940–1 a belief in being ultimately indefatigable acquired as a rider a conviction that their armed forces were somehow immune to surprise attack.[20] This geographic non sequitur was brutally exposed at the start of Barbarossa.

It has been suggested that western aid added 5 per cent to Soviet economic output in 1942 and as much as ten per cent by 1944.[21] Certainly it will have been quite contributive. But to explain at all adequately why the USSR proved more resilient in adversity than had the Russia of Tsar Nicholas II, one needs introduce additional factors. One would be that by definition a "totalitarian" regime may be better able to retain social control than one which is merely "authoritarian", to employ the Jeanne Kirkpatrick distinction. Another could be the Marxian shibboleth that anything purported to be a proletarian revolution was irreversible from within or without. Even half believing something can happen sometimes help it come true.

Second Front Now?

An Aid to Russia public campaign was launched on the British Left shortly after the USSR was invaded. By 1942 it had evolved into a demand for a Second Front Now. That September, this drew a crowd estimated at 35,000 to a Trafalgar Square rally. Its lead orator was Aneurin Bevan, an MP from a Welsh coal-mining valley and an emergent doyen of the Labour Left. Calling for Churchill's resignation in this context, he ruptured intermittently the wartime parliamentary consensus.

Through the middle of 1942 at any rate, the option of an early Second Front in France attracted some weighty support within the Washington administration. Correspondingly, the British government paid more lip service to a loose interpretation of this aim than otherwise it would have. Meanwhile, those who backed it more positively saw it as a genuine alternative to what Bevan depicted as a strategic bombing offensive belonging to "the realm of rhetoric".[22] They also saw it as more germane than any Mediterranean strategy. An invasion of Italy, in particular, they deemed unlikely to be impacting. Sometimes this was because it was considered too easy. At other times it was said to be too difficult.

However, there is no doubt that the interaction between time lines and geographical constraints presented no viable alternative to any full-blooded D-Day being in mid-1944. Britain did not even have sufficient sea control through 1942. That February the Royal Navy failed to prevent two German battlecruisers — the *Scharnhorst* and the *Gneisnau* — steaming from Brest up the Channel to their home waters. In August an Anglo-Canadian attempt to seize and hold for some days the port of Dieppe was summarily repulsed — an episode represented by Berlin as a failed Second Front.

More particularly, this Dieppe raid confirmed it was necessary initially to invade across open beaches; and then to be prepared to resupply and reinforce via that routeway for some weeks. This prospect was one which underlined an unavoidable dependence on special facilities (advanced minesweepers, landing craft, prefabricated Mulberry harbours . . .) liable to take up to four years from 1940 to develop, adequately produce and render operational. A much wider margin of air supremacy was also needed.

The Second Front Now campaign never, in Britain or anywhere else, reached a pitch at which it threatened to sunder the war-winning consensus. Nevertheless, it may stand as another manifestation of an overly rhapsodic attitude to the Soviet experiment on the part of many of the West's intelligentsia in the 1930 to 1955 era. Of course, in this instance the empathy derived considerably from a vivid awareness of the USSR's military resilience exemplified by how Leningrad, Moscow and then Stalingrad had all got terribly close to being overrun yet none were.

The negative side was ignorance of the Soviet military culture — meaning, above all, little knowledge of or concern with how roughly the lower ranks were so often treated. The lot of the common soldiery was certainly worse than in the Wehrmacht, and likely worse than in the army of Imperial Japan. Witness the thousands of allegedly fainter hearts executed during the Battle of Stalingrad and left in unmarked graves.[23] Witness, too, the deployment of punishment battalions to use their own bodies to blast a path through minefields. Witness the many "human wave" attacks, an old-fashioned tactic that raised casualty rates appreciably. Life in the wartime Red Army must have been hell for many squaddies.

The Security of Obscurity

Unfortunately, the USSR's sheer opacity left too much room for self-deception by progressives elsewhere, people often made lonely and rootless by anxiety about the state of the world. Might Moscow serve as a philosophic anchor? One is reminded again of the *Annales* account of the oneiric attitude of medieval Europe towards an Indian Ocean about which it knew veritably less than nothing. This supposedly enclosed sea became for "the medieval West, the place where its dreams freed themselves from repression".[24] **(See 3 (2))**.

Take a book about Soviet convict camps (*gulags*) which circulated quite extensively in Britain from 1938. Writing of a sub-polar site, its author has the guards promise the prisoners that if they and their families "behave well", the latter might join them. He opines how "Convict colonies have been known to become prosperous new countries before. There is nothing unique in this Arctic convict camp. What I found new was the great and sincere belief of the young administrators that they were really pioneers of the soul in the wilderness of these ruffians minds".[25]

That roseate and snide encomium can be tested against the overview afforded by Alexander Solzhenitsyn drawing on his inmate experience. Horrible enough were the workaday basics: "inept roll calls . . . long waiting under the beating sun or autumn drizzle; the still longer body searches . . . foul-smelling toilets . . . perpetually crowded, nearly always dark, wet cells . . . the wet, almost liquid, bread; the gruel cooked from what seems to be silage".[26] And, of course, outright brutality.

Stalin's first grand project within the *gulag* regime was the Baltic-to-White Sea or Belomor canal (1931–3). Putting a Wittfogel spin on, Solzhenitsyn observed that "In his favourite slave-owning Orient — from which Stalin derived almost everything in his life — they loved to build great 'canals'."[27] Pressure to complete in 20 months was frenetic throughout. The technology literally to hand was little more than pickaxe, spade and barrow. Reportedly, 100,000 died the first winter. The notion that it would allow naval forces to switch was as fatuous as with the Volga–Don. **(See 5)**. Yet it was advanced to justify deepening the canal immediately upon its initial completion.

As noted, gratitude for Soviet resistance to Hitler played a decade or more. But by the sixties, indulgence had assumed a different mien. The Neo-Marxism waxing strong within the West's intelligentsia proved adept at detaching distaste for the Kremlin from its own world view. Seeking to set the May 1968 turbulence in Europe within a broad historicist sweep, the Cohn-Bendit brothers revived a theme the forthright Polish Jewish Marxist, Rosa Luxembourg, had fierily articulated hard upon the 1917 revolution: "No matter what Trotskyist historiographers may tell us today, it was not in 1927 . . . but in 1918 and under the personal leadership of Trotsky and Lenin that the social revolution became perverted . . . "[28]

One might have expected that a Bolshevism skewed since its very inception, half a century ago, was enough of a menace to warrant collective security arrangements.

Yet that summer, some 25 of Britain's New Leftists essayed an overview of the world scene. A chapter on "The New Imperialism" portrays the Polaris strategic missile as a means of dominating the newly developing world. The Soviet superpower might not have existed.[29] Nor did phantasy end there. Nor was it confined just to the Marxian Left. Thus A. J. P. Taylor, a ranking Oxford historian and avowed non-Marxist Leftist, reverted in 1976 to the canard that it would have demanded resources the Russians had not then to hand to move on Warsaw in support of the Polish Home Army rising in the summer of 1944. The Red Army was said to have outrun its supplies, partly because of a need to convert railways to broad gauge.[30]

Surely to goodness, Soviet troops holding the Vistula's east bank could have done other than bathe in the river while the dreadful battle raged. Nor need airfield facilities have been refused to aircraft from the West (Polish, South African . . .) paradropping supplies. A modest pro-active response could have contributed mightily to Wehrmacht discomfiture and Allied solidarity. Also Alan Taylor was being too much of a geopolitician of the old illiberal school when he went on to opine that "A Poland independent of both Germany and Russia perished for ever in September 1939". Recent history gainsays this rather callow geopolitical determinism.

Still, oneiric escapism in the direction of Moscow is today a thing of the past. So now we can focus more steadily on fundamental issues. How could someone such as Stalin, very nasty and pretty ignorant, so long exercise such unbridled power over so many people and so much space? And could such a circumstance arise again?

An Approaching Soviet Crisis

After Stalin's death, in shadowy circumstances, in March 1953, speculation was rife about political "convergence" between the USSR and the West, this putatively in line with broadly parallel material change. Well to the fore exploring this theme was another dissentient Polish Jewish Marxist — this time someone reaching out to social democracy. Isaac Deutscher averred that the imperatives of technocratic modernization made freedom of thought and expression ultimately unavoidable: "The aircraft designers must be let out of the prisons literally and metaphorically if Russian aircraft design is to meet the demands which the international armament race, to mention only this, makes on it."[31] Never mind that only the previous decade, German designers, working for tyranny, had been chillingly innovative in aeronautics, rocketry and much else of military relevance.

However, the Soviet ethos and institutions found especial difficulty adapting to the agrarian sector — this being susceptible to weather and climate cycles and specific local circumstances. So in 1954, Nikita Khruschev plumped for the external margins. He launched a Virgin and Idle Lands programme whereby the acreage ploughed would increase by 25,000,000 acres in Transcaucasia, Kazakhstan, the Volga basin and Siberia. Leonid Brezhnev, involved in the launch as the Party Secretary in Kazakhstan, dared claim in 1956–7 that "the epic of the virgin lands . . . has become a symbol of selfless service to the homeland". He said it also acted as a pump primer for the whole Kazakhstan economy. He cited the installation of a fast-breeder reactor.[32]

Certain members of the Praesidium of the Central Committee had been sceptical from the start. Georgi Malenkov, Chairman of the Council of Ministers since Stalin's death, intimated that, given Soviet geography, priority should have been accorded the intensive margins instead. Recorded grain production union-wide did rise remarkably from 82 million tons in 1953 to 140 million tons in 1958 but by 1963 had fallen back to 107 million tons. "An investment of 30 billion roubles, which could have achieved much in drainage in the west or irrigation in the south east, had largely been dissipated".[33] A collateral programme to boost maize output *per se* also faltered. In 1972 and again in 1975, the USSR imported over 20 million tons of grain, much of it surreptitiously from the USA.

The Consumption Function

Still, the most basic error apostles of convergence made was to concentrate too heavily on the production function as opposed to the consumption. More affluent life styles are liable to engender political aspirations. Somebody with a car or even a camera is bound to want freedom of travel. Evidently, too, suburban dwellers will seek to air a wider range of opinions than the denizens of remote villages. Between 1968 and 1974, the recorded percentage of Soviet urban dwellers rose from 55.1 to 60.1.[34]

The Nationality Contradiction

Anywhere, a tension between production and consumption would pose difficulties for a "dictatorship of the proletariat". But in the USSR an extra rub was that greater freedom of expression for all Soviet citizens would encourage the germination of national secessionist movements, ultimately likely to split the union asunder. Imagine an open debate on issues like devolved decision-making, decollectivization of agriculture, religious toleration, the defence budget, the price mechanism, and — not least — the individual and the law. Attitudes could soon differentiate along nationality lines, with most minorities adopting a liberal pitch initially. Moreover, some of the most restive minorities would be ones concentrated in well-defined territories just inside the western or southern borders of the USSR. They would therefore be well-placed to secede as the Soviet constitution formally allowed.

In the vanguard of the devolutionist revival in train by 1968 was the Jewish community of some two million, distributed across many towns and cities though with a quarter dwelling in Moscow. The percentage citing Yiddish as their mother tongue had declined from 70 to 17 since 1926. Now but 50 synagogues were open as against 450 as late as 1956. Emigration to Israel had been a much more overt theme since 1967.

The crux of the whole matter was that, whilst the Kremlin could not afford to liberalize the Soviet polity, it could not afford not to either. Referring to the nationality question in 1968, Zbigniew Brzezinski regretted the disposition of "many western scholars of Soviet affairs to minimize what I fear may be a very explosive issue in the Soviet polity".[35] By 1975, his foreboding did seem on course to confirmation by 1990.[36]

Homeland Security

One way to defer Western "consumerism" or, if one preferred, "socialist abundance" and all the attendant perils was to reaffirm the long tradition of all round defence of the homeland. Above all, retain a large navy, albeit one divided between several constricted seaboards. In 1914, for instance, 200 submarines were in the Tzar's service. Likewise, his army had at least as many artillery guns as did his Eastern Front adversaries. The rub was that much of his ordnance was tied up in fortresses, these mostly by a frontier or coastline.[37]

No doubt this disposition to give the national bounds such martial definition derives from a continuing sense of exclusion. After all, Russia had remained outside the diversifying influences of sixteenth-century Europe, the individualism the Renaissance extolled and the pluralism the Reformation engendered. Meanwhile Moscow was only the fifth city in formal precedence within Eastern Orthodoxy — below, that is to say, Constantinople, Alexandria, Antioch and Jerusalem. Under the early Bolsheviks a sense of "the other without" intensified. Hence the shift of the seat of government back to Moscow from Petrograd (alias St Petersburg to 1914; and Leningrad from 1924). All throughout, "old believers" within the Communist party will have relished the disciplined puritanism Soviet separateness encouraged.

A Strategic Culture Contrived

So it was that, through the 1950s, the Soviet Navy acquired 600 submarines, nearly twice what Nazi Germany maximally possessed. No matter the planners were bound to have realized that many of these boats were too small for Atlantic or Pacific patrols. Nor that declared Soviet doctrine was by then that a general war would go all out nuclear either immediately or very soon. Clearly once this had happened, the critical Allied problem would be not a shortage of ships but a lack of usable ports for them to ply between. Similarly, the Soviet Navy in 1990 retained in service no fewer than 350 minesweepers, many designed for inshore or coastal service against a threat hardly worth presenting in waters so landlocked and ice-locked.

Yet nowhere was this military–industrial application of resources more otiose than in the effort diverted to continental air defence, this largely under the rubric of *Protivo-Vozdushnaya Oberona Strany*, an elite force dedicated to ensuring the strategic bomber could never get through. Early in 1941, this PVO-Strany had been accorded direct representation on the high command; and in 1948 became an independent service. A decade or so later, the Soviets were reckoned to be spending three times as much on continental air defence as were the Americans.[38]

As the "9/11" crisis was to show in a different context, the Pentagon was not incapable of underrating the importance of homeland air defence. As is indicated in Chapter 11, however, the existence of so large and distinct a PVO-Strany could have compromised the flexibility classically regarded as the supreme attribute of air power. It was also going to prejudice the resource reallocation required as one moved from the strategic bomber to one dominated by the strategic missile.

Operational Purposes

What nobody would deny, however, is that the Soviet armed forces did have substantial commitments to address throughout the post-Stalin decades. When all was said and done, strategic nuclear deterrence remained a real enough requirement. So did external and internal security concerns within what Leonid Brezhnev entitled the "Socialist Commonwealth", the states of Eastern Europe. So was preserving an acceptable military balance between the Warsaw Pact and NATO, especially in and around the two Germanies. So was exerting leverage on China. So was supporting friendly influences further afield.

China's "cultural revolution" was especially vexing. In its wake, there was talk in the West (from Charles de Gaulle among others) about demographic overspill from China into the USSR, a possibility which if envisaged large-scale fitted ill (a) the idiom of the age and (b) the character of the 7000-mile border and the respective hinterlands. But in any case, Moscow was desirous of constraining somehow any continuation or repetition of a social mania that (a) undeniably had very ugly aspects and (b) was so impertinent as to envision a road to utopia that had eluded them (i.e. the Soviets) for half a century.

Preparations for the missile defence of Moscow were initiated when the Cultural Revolution began in 1966. **(See 11)**. Soon a transfer was effected of general military capability to the Chinese border zone. It brought the land order of battle there (from medium or intermediate range strategic ballistic missiles to riflemen) to a third of the Soviet total.[39] Some limited border skirmishing did occur in 1969. And at one point Brezhnev aired with Henry Kissinger the possibility of an attack on China's military nuclear installations.[40]

The requirement to support friendly states and insurgent movements further afield was met considerably by non-military leverage. However, military missions often figured. Naval presence, too, was valued. Until 1958, the Soviet navy had a limited base facility in Albania. But as the Moscow–Beijing cold war developed, Tirana allied with Beijing if only because Belgrade did, albeit circumspectly, with Moscow.

Then in 1964, a Soviet Mediterranean squadron was re-established regardless of Tirana. It proved to be the *entrée* to a much wider naval spread, intended not to "counter" Polaris (as was sometimes suggested) but to gain general influence. From 1967, two 19000–ton carriers — the *Moskva* and the *Leningrad*— were commissioned, each bearing 20 Anti-Submarine Warfare (ASW) helicopters. Warsaw Pact naval manoeuvres — Exercise North — were conspicuously held in the Baltic and North Atlantic several weeks prior to the invasion of Czechoslovakia. Earlier that year a Soviet naval presence had been established in the Indian Ocean. Soviet naval task forces were visiting Cuban ports from 1970.

A lack of an adequate air umbrella was a source of vulnerability, not least for those ASW carriers and the stately, too stately, *Sverdlov* cruisers. To help offset this, considerable store was placed by installing cruise missiles afloat, especially one of 300 miles range codenamed by NATO the Shaddock.

Also some saw a quest for contingent facilities from ashore in aspects of Moscow's foreign policy. Take the 1960 *Statement of the Conference of 81 Communist and Workers' Parties*. This introduced the concept of an "independent national democracy" : a polity not rated a "people's democracy" *à la* Eastern Europe and Cuba but which was deemed to fight against economic neo-colonialism and such more formal expressions of Western imperialism as alliance membership and foreign military bases. Among those duly said to qualify were Angola, Burma, Syria, Tanzania and land-locked Afghanistan. Their commonest denominators were small to medium size, compact shape and helpfully strategic location.

The Socialist Commonwealth

As regards Communist Eastern Europe, Brezhnev made it clear in 1968 that the sovereignty of member states was strictly "limited" in respect of internal as well as external affairs. Hence the Warsaw Pact invasion of Czechoslovakia. Khrushchev had applied the same principle not expressly but just as determinedly in 1956. Hence the Soviet invasion of Hungary.

Of the two subjugations, that of Hungary was operationally the uglier. It took three weeks effectively to complete not least because, even after receiving adequate infantry support, Soviet armour had much difficulty in coping with *ad hoc* groups of freedom fighters sometimes equipped with anti-tank ordnance though in the main with small arms, Molotov cocktails, dummy mines and oils intended to compromise tank tracks. Several thousand Hungarians died; 720 Soviet troops were killed or missing.

Conversely, the reduction of Czechslovakia was a set piece *coup de main*. It followed several months of exploring a diplomatic way through the challenge posed by "Prague Spring" liberalization. Come early August, President Johnson retired to his Texas ranch for several weeks. Meanwhile, about half a million Warsaw Pact troops were ostensibly returning to barracks after large-scale manoeuvres in the immediate region. From them were drawn the soldiery who conducted the tightly synchronized invasion on 20 August. The Soviet standing army in East Germany did not participate. Columns advanced into Czechoslovakia from north, east and south. Big landings on civil airfields were made, these involving Aeroflot considerably. The Czech government under Alexander Dubcek ordered the Czech forces (250,000 strong) not to resist. The border with NATO was secured within 24 hours. A virtually bloodless counter-revolution had been effected.[41]

The Crisis Looms

Leonid Brezhnev (1906–82) died in office as Chairman of the Praesidium and General Secretary of the Communist Party. His lugubrious style was in sharp contrast to the effusiveness of Nikita Khrushchev. Brezhnev could never have advised farmers in the American Corn Belt that their "grandchildren will live under Socialism".

Yet the difference was more stylistic than substantial, a difference in part between the world of the seventies and that of the fifties. On the control of political expression, control of information, Eastern Europe and China, the correspondences are close. Granted, too, both men were far less disagreeable than Stalin. But in the ultimate, each was so anxious to uphold stability that he slid into inertia. Brezhnev, in particular, came to symbolize two wasted decades. Crisis duly moved closer.

An article in 1981 in *Foreign Affairs*, the house journal of the US Council on Foreign Relations, predicted Soviet economic growth would slow during the eighties, probably enough to level out the living standards of working people. Any attempt to head this off would likely be vitiated by how heavily the existing infrastructure was concentrated in the European USSR while untapped natural resources were largely located elsewhere.

In principle, this could "be a period ripe for ferment . . . and for opening up the political process". The decade would see the effective takeover of many leadership cadres by a new generation less imbued with "rooted suspicion of the outside world". But it could well see, too, a "coincidence of outbursts among increasingly restive elites in Eastern Europe". Yet the USSR "remains unshakeably committed to controlling its East European empire" through intimidation or, indeed, force. Also, Slavic and indigenous elites could come into conflict in Soviet Central Asia.

Something the study did not mention is pollution. Something it pointedly did was the possibility that Soviet leaders would, after all, "seek to recreate the atmosphere of a besieged fortress, to rally round the theme of external enemies; and to foster public xenophobia".[42] Since when, there have been low profile tendencies in this general direction. In 2000, the added emphasis on theatre nuclear weapons in a new Soviet national security concept was adjudged by American analysts to be a contingent corrective to the overstretching of military resources.[43] In 2006, a Kremlin spokesman saw the interest Georgia and the Ukraine had expressed in NATO membership as part of an attempt to use that alliance to surround his country.[44] In 2007, Putin similarly interpreted American plans to position Anti-Ballistic Missiles in Eastern Europe. About that time, Moscow's renewed emphasis on long-distance maritime air patrols was accented.[45]

The Strategic Revolution[46]

All things considered, the availability of Mikhail Gorbachev to assume Kremlin leadership in March 1985 still rates as great good fortune for humankind. He sought to address, in the nick of historical time, the worst deficiencies in the Soviet system. And early on, he turned to his adversaries in the West and conceded their victory in the Cold War with grace and even panache. Few statesmen anywhere ever have been possessed of the stature such responses required.

History may well see him as having altered the structure and complexion of Christendom more profoundly than anyone since Martin Luther. After all, Luther had ended up focal to a Reformation which had become, in essence, the sharp and quite even cleavage of Europe — philosophically and geopolitically — between a sunbathed south and a sombre north.

Gorbachev was destined to go the other way about. He inherited a well-defined and roughly equal division of Europe between a Moscow-oriented East and an ocean-oriented West. He found himself setting off the dominos of barrier removal (and, most conspicuously, of the Berlin Wall in 1989) that ended this dichotomy. He, too, acted at a time of unprecedented advance in information exchange. Transistors, VCRs, computers, fax machines and photocopiers were undercutting the protective isolation of the Soviet bloc.[47] Both Gorbachev and Luther set in train changes far more radical than what they had envisaged. Each thereby ushered in a Strategic Revolution.

As to what bestirred Gorbachev to act quite the way he did, one signal influence must have been the sharp deterioration in the situation in Afghanistan as the West started to supply the Taliban with hand-held surface-to-air missile launchers. But another does appear to have been that he and other seniors in party and state were somewhat awestruck by the Strategic Defense Initiative (SDI) of 1983, alias "Star Wars". The panache it bespoke, their side lacked the resources and ethos to respond to. Never mind the critiques of SDI emanating from their own Academy of Sciences,[48] critiques a sight more factual and incisive than previous Soviet disquisitions along such lines.

Raymond Garthoff, as Brookings analyst of Soviet military affairs, inferred that central to Moscow's concern over SDI was its possibly acting as a fount of technology spin-offs across the panoply of theatre war.[49] Meanwhile, westerners in contact with Gorbachev and his entourage, notably at the Reykjavik summit that October, were getting intimations that SDI had persuaded the Kremlin the Cold War was unwinnable. Later Margaret Thatcher said this was what Ronald Reagan had predicted two years before.[50] Furthermore, this interpretation was confirmed at Oxford in the Spring of 1992 by Roald Sagdeev who, through the middle eighties, had headed the Institute for Space Research at the Soviet Academy of Sciences.[51] He came across as eminently trustworthy as well as enlightened.

In April 1986, however, another searing crisis had broken — the explosion of the Chernobyl nuclear reactor in the Ukraine. Other countries affected badly by fall-out included Georgia, Belarus and Russia within the USSR; and — elsewhere — Turkey, Poland, Germany and so on. In Finland, Sweden and Norway, carcasses of reindeer that had grazed on contaminated vegetation were destroyed.

Twenty years later, debate continued as to how many people would eventually die as a consequence of this event. One analysis indicated 9000 in Ukraine, Belarus and Russia. Another pointed to between 30 to 60 thousand worldwide.[52] Anger across Europe rekindled long-standing resentment at the way Soviet-style economic management in Eastern Europe had regularly been causing pollution in Lapland and the High Arctic.[53]

Post-Reykjavik Crises

Soon after Reykjavik, however, a succession of other issues turned acute in ways which Gorbachev, for all his virtuosity, proved unable so to modulate as to retain control of his own revolution. One was Afghanistan which went quite unmanageable with the introduction to the Taliban from December 1986 of light but sophisticated anti-aircraft weapons deadly against helicopters and often against monoplanes. These were principally American-made Stinger hand-held missile launchers.

In 1989, the Red Army withdrew completely from Afghanistan, leaving behind millions of land mines. According to the Soviet-supported President, Mohammed Najibullah, former head of the notorious Afghan secret service, over a million of his country's 17 million had been killed or injured, including 300,000 children with lost limbs. Five million Afghans were living abroad while another five had been displaced in-country.[54]

In May 1987, a young German peace activist, Mathias Rust, illicitly flew a light aircraft at low level from Helsinki to land on a bridge in central Moscow. He attracted worldwide attention and precipitated a big purge of the Soviet military, especially PVO-Strany.[55] Later on, the Soviet Navy's inability to essay the rescue of the crew of a sunken nuclear submarine caused further aggravation and anguish.

In September 1990 came another incident involving a nuclear installation, far less dreadful than Chernobyl yet still bad enough. Perhaps 100,000 people in Kazakhstan were at some risk from a toxic gas vent from a military facility for nuclear reprocessing at Ulba. This event underscored a Kazakh sense of grievance about how much and how carelessly their land had been used for nuclear weapons development. Meanwhile the disposal of nuclear and other waste in the Soviet Arctic had entered the controversy frame.

Frustration and low morale throughout the population was reflected in the standard indicators of malaise attracting much more attention. Alcohol assumed prominence. Between 1955 and 1984 consumption nearly trebled; and by 1985 one in twelve of the entire population was unduly dependent. That year, Mikhail Gorbachev launched an anti-drink campaign. By 1988, it was being cut right back as a virtual failure. Meanwhile crime seemed to be rapidly rising. Also, hundreds of strikes occurred in 1989 and again in 1990, ethnic tensions often being interwoven with material grievances. Inflation was accelerating to 20 per cent and more.

Big Departures

Two initiatives Gorbachev took early on vividly betokened a concern to start anew. In August 1986, a joint meeting of the CPSU Central Committee and the Council of Ministers cancelled plans to divert water from (a) northern Russia to the Volga then Caspian and (b) Siberia to Soviet central Asia. The latter, the most remarked of these two schemes, had first been enunciated in 1949 by Leningrad hydrologist, M. M. Davydoy.[56] Its central feature was to be a damming of the Irtysh river to create a "Siberian Sea". The declared intent was to effect "socialist transformation" by "correcting the mistakes of Nature". The annual discharge of water to central Asia would approximate to the pre-existing flow from the Ob to the Arctic.

By 1965, however, this vision was coming under stringent scrutiny, not least *vis-à-vis* possible inundations of Siberian hydrocarbon deposits. A decade further on, Moscow was backing proposals involving less water diversion; more attention to water conservation in central Asia; and more preparatory research, not least into environmental impact. Then in the early 1980s, more recourse within the debate to econometrics served to uncover massive corruption in Uzbekistan, this effected by inflating the returns for cotton production. There was also criticism of Soviet macromanagement for prioritizing new investment as opposed to renovation. Meanwhile, Russian literati were warning that the proposed European diversion, in particular, would irrevocably alter the "cultural heartland" of Russia. And Siberians were protesting too.

Contrariwise, early 1985 also witnessed a backlash by Communist Party "old believers". They insisted that the debate about basics was concluded, that work on site was already beginning. Given that this programme was so inspirational to some, Gorbachev needed courage and aplomb to curtail it.[57]

More immediately impacting, however, was the curtailment, as from January 1987, of Soviet jamming of foreign broadcasts, an activity on which several billion dollars equivalent a year was still being spent. Some 81 local jamming stations of up to 500 kw power (and, jamming on short wave, fairly effective up to 30 km) had been installed, principally in or near big towns and cities. Sky-wave jammers with slant ranges to 2500 km were also used, their utility lying in their weakening of incoming microwaves rather than actually negating them. Television was always harder to jam because its signals were some orders of magnitude less divergent. (See Appedix A).

The Soviet jamming programme effectively began in 1948. After 1960, Moscow episodically acknowledged its existence. After 1970 or thereabouts, it was no longer too incriminating to be caught listening in. Throughout, penetration of peripheral regions was considerable and of hinterland locales appreciable. There was a fair amount of audience feedback.

Glasnost *and* Perestroika

Yet gratifying as it must have been for millions to see those jamming transmitters redirected to propagation, this switch highlighted a critical weakness in Gorbachev's reform strategy. *Glasnost*, the new openness he was engendering in Soviet life, was moving far ahead of *perestroika*, the institutional restructuring he was collaterally seeking. The latter was always going to be the harder to effect. But the institutional inertia was compounded by Gorbachev himself remaining reluctant to step outside the Communist Party — the dictatorship of the proletariat.

Shortly after his accession, he had squashed the concept of "developed socialism" expounded by his immediate predecessor, Konstantin Chernenko. This spoke of the USSR being poised to enter a phase in which "a reorientation of social consciousness" would be achieved by "the wider flow of information through society". But having jettisoned this mumbo jumbo, he himself failed to enunciate an alternative paradigm, simply relapsing into equivocal talk about advance "within a framework of socialist choice". Even the decollectivization of agriculture (undertaken long since by Communist regimes in Yugoslavia and Poland) was eschewed.

Yet popular expectations were on the rise, encouraged further by a strong surge towards pluralist democracy in Eastern Europe, 1989–90. On 24 July 1991, Mikhail Gorbachev felt able to announce that the nine Soviet republics had agreed on a new relationship, essentially confederal in character. But that development was swiftly overtaken by an attempted coup led by a column of tanks in Moscow. This coup was thwarted, most conspicuously by Gorbachev's arch-critic, Boris Yeltsin. By the end of the year, all the constituent republics bar the Russian federation had adopted declarations of complete independence. The USSR was no more. The Commonwealth of Independent States (CIS) that succeeded it in December provided, in principle, for the unitary control of nuclear arms, a single currency and a "single economic space". In practice, it was to amount to little or nothing. The Cold War was over. Moscow had lost nearly all the coastline over which she had exercised sovereignty on the Baltic and much of that on the Black Sea. She was now hemmed in by land and polar sea except for her Pacific coast, much of it possessed of the most barren hinterland in the world. Her "Socialist Commonwealth" of states in Eastern Europe had disappeared. The dominos had fallen.

The Eastern Question had therefore dissolved into the Western Question. Could the West use its new-found paramountcy wisely enough to build a lasting peace, however that be defined?

A Retrospect on Stalinism

Khrushchev, Brezhnev and, of course, Gorbachev were among various Soviet leaders for whom one can have respect. This makes it all the more ironic and sad that for three decades the world's largest sovereignty had been ruled by an utterly cracked personality, one warped almost as much as Hitler's was by manic paternal beatings. Stalin was driven by an insecurity that required the tea he drank to be "taken from sealed packets opened only by a special servant. Even the air in his apartment in the Kremlin had to be tested for toxic particles".[58] How he made the Soviet experiment sustain his psychosis is hard for either Psychology or Political Philosophy to explain unaided. So is how the USSR prevailed against the Third Reich. Geography certainly bears on the second question and may contribute to answering the first.

Notes

1 Paul Kennedy, *The Rise and Fall of the Great Powers* (London: Unwin Hyman, 1988), pp. 278–9.
2 Richard Storry, *A History of Modern Japan* (Harmondsworth: Penguin, 1967), pp.158–9.
3 Malcolm Mackintosh, *Juggernaut* (London: Secker and Warburg, 1967), Chapter 2.
4 Ian Cummins, *Marx, Engels and National Movements* (London: Croom Helm, 1980), Chapter 6.
5 Robin Milner-Gulland with Nikolai Dejevsky, *Cultural Atlas of Russia and the Soviet Union* (Oxford: Equinox, 1989), p. 153.
6 W. H. Parker, *An Historical Geography of Russia* (London: University of London Press, 1968), pp. 290–1.
7 Lionel Kochan, *The Making of Modern Russia* (Harmondsworth, Penguin, 1963, p. 289.
8 Robert Conquest, *The Great Terror* (London: Macmillan, 1968), pp. 62–81.
9 Stephen B. Wheatcroft, "Agency and Mass Terror: Evdokimov and Mass Killing in Stalin's Great Terror", *Australian Journal of Politics and History* 53, 1 (2007): 20–43.
10 Conquest, *The Great Terror*, pp. 236 and 463.
11 For example, John Erickson, *The Soviet High Command* (London: Macmillan, 1962), pp. 536–7.
12 *Scandinavian Economic History Review* 53, 3 (2005): 109–11.
13 Mackintosh, *Juggernaut*, p. 133.
14 Erickson, *The Soviet High Command*, pp. 406 and 576.
15 Mackintosh, *Juggernaut*, p. 131.
16 Evan Mawdsley, "Crossing the Rubicon: Soviet Plans for Offensive War in 1940–1", *International History Review* XXV, 4 (December 2003): 818–65.
17 Martin Van Creveld, *Supplying War* (Cambridge: Cambridge University Press, 1977), pp. 175–80.
18 Mark Harrison, "The USSR and Total War. Why Didn't the Soviet Economy Collapse in 1942?" in Roger Chickering, Stig Förster, Bernd Greiner (eds.), *A World at Total War? Global Conflict and the Politics of Destruction, 1937–45* (Cambridge: Cambridge University Press, 2005), (Chapter 7).
19 Parker, *An Historical Geography of Russia*, Fig. 91.
20 Mackintosh, *Juggernaut*, p. 133.
21 Harrison, *detail needed from note 18*, p. 144.
22 Michael Foot, *Aneurin Bevan, 1897–60* (London: Victor Gollancz, 1997), p. 180.
23 Antony Beevor, *Stalingrad* (London: Viking (1998), p. 431.
24 Jacques le Goff (Arthur Goldhammer, trans.), *Time, Work and Culture in the Middle Ages* (Chicago: University of Chicago Press, 1980, pp. 189–200.
25 P. S. Smolka, *Forty Thousand Against the Arctic* (London: Hutchinson, 1938), pp. 161–2.
26 Alexander Solzhenitsyn, *The Gulag Archipelago, 1918–56*, 3 Vols. (London: Collins, 1975), Vol. 3, p. 533.
27 *Ibid.*, Vol. 1, p. 86.
28 Gabriel and Daniel Cohn-Bendit, *Obsolete Communism, the Left Wing Alternative* (London: Penguin, 1968), p. 244.
29 Raymond Williams (ed.), *May Day Manifesto* (London: Penguin, 1968), Chapter 18.
30 A. J. P. Taylor, *The Second World War* (London: Hamish Hamilton, 1975), pp. 206–7.
31 Isaac Deutscher, *Heretics and Renegades* (London: Hamish Hamilton, 1955), p. 215.
32 L. I. Brezhnev, *Virgin Lands* (Oxford: Pergamon Press, 1972), pp. 99–100.
33 W. H. Parker, *An Historical Geography of Russia* (London: University of London Press, 1968), p. 348.
34 *The 1974 Demographic Yearbook* (New York: United Nations, 1975), p. 148.
35 *Problems of Communism* XVII, 3 (May–June 1968): 47.

36 See the author's *Future Global Challenge, 1977–90* (London: Royal United Services Institute, 1977), Chapter 14, "The Approaching Soviet Crisis".

37 Norman Stone, *The Eastern Front, 1914–7* (London: Hodder and Stoughton, 1975), pp. 23–8.

38 Herman Kahn, *On Thermonuclear War* (Princeton: Princeton University Press, 1960), Chapter 4.

39 See the author's, "The Myth of an Asian Diversion", *Journal of the Royal United Services Institute for Defence Studies* 118, 3 (September 1973): 48–51. For the definition of "medium" and "intermediate" in this context, see Chapter 11, Footnote 4.

40 *The Economist* 383, 8529 (19 May 2007): 91.

41 P. H. Vigor, *Soviet Blitzkrieg Theory* (London: Macmillan, 1963), Chapter 11.

42 Seweryn Bialer, "The Harsh Decade: Soviet policies in the 1980s", *Foreign Affairs* 59, 2 (Summer 1981): 998–1020.

43 Paul Mann, "Russia Waxes Assertive, Fearing "Encirclement", *Aviation Week and Space Technology* 152, 19 (8 May 2000): 54–5.

44 "Surrounding Russia", *The Economist* 379, 3482 (17 June 2006): 43–4.

45 "Russian bombers resume long-range patrol flights", *International Herald Tribune*, 10 August 2007.

46 See the author's *The Strategic Revolution* (London: Brassey's, 1992), pp. 4–5.

47 Robert G. Kaiser in Nicolas X. Rizopoulos (ed.), *Sea Changes* (New York: Council on Foreign Relations, 1990, p. 27.

48 E.g. Yevgeni Velikov *et al.*, *Weaponry in Space. The Dilemma of Security* (Moscow: Mir Publishing House, 1986).

49 *Scientific American* 261, 6 (December 1986): 64.

50 Margaret Thatcher, *The Downing Street Years* (London: HarperCollins, 1993), p. 467.

51 The Tanner Lectures, Brasenose College, March 1992.

52 Rob Edwards, "How many more lives will Chernobyl claim?", *New Scientist* 190, 2546 (8 April 2006): 11.

53 "Reds Sound Green Alert", *The Observer*, 28 May 1985.

54 Alan J. Day (ed.), *The Annual Register of World Events, 1989*, Vol. 231 (Harlow: Longmans, 1990), p. 300.

55 "What Happened Next?", *The Observer*, 27 October 2002.

56 M. M. Davydoy, "The Ob will enter the Caspian: the Yenisey-Ob' and Caspian Water Connection and the Energy Problem", *Soviet Geography: Review and Translation* XIII, 9 (November 1972): 303–17.

57 See a paper distributed in 1997 by David Duke of Acadia University, Nova Scotia. It is entitled "Seizing Favours from Nature: the Rise and Fall of Siberian River Diversions".

58 Vivian Green, *The Madness of Kings* (Stroud: Alan Sutton, 1993), p. 282.

9 | Peripheral Wars

I A Red Acropolis?

The geopolitical background to the civil wars in Greece (1944–9) has been considered above. However, there may be lessons of general import in how things developed in-country. Greece had a chronic monarchy problem. How close should the king be to the political process? Should his family ties be with Germany? Should he be around at all? Unrequited irredentism also persisted regarding Cyprus, the Dodecanese and Epirus. Macedonian nationalism was abhorred by Greek conservatives. Exacerbating these tensions was economic backwardness. The per capita income immediately pre-war of $75 annually was low by European standards.

Yet despite everything, the Greeks had responded with inspirational solidarity to naked aggression by Mussolini. His 1940 invasion from Albania was expeditiously turned into a slow but sustained Greek offensive into a wintry Albania, roughly a third of its area being captured by the time the Führer came to the Duce's rescue in April 1941. But the ensuing occupation by the Germans, Italians and — quite the worst — Bulgarians had left Greece all but ungovernable come the liberation in Autumn 1944. Nearly one Greek in every 15 had died from war-related causes. Destruction had been "on an enormous scale and every sector was left utterly devastated". Hundreds of villages had been "totally burned down". Industry was at a standstill. Likewise, mechanized transport. Two-thirds of the prized mercantile marine had been lost in allied service. To cap it all, inflation was rampant. The Athens cost of living index (1940 = 100) rose from 2.21 billion end September to 18,850 billion on 10 November.[1] Not for several years would price rises be adequately constrained.

Nor was economic trauma the only factor militating against a rebuilding of Greek democracy. After the heroic unity evinced in 1940–1, a guerrilla campaign was begun which was robust but ideologically riven. There is solid agreement that, come liberation, quite the largest resistance movement (especially across the central mainland) was the broad left National Liberation Front (EAM), ELAS being its military wing and the KKE its Communist Party hard core. The weaponry of EAM/ELAS was to be much augmented by Italy's surrender to the Allies hard upon Mussolini's being deposed in 1943. All in all, the situation looked *prima facie* ripe for a Communist takeover.

However, the attempt by perhaps 40,000 ELAS guerrillas to seize Athens late in 1944 brought forth a rapid British intervention, at divisional level plus air support, undertaken with Stalin's compliance. ELAS took heavy casualties. Besides which, the battle for Athens accentuated a national shift towards a zealous and unscrupulous authoritarian Right within the gendarmerie, police and civil service. This reality helped to vitiate the Varkisa peace agreement of February 1945 whereby a now crumbling EAM was invited to engage in politics subject to surrendering their arms. Over 40,000 rifles were handed in along with 2000 machine guns, 160 mortars and 100 pieces of artillery. But these "were all old weapons; newer German weapons were cached".[2]

British support enabled Themistocles Sophoulis, a spry though not incorruptible octogenarian with radical and republican proclivities, to become head of the Greek provisional government in November 1945. His immediate task, preparation for a general election, was hampered by the rise of the hard Right within the public service. Sophoulis had hoped to correct this imbalance by studiedly appointing liberals to key positions. However, he was dissuaded from this by British advisers. These people probably erred again by insisting on elections being held as early as March 1946. Reciprocal and often vicious intimidation had become widespread in the countryside as had official manipulation of the electoral roll. The KKE decided, in the event, to boycott the ballot.

Nevertheless, 60 per cent of those registered did vote in a poll which, in the opinion of an Anglo-French-American observer team, was conducted correctly on the day. The outcome was victory for the Populists, a triumph the Right capped in September with a large plebiscite majority favouring King George's return. The Communist armed forces, now the DSE, made ready to resume hostilities. This time round, they fought with their backs to the northern border, figuratively speaking, seeking to exploit the South Slav Federation theme. A varied panoply of Soviet origin was soon coming over the border as were, most importantly, copious supplies of ammunition. The heavier weapons included mortars, flame-throwers, field guns up to 105mm calibre, and anti-aircraft guns. Yugoslavia was the chief supplier but Bulgaria and Albania were involved too. No cross-border clashes occurred.

For some months, 10,000 British troops and a deal of sterling currency stiffened the government's resolve. Then, stressed acutely by the fuel crisis brought on by the bitter winter of 1946–7, Britain peremptorily indicated it could not actively support counter-insurgency in Greece beyond 31 March. London thereby obliged Washington to re-examine urgently the opposition to forceful intervention in Greece, previously expressed so adamantly. In doing just this, President Truman questionably linked the Greek situation with overtures Moscow had made to Ankara in 1945–6 about the future status of the Turkish Straits. He was less mindful of other signs apropos China, Persia and, of course, in 1944–5 Greece itself that Stalin was turning towards retrenchment abroad because he was apprehensive about the American nuclear monopoly and wished, in any case, to draw *les limites naturelles*. On 22 March, exactly a month after the British had first indicated their intentions, the President asked Congress to authorize aid to Greece and Turkey. In doing so, he enunciated the new Truman Doctrine. Its gist was that the USA must "support free peoples who are resisting attempted subjugation by armed minorities or by outside pressures". The concept of "indirect aggression" was soon part and parcel.

A central plank of the support for Greece was a large and proactive military mission. By 1949, the USA had 400 commissioned officers on advisory attachments "in country" and her post-war aid bill, including originally contributions through the United Nations Relief and Rehabilitation Administration (UNRRA), had topped $800,000,000.

The influence this conferred on its donor was used without inhibition. Counsel was firmly given on such matters as budget details and army staff appointments. Above all, in August 1947, the recall of Sophoulis was effected. This ensured, among other things, that civil liberties were not entirely eroded by the exigencies of war. Though the leading Communist newspapers were ordered to close that October, the administration was still subject to much press criticism.

By 1947, the Greek National Army was 180,000 strong, albeit considerably dispersed into local garrisons. The DSE field force was an estimated 23,000 strong, while its YIAFAKA underground movement was put at 50,000. The DSE was little in evidence militarily in the islands and weak in the Peloponnese. In December 1947 it decided to organize a "Free Democratic" zone built around the Vitsi and Grammos massifs in the north-west.

But herein lay a fateful paradox. The Communist side in the Battle of Athens had considerably comprised sturdy peasant stock of the ilk who had defeated Mussolini so roundly in 1940–1. Those who sustained the rural-based insurgency from 1946 were more city-dwelling progressives who did not connect easily with the peasantry. Nor did the abduction across the northern border in 1948 of Greek children (over 10,000 below the age of 10) enhance DSE popularity abroad or at home. Instead, of course, it reduced the inhibitions the government side was subject to. Nevertheless, through that year the tide of war looked equal.

Come 1949, however, the balance shifted dramatically. Belgrade's rift with Moscow was now so deep that Tito was being obliged to become what he later termed "Ideologically on the side of the East but strategically on the side of the West". So in July he forbade the rebels to operate from Yugoslav territory. In February, Markos Vafiades, the DSE leader, had been relieved of his command because of failing health and, it seems, discord over Macedonia. His successor, Nikos Zachariades, sought to sustain positional defence but proved unable to do so, not least because the Royal Hellenic Air Force had lately acquired Curtiss dive-bombers. In early 1949, the Peloponnese was finally pacified; by mid-summer the central mainland was too. The Vitsi and then the Grammos natural redoubts fell in August. The war was over. Perhaps 38,000 Communists had been killed, while 40,000 had surrendered. Greek army killed and missing exceeded 17,000.

The Alternative Future

How Greece would have fared had the DSE gained power by 1950 remains conjecture. But one could apprehend that the archipelagos and islands which made Classical Greece the mother of pluralist democracy could then have worked in a contrary sense. After all, city state pluralism was badly undermined by the Peloponnesian War (431–404 BC) between Athens and Sparta. Athens depended on maritime trade and a big navy; and it duly became the head of a locally extensive maritime empire. The factionalism that can so readily be the downside of a disparate structure compromised its war effort. Sparta headed an alliance of land states plus Corinth and Syracuse. It had kept factionalism at bay by being ultra-totalitarian.[3] This gained it victory but almost flaunted the decline of the city state ideal.

It is hard to see how an outright DSE victory could have brought true democracy to Greece even allowing for there being influences within the Left that would have favoured it. On the other hand, a stalemated war might have led to a broad democratic coalition in a way it did not in rectilinear Korea. But in this Greek conflict, a definable stalemate was improbable. The geopolitical setting was simply too convoluted. Perhaps, too, the Classical heritage overly encouraged the politics of absolute principle. Moreover, the army officer corps, unlike its Turkish counterpart, had little rapport with the urban intelligentsia.

The Colonels

As it was, parliamentary democracy did not progress smoothly in Greece. A coup in 1964 brought "the colonels" to power. Then perhaps the first sharp check on this military regime came from an hellenic outer island, namely Cyprus. There in the 1950s EOKA insurgents had fought the British in pursuit of Enosis, union with mother Greece, regardless of the concerns of the Turkish Cypriot minority or anybody else. In 1967, the colonels sought to encourage an EOKA resurrection. But they were thwarted by Archbishop Makarios III, Ethnarch and President. He outmanoeuvred EOKA by insistently drawing a distinction between "Organic Enosis" and "Genuine Enosis". Neither was precisely defined but the understanding put across was that the one was shot-gun integration, the other a link more voluntary and informal. Given the latter's adoption, the "organic" alternative could hopefully be postponed to the Greek Kalends. This formula proved unable to ensure the continuation of an undivided Cyprus but it did rein in the pretensions of the colonels and an EOKA now subordinate to them.[4]

One question to ponder is how far a Byronic preoccupation with Greek liberty conditioned British strategy. It surely comes through in 1944 with Churchill's obsession with liberating the Greek islands. Come October, this had boiled down to a desire to gain historic Rhodes. It was still under the Italian military despite Italy's formal capitulation in September 1943. Yet strategically it stood in limbo. On the 7th, Field Marshal Alanbrooke, Churchill's Chief of Staff since 1940, wrote in his diary of "Another day of Rhodes madness!". On the 8th, the entry states, "Winston is becoming less and less balanced! I can control him no more. He has worked himself into a frenzy of excitement about the Rhodes attack, has magnified its importance so that he can no longer see anything else . . . "[5]

In the event, there was no Rhodes attack. Instead, the Prime Minister's attention turned to the Athens situation. So Alanbrooke did retain influence after all. So, one can presume, did the Deputy Prime Minister, Clement Attlee. For his first several months in the premiership, from July 1945, this quietly proud Gallipoli veteran was to evince no taste for a continuing Mediterranean commitment.[6] This was in line with antipathy on the Labour Left to the Greek involvement, 1944–7. It was not in line with burgeoning Labour concern about Palestine.

2 Communism's Greatest Gain

China had emerged from World War Two less damaged materially than might have been expected. But a state of generalized disarray was compounded by how the threat to Kuomintang (KMT) authority posed by the Communists had grown. Granted, Chiang Kai-shek had 2,000,000 under arms, perhaps a quarter of them in his Central Army which was American-equipped and trained. But Mao Tse-tung now boasted a field army of 900,000 plus two million militia. Over half the former are said to have been in north China and more than a third in the centre. A weakness still was an acute shortage of small arms and a desperate one of heavier weapons.

Nevertheless, Mao could claim his forces controlled *in toto* land lived on by 100,000,000 people, nearly a quarter of the national total. Beyond that, it was and is difficult to go, as so often in revolutionary situations. Facts are in short supply and hard to check. Their interpretation is subject to much ambiguity. The notion of control tends to meld into that of claiming genuine allegiance. In this instance, matters were further complicated by the roots of Communism in China having originally been sunk not in Yenan but in the maritime South East. Marxism was, after all, an import from the West.

Great Power recognition of the KMT Nationalist administration as the interim government of post-war China imposed on it the triple responsibility of accepting in the field the Japanese surrender, setting in motion economic revival, and extending itself into a reasonably broad coalition. So demanding an agenda was beyond it, in terms of geographic positioning but also of more intrinsic respects. Thus while Chiang himself may not have been tainted too directly by corruption, most of his relations and key officials were deep into peculation and practices bordering on warlordism.

Kremlin Ambiguities

The treaty signed with the USSR in August 1945 appeared to guarantee Chiang the collaboration of the 300,000–strong Soviet army which, under Marshal Rodin Malinovsky, had lately liberated Manchuria: a territory the Japanese had made the most industrialized of the Chinese provinces and one where the Chinese Communist Party (CCP) had become operationally strong, in the countryside at least. However, that apparent guarantee was laced with ambiguity. Chu Teh, the CCP army commander, felt able to reject with scorn an injunction from the KMT authorities to have his People's Liberation Army (PLA) stand fast while they, the internationally legitimated government, disarmed all Japanese troops. In the event, large quantities of weapons from Japan's crack Kwantung army in Manchuria (including many of its 500,000 small arms and 2700 artillery pieces) passed to the PLA. Both the CCP and KMT later said as much.

Central to any rounded interpretation, yet hard to arrive at, is an elucidation of how big was the part Moscow played in this. The Soviet withdrawal from Manchuria was delayed till March 1946. This could have been to inhibit a Communist takeover of the North, an aim akin to that of the 50,000 US Marines who (for a year from October 1945) maintained an expeditionary enclave embracing Beijing and Tientsin. Such a Soviet pitch would have been consistent with (a) Stalin's jealous resentment of Chinese Communism and (b) a Soviet disposition during 1946 towards global retrenchment.

Alternatively, the Soviet Far East Command could, under the forceful Malinovsky, have been evolving an external strategy of its own, one which would probably have involved allowing Kwantung weaponry quietly to slide into PLA hands. The other side of this coin will have been mollifying the Kremlin by the pillaging as "war reparations" of what were huge quantities of industrial equipment. Without postulating some such machination, it is hard to explain how the PLA so suddenly became comparatively rich in weaponry and maybe ammunition. One needs bear in mind the Soviets' claim that — during Malinovsky's offensive — their troops took prisoner 610,000 of the 760,000 troops then in the Kwantung army.[7]

KMT Unilateralism

A joint Chiang–Mao communiqué, released on 10th October (China's national day) in 1945 ostensibly bespoke a fair measure of agreement. Pending the framing of a constitution, Chiang was to appoint an advisory council, not more than half of which were to come from his own party; and a council majority of three-to-two could override any Presidential veto. But among the problems unresolved were the actual modes of integration of the two armies; the future of various local Soviets; and pressures within Manchuria for a fair degree of provincial autonomy. Meantime, a civil war was ramifying. An attempt by the Nationalists to reach Kalgan alias Changkiakow (a communications hub close by the Great Wall in Hopei province) failed though they did prevail against the PLA in the lower Yangtse basin. Then in November, the Nationalists began to dispatch 250,000 troops, mainly from the Central Army, to Manchuria.

This deployment disconcerted the Americans on two counts. Might it finally dash the prospects for a compromise peace? And, in any case, was not the KMT overextending itself militarily? President Truman was to write of its having a "walled city complex", Chiang's presumption apparently being that everything hinged on holding the big towns plus the desperately vulnerable railways and roads linking them. To these anxieties was added concern about how the KMT soldiery (many of them from middle and South China) looted without compunction.

In December 1945, General George Marshall arrived on an extended peace mission, representing the White House. However, the national truce he negotiated with some expedition was woefully ineffectual, especially in Manchuria. There the Sungari river would soon become a litmus test of apparent territorial advantage. Having defeated Lin Piao's PLA forces severely in fighting south of the Sungari in May 1946, the Nationalists crossed the river in June. Concurrently, however, many townships to southwards fell back into PLA hands, various garrisons having exhibited an ominous lack of martial resolve when facing moderate odds. Meanwhile, fighting resumed widely within China Proper with the Nationalists making some insubstantial gains.

Inside the KMT, illiberalism now waxed strong. Anti-American demonstrations were organized; secret police activity was greatly extended; and two leaders of the intellectually progressive Democratic League were assassinated. Then in November, the Nationalists convened unilaterally a National Assembly, the elections for which only one voter in 200 participated in. Forthwith, Chou En-lai abandoned his assigned role as Communist representative at the Nationalists' resumed capital, Nanking. Several weeks later, Marshall likewise departed. Thenceforward American aid to the KMT was just enough to tarnish the Truman administration with blame through association, nothing like enough to alter the outcome.

Taiwan and Manchuria

The island of Taiwan (alias Formosa) had been ceded to Japan "in perpetuity" following her defeat of China in the 1894–5 war. In technocratic terms, Tokyo's rule was to be modernizing throughout. Early on came the suppression of piracy in local waters, this underpinned by an efficient coastguard and well-lit coasts. Later came widespread electrification (much of it hydro generated); excellent railroads and highways; and 80 per cent literacy.

At the Cairo summit conference of Allied leaders in 1943, Taiwan was casually reallocated to China. But post-1945 there was a native backlash against a heavy influx of mainlanders, especially into administration and teaching. Things turned critical after 15,000 Nationalist troops had stormed ashore early in 1947. Hundreds or thousands of Taiwanese were killed in a repression still notorious for its wantonness.[8]

Meanwhile in Manchuria, probes by Communist columns of at least divisional strength were being made across a frozen Sungari. By June the Nationalist garrisons at Changchun and Kirin were surrounded. Heavy Nationalist reinforcements were duly dispatched to the province. But by November Mukden's rail links were cut.

Currency Catastrophe

In 1945, the Nationalists had drastically overvalued their *fapi* — as their standard unit of paper currency was called — in relation to notes redeemed from former Japanese-held areas. This error was made through taking cognizance of a Shanghai black market rate driven unnaturally high by a flush of post-war optimism and a temporary dearth of *fapi* in liberated areas. Being officially overvalued, the *fapi* flooded into those parts causing inflation within them; elsewhere a currency dearth was remediable only by more printing. Soon the government was printing still more notes to correct budget deficits caused by rising prices, incorrigibly bad tax collection, and soaring military appropriations. Thus began an inflationary spiral which, during the first three years of peace, forced up Shanghai price indices by a third each month.

In the summer of 1948, a new gold yuan currency was introduced. To support it, rigid price controls were introduced. To be effective nationally, these had to be so in Shanghai. Chiang's son, "Tiger" Chiang Ching-kuo, was in charge of this megalopolis, now characterized by rice riots and the deaths of thousands of untended refugees. He managed well until he tried to take action against the mighty Tu Yu-shan — gangster, dope pedlar and philanthropist. Years ago, Tu, then an Elder in the Green Shirts secret society, had met a young novitiate by the name of Chiang Kai-shek. Tu now used this link to ensure for his agents just nominal sanctions. Next, Tiger's mother made him go easy on a Yangtse black market ring controlled by her brother-in-law. By October, Tiger Chiang had resigned and raging inflation resumed.

The Final Phase

Systemic inflation and corruption undermined Nationalist morale terribly. No matter that, in June 1948, Nanking reckoned each side had a million riflemen and 2000 artillery guns. Nor that the Nationalists had a monopoly of air power, this in the form of 400 warplanes. These saw little action.

In the event, the Mukden garrison of 200,000 men surrendered in November

1948. In January 1949, the North China army (from which, for months, arms had been passing to the PLA) capitulated. This enabled a North China People's government to be established in Beijing. The consolidation regionally of control over civil society will have followed from that.

Faced now with an incipient disintegration of his high command and a crescendo of criticism from within the legislature, Chiang temporarily handed the presidency to the radical Li Tsung-jen who tried to negotiate with Mao about forming a progressive coalition. But his functional inability to disband the secret police or have political prisoners released convinced the Communists his progressivism had shallow roots.

Two days after peace talks finally collapsed in April, a strong Communist column easily crossed the Yangtse — a natural defence line which, stoutly held, might have claimed a million casualties. Then hopes that the Communist forces would be distracted and corrupted by Yangtse city life proved vain. In December, Chiang Kai-shek led a rump Kuomintang to the comparative security of Taiwan. The civil war on the mainland was virtually over. Belief in Communism as the wave of the future had received a big though not quite decisive boost.

Railways

China between 1931 and 1958 may stand out in future histories as being the last theatre of war or warlike crises in which railways are seen as the key means of strategic mobility, the linkages the combatants were especially concerned to protect or disrupt. The Japanese take-over of Manchuria from 1931 began with a staged railway incident. The Soviet sale to Japan, agreed in 1935, of the Chinese Eastern Railway through Harbin to Vladivostok represented the disposal (as a placatory gesture and measure of consolidation) of an asset that was valuable but hopelessly exposed. Completion around that time of a double-tracking of the Trans-Siberian railway made any Far East military build-up rather less difficult. Then, in late 1941, it facilitated a critical switch to the Moscow front.

The completion, in November 1951, of a railway to a Vietminh-controlled sector of the Indo-China border was an important precursor to Giap's investment of Dien Bien Phu. Railway construction likewise preceded the furious bombardment in 1958 of Quemoy and Matsu, two islands the Nationalists had retained just off the mainland near Amoy. The declared purpose of the new rail link was to export iron ore, timber and figs. But immediately upon completion it was closed till January 1958 for "experimental operation". The bombardment began that summer.

In 1949, the Communists calculated there were 21,700 km of railroad nationwide. By end 1957, another 5000 km had been laid. Neither figure bears comparison with the 160,000 km Sun Yat-Sen had called for in 1912. But the heavy concentration in the north-east (Manchuria included) helps one understand the importance both sides attached to that region in the 1946–9 civil war.[9]

3 The Limited War[10]

The Korean peninsula was and, indeed, has remained pivotal regionally, pivotal yet discrete. Though far from isolated, it is isolatable. So it was that three years' savage conflict saw no hostilities against neighbouring territories. Britain's apprehension that its involvement might precipitate an attack on Hong Kong was not borne out. Nor were allied ships attacked, except close inshore to hostile coastlines.

But neither did Harry Truman air publicly the question of Soviet involvement in military operations. Nor did he countenance any out-of-country escalation by the USA or UN. Much would be written the next 20 to 30 years about "limited wars" (regular warfare, geographically confined). David Rees did well to typecast this as *the* example of such a conflict.[11]

Korea had been split in 1945 between a Soviet-occupied North and an American-occupied South, this along the 38°N parallel of latitude. As a straight line across the map, it did supposedly connote a mutual commitment to organic reunification but was not conducive to military stability and political coherence. Moreover, the atmosphere was soured the more by the legacy of a Japanese colonialism (1895–1945) which encouraged industrialization but was otherwise repressive. Korea was deemed part of *gaichi*, the outside world not *naichi*, the "inner world" or homeland extending to Okinawa and Taiwan. The enforced induction into military brothels abroad of tens of thousands of Korean women had betokened this brutally. Korean insistence on the swift exit of Japanese administrators (one for every 250 Koreans) was an immediate postwar backlash against the whole colonial experience.

The 38° partition disadvantaged the South asymmetrically. On its west side, it proffered no military defence in depth. As regards economic imbalance, the South comprised 43 per cent of the country's area but in it lived, in 1945, 64 per cent of the 25 million Koreans. Over half the electricity was generated in the North (overwhelmingly from Yalu river dams) but industry was largely in the South. Without hydropower and Japanese technocrats, output there slumped initially by four-fifths.

Under Kim Il Sung (1910–94), the North's regime, based on Pyongyang, evolved a singular brand of Communist self-reliance, *chuch'e*. It blended the perspectives of cadres who had been exiles in Moscow with some back from Yenan. From 1946 it was underpinned by land redistribution. That year five million acres were shared among 750,000 tenant families, the maximum holding being set at 12 acres. The South made no moves at all in this direction until 1948–9.

All the same, two million refugees (some of them Communist insurgents?) had reportedly moved to the South by 1950. Overall, the South was the slower to achieve stability. This was due partly to the human geography but perhaps more to the reactionary bitterness of Syngman Rhee (born 1886). He was initially encouraged by the Americans to assume a lead role after 25 lonely years abroad campaigning for independence. A big rebellion erupted in the Yosu district in October 1948. Twenty thousand were "convicted" and many executed by Rhee's National Constabulary, a huge force known as immoderate, corrupt, inefficient and factional.[12] Thenceforward insurgency as such diminished while clashes along the parallel increased.

By 1948, both Superpowers were expressly giving up on early reunification. Instead, each declared its fief independent that summer; and each ended its occupation in 1948–9. In the North the USSR was masterminding the build-up of a four-division, tank-rich North Korean army complete with air support.

Dissuasion through Distancing

The South (now the Republic of Korea) was creating an army of comparable size but without tanks or effective anti-tank guns. The immediate American concern was

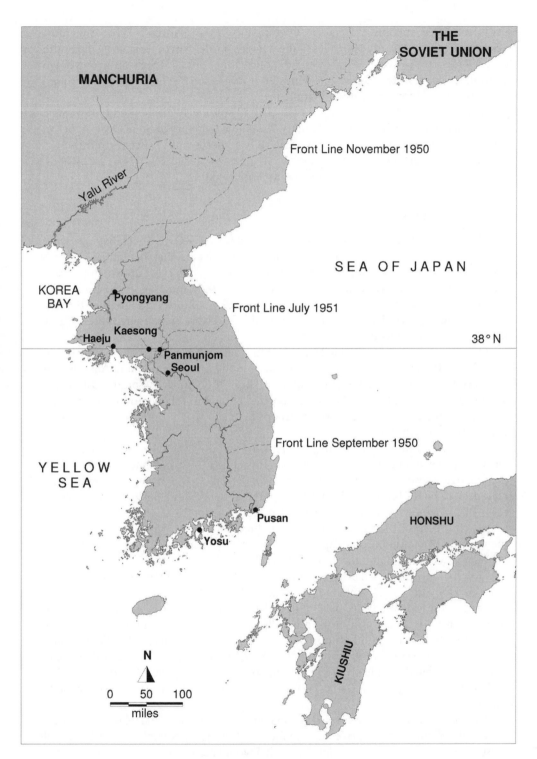

The Korean War

President Rhee and his threats to "liberate the lost provinces". What one must acknowledge is that nearly 100 miles of the western border was defensible by the South only via recourse to instant escalation. Still, one cannot put all Rhee's posturing down to arbitrary geography. After all, the Minister of Education he chose had been through the Nazi school of philosophy at Mannheim.

In 1947, MacArthur and the Joint Chiefs had studiously discounted the strategic importance of Korea. Then on 12 January 1950, US Secretary of State Dean Acheson publicly excluded the Republic of Korea (ROK), along with Taiwan, from the US defence perimeter, telling them to look to themselves and to the UN. That March, Pyongyang activated a fifth division; and in April a sixth.

Guerre à l'Outrance?

On 25 June, full-scale war was launched against the ROK, Stalin having given grudging approval. Several early encounters were along that benighted western border, the expectation no doubt being it would be a good sector on which to crack ROK resolve. However, Seoul the capital fell on the 28th, much of its covering force having been trapped north of the Han river by the premature demolition of the one or two bridges. Nevertheless, ROK troops in the field were responding more toughly than some on the general staff.

On 27 June, a Security Council with the USSR absent resolved that the UN should furnish the ROK with enough armed assistance to thwart the aggression. Already that day Truman had ordered US planes and ships to bombard Communist forces advancing on Pusan. On the 29th he authorized air and naval operations north of the parallel and the commitment of ground troops to its south. The 24th Infantry Division began arriving next day.

Through the third week of July, several thousand of the 24th stoutly defended Taejon, thus enabling the UN to form the "Pusan box" — a rectilinear defence zone delimited to westward by the Naktong river. Its fall would have rounded off the conquest of the ROK; and until about 8 September the box did look decidedly vulnerable. But whether the North Koreans were gunning for its outright capitulation is uncertain. Much of their intense effort was directed to destroying artillery emplacements.[13] The intention could have been so to enfeeble the UN as to induce its negotiated withdrawal, humbled at every level.

However, events took a different turn. On 15 September, Douglas MacArthur — by then UN Supreme Commander — took what he termed a 5000 to 1 chance. Flouting professional advice, he dispatched the US 10th Corps across the 40-foot tides, extensive mudflats and strong sea walls of Inchon, to the port for Seoul. Twelve days later the Corps stood triumphant in shattered Seoul itself. Outflanked and coming now under pressure from the 8th Army within the box, the North Korean army disintegrated. Inside six weeks, the prisoners of war totalled 135,000; and 240 T-34s (Soviet-design Main Battle Tanks) had been captured in varying serviceability states.

But the success was overweening. China imported North Korean electricity generated on the Yalu, the border river. So did the USSR. On 25 September, the PLA Chief of Staff told the Indian ambassador that China would resist a push to the Yalu, atomic bombs notwithstanding. On 1 October, ROK troops crossed the parallel. Six days later, the US 1st Cavalry was ordered to, heedless of further warnings. By the 15th,

PLA "volunteers" were swarming into North Korea. Moving largely on foot, they avoided the roads except when ambushing. Meanwhile, the North Korean capital, Pyongyang, fell to the 1st Cavalry; and Rhee, flatly defying the UN General Assembly, extended his jurisdiction north of the parallel, introducing there a characteristic reign of terror.[14] Trashing land reform was part and parcel.

Northwards from Pyongyang the front widened dramatically; and there were no lateral roads, only gravel tracks. Soon a yawning gap opened between the 8th Army on the left and the 10th Corps now on the right. Come early December command and control in the former neared collapse. The onset of the quasi-Siberian Korean winter much exacerbated a tough situation.[15] By early 1951, the Communists had regained everywhere down to a line generally 50 miles or so south of the parallel. The PLA was proving well able to plan and execute frontal offensives at divisional level.

To cap everything, Truman felt obliged in April to dismiss MacArthur. This was partly because of his flaunted contempt for the President (his Commander-in-Chief) though also on account of his constantly pressing for a wider war built around a grand offensive to the Yalu if not beyond. Still, the situation overall had its positives. Since December, MacArthur's successor-to-be, General Matthew Ridgway, had been commanding the 8th Army with 10th Corps now incorporated. He, too, was a tough veteran but one more pragmatic in the given circumstances than MacArthur had proved. Furthermore, he was then managing a front not half the width and more joined up than in the autumn. Meantime, recognition was burgeoning on both sides of the need to seek a "no winner, no loser" peace.

In March, the UN retrieved Seoul yet again and were to hold it against counterattacks in April and May. Then a renewed UN offensive threw the Chinese into some confusion, 17000 prisoners being taken. Pyongyang was seized though not held, the object being to create a situation conducive to bargaining. Formal armistice talks began in July. These were to drag on, first at Kaesong and then at Panmunjom, a full two years.

The New Phase

On both sides, the forward areas remained heavily manned. As noted above, the UN had a division deployed every eight miles or so along what it now designated the Main Line of Resistance (MLR), albeit with little structured defence in depth and few mobile reserves. The Communists had thrice the number of divisions close by the MLR, these troops manning robust defence works some 14 miles deep. These were largely completed in the quiescent weeks prior to the Kaesong talks.

Much artillery was in position each side. However, the UN enjoyed complete air suspremacy over the battlefield. Airfield construction was often difficult, given the topography and soil structures. But warplanes also sortied from Japan or off carriers. Meanwhile, UN forces on the MLR were ordered to limit offensive operations to battalion level. The MLR signified little net gain of territory either way, since June 1950.

Yet despite the constraints, much ground action continued. UN casualties the last two years combined were not so far short of those during the first. Generalized exchanges of fire were punctuated by more organized encounters, often for a prominent hill. Witness the ferocious battles in the Spring of 1953 to retain Pork Chop then the Hook.

As F. W. Lanchester would have anticipated, however, the conflict these last two years considerably hinged on UN interdiction of Communist resupply and reinforcement. The PLA and North Korean armies were usually able to hold in the field between 36 and 60 average day's consumption of supplies,[16] thanks to skilful deception and concealment. However, average daily usage of ammunition would be quite a minor fraction of what a day's heavy fighting might demand. Therefore one can reasonably infer that the interdiction campaign waged from mid-1951 against Communist logistics did preclude still bloodier fighting. Still, the interdiction never truly became Operation Strangle, the code name the USAF assigned it in 1951. Frustration built up over this and, of course, the halting progress of the peace talks.

So the target list was extended in June 1952 to include *inter alia* transformers drawing electricity from the Yalu dams; and in May–June 1953 a quarter of the dams in the Haeju area, the "rice bowl" of North Korea — these attacks coming immediately after rice transplanting. The professional claim was that this escalation must have been what induced the Chinese and North Koreans to abandon, between 7 May and 4 June, a precept they had adamantly insisted on: namely, the repatriation of all prisoners of war, regardless of their personal wishes.[17] Still, this pressure complemented a warning conveyed covertly (via Panmunjom, New Delhi, etc.). It was that, unless the deadlock was broken by the requisite Communist concession, the USA would widen the war geographically and possibly resort to nuclear weapons as well.[18]

Air Losses: Chinese and Soviet?

A facet of the air war which particularly invites discussion is the claim that, in battles high over the Yalu approaches, the F-86 Sabre (the USAF's interceptor mainstay) destroyed 818 MiG-15s while losing but 58 of its own number. Doubts arise about this claim and the related view that total Chinese losses in fighter aircraft exceeded 2000, albeit largely through training accidents.[19]

Seeing that Chinese Communist air power had built up from scratch since 1949, heavy losses in action or otherwise seem retrospectively predictable. Nevertheless, certain considerations point the other way. The three-fold advantage they often enjoyed in numbers of planes committed will have favoured them *vis-à-vis* the Lanchester square law. Also their MiGs were directed by ground control centres located just inside Manchuria. The MiGs were superior to the Sabres in some aerodynamic respects and had stronger armament.

Above all, if Chinese losses had run into the low thousands, the interceptor inventory of this fledgling air arm could hardly have risen the way it appeared to, perhaps to 3000 by 1953. In air combat *en masse*, guestimates of enemy losses have often been too high historically. Thus 15 September 1940 was a day of truly decisive ascendancy for the RAF in the Battle of Britain. Yet the fact remains it initially exaggerated Luftwaffe losses three-fold.

On the other side of the argument, shortfalls in advanced training forced the Chinese to rely excessively on preserving close formation in actual combat. Then again, just the F-86 had a radar-directed gunsight, if only one too unreliable and hard to service to utilize much.[20] All in all, a revisionist estimate put the MiG loss rate at 1.7 per 1000 sorties as against 1.3 for the Sabre.[21] This differential probably implies an aggregate loss ratio of three or four to one. Much the same would be registered in 1958 during Taiwan Straits aerial encounters.[22]

However, there is to all this another dimension. UN pilots passing close to "Chinese" aircraft not infrequently noticed how caucasian their opposite numbers appeared. Now it is officially acknowledged in Moscow that several Soviet air defence divisions were deployed in Manchuria during this war. We are told they lost 335 planes and 120 pilots. Another 179 Soviet military personnel were killed on staff duties in North Korea.[23] *Prima facie* those figures could suggest few Sabres were shot down by the Chinese themselves.

The Outcome

During its first year especially, this was a grim conflict, exceptionally frightening and dispiriting. For the UN soldiers the horror was compounded by remoteness from home and the unattractiveness of both the local regimes. Nevertheless, in the circumstances then obtaining, the United Nations involvement can be rated a strategic success despite much being left unresolved. It is not too difficult to imagine how the seizure of South Korea might have strengthened immoderation in many countries — not least China, the USSR, the USA and Japan.

Fourteen nations eventually sent ground troops to Korea; and the total strength of UN military personnel in the theatre, other than Americans and South Koreans, approached 50,000. The forces in country by the first Christmas included Australian, British, Filipino, French, Netherlands, South African, Thai and Turkish combat formations, plus Indian and Swedish medical units. The ROK had an estimated 400,000 killed, civilians included; the USA had 54,000. The next longest roll of honour was the Turkish, 717 dead.[24] Altogether several million Koreans, South and North, were killed, wounded or made homeless.

4 No Neutral, The Jungle

Some bestirring books have been written about the Indo-China wars (1946–75), notably by journalists with extensive in-country experience. If they have a generic fault, it may be that they take too much for granted the part landscapes play in the sequence of events.

Viewed broadbrush and from above, Indo-China resembles an involuted egg-timer. The mountain spine separating the Red River and the Mekong often rises several thousand feet, very generally enveloped to its peaks and ridges in rain forest. Still, the insurgency campaigns were ultimately for control of the hamlets of the riverine paddy. "Hamlets" was the term locally preferred for small rural settlements. A "village" usually meant a cluster of hamlets. Note should also be taken of "transition zone" rubber plantations and the like, a semi-developed ambience subject to less war than one might have apprehended.

Cultural Geography

In a seminal 1955 article, a distinguished Far East editor of the *Times* saw Indo-China as transected by a deep cultural divide between the Vietnamese to the East and the Lao and Cambodians to the West and South. The Vietnamese were "the most active and best educated", the South Asians more quietist.[25] Granted, the Khmer Rouge's diabolical reign of terror in Cambodia from 1975 (see below) was to fit ill this dichotomy. But that does not mean it never had validity. The temple states herein mentioned (in connection with "hydraulic despotism" but also as marking the limits of Mongol imperialism) were archetypically South Asian. In the sixties some in South East Asia (among them Prime Minister Lee Kuan Yew of Singapore) would tell you how in Vietnam the Americans were up against Zen, broadly defined. Historically, Mahayana ("big wheel") Buddhism had progressed through from India via China to impact on South Asia in Indo-China. The Hinayana ("small wheel") Buddhism of Cambodia, Laos, Thailand . . . had remained South Asian throughout. It was less proactive, believing ordinary mortals could never attain full buddhahood.[26]

Alternatively or additionally, Vietminh/Vietcong zeal is open to a more nationalist interpretation. Between 208 BC and AD 1428, imperial China had been a perennial threat, one leading to longish spells of harsh occupation. The most celebrated of a succession of uprisings climaxed in the defeat of Kublai Khan's imperial army in 1287. Then, as the Chinese threat unsteadily receded, the Vietnamese turned in upon themselves: "the war between North and South Vietnam after 1954 largely expressed ancient regional animosities only newly overlaid with an ideological veneer".[27] Another Vietnamese inclination was to advance their southern frontier, political and demographic. In 1471 they attacked Champa, the Indianized capital of the central east coast of IndoChina. By 1750 they had reached the Gulf of Siam, this at the expense of the once entrenched Khmers.

As a frontier people, the southern Vietnamese acquired a fractured cultural geography. Latterly this state of affairs had been accented by Japanese "divide and rule" (1941–5). Prominent among the sects were the Cao Dai ("High Tower") founded in 1926 and centred on Tay Ninh, a holy city in the Mekong basin 10 to 15 miles from a jungly Cambodian border. Its social ethic could be characterized as Confucian; and its core belief, karma and rebirth. Among its many deities are or have been Joan of Arc, Gautama Buddha, Winston Churchill, Mahatma Gandhi, Vladimir Lenin and William Shakespeare. Syncretism *à l'outrance*.

By 1950, it was claiming a million adherents but still felt constrained to upstage in patriotic militancy the Vietminh: the Communist-led front Ho Chi Minh, Vo Nguyen Giap and others had inaugurated in a desolate cave in 1941. Come 1955, the Cao Dai was suppressed by President Diem of South Vietnam. Then after the Communist victory in 1975, it was suppressed more thoroughly throughout Vietnam. Today it survives elsewhere, particularly in Australia and, the more so, Canada. As of 2003, it was very dubiously claiming six million adherents worldwide.[28]

Another component of the human geography were the montagnards, the plurality of aboriginal clans driven out of the fertile lowlands historically by Vietnamese or sometimes Lao. Their counterparts elsewhere (e.g. the Dyaks in Borneo and the Nagas in Assam) looked to European imperialism to safeguard them against the indigenous majority. Thus in Borneo (1963–7), Commonwealth forces, mainly British, beat back endeavours by President Sukarno of Indonesia to destabilize Eastern Malaysia by means of insurgent infiltration. Hoping thereby to persuade the British and Australians to stay around to shield them from the Malays (Indonesian but also Malaysian), Dyak tribesmen provided invaluable intelligence about Indonesian movements. This could enable a single infantry battalion (*c.* 750 men) to screen a 100-mile sector of rain forest.[29] Similarly, when the Communist insurgency in Malaya (1948–60) was into its final phase, the Senoi Pra'ak (aborigines numbering not more

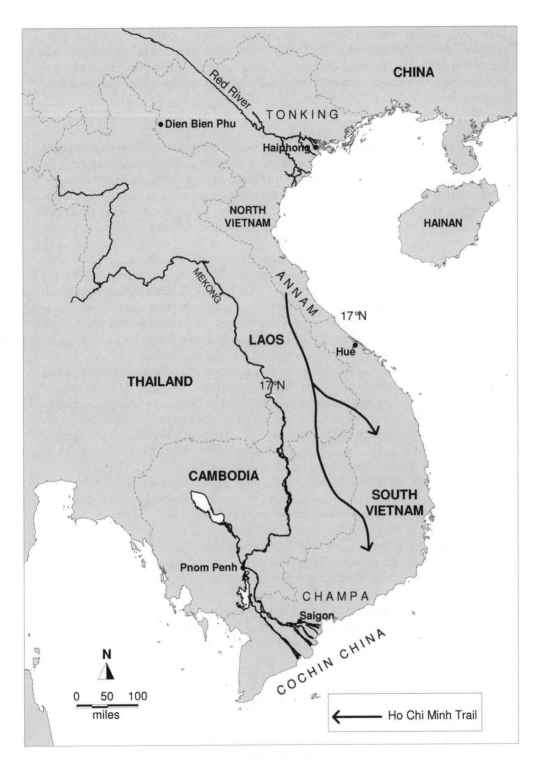

Indo-China

than 300, many armed just with blowpipes) killed more guerrillas than all the other security forces.[30] At the start of the second Indo-China War, there was considerable speculation about the montagnard alignment. In fact, the conflict raged so fearsomely that they had to concern themselves with daily survival, not long-term commitment.

The First War

Early in 1945, the Japanese occupation regime induced Bao Dai as Emperor of Annam to declare his country independent under their protection. Upon their surrender, he abdicated in favour of Ho Chi Minh. At that stage, the Vietminh broad left front laid claim to 50,000 members, a tenth of them avowed Communists. It had conducted low intensity guerrilla activity towards the Pacific War's end, duly receiving American arms and advisers. One factor in play of late had been Roosevelt's adamant opposition to any return to French colonization.

Come Japan's surrender, Kuomintang forces entered Tongking while French and British landed in the south. The French and the Vietminh clashed in the ensuing months. But in March 1946, it was agreed that 15,000 French soldiers should join with 10,000 Vietnamese to keep order in Tongking. Also Annam and Tongking plus perhaps Cochin China would join an Indo-Chinese federation within the French Union. Ho Chi Minh wanted the French in Tongking awhile to preclude Kuomintang imperialism. Soon, the Chinese troops withdrew as did the British.

The March accord was short-lived. Using Japanese and Chinese instructors and materials, the Vietminh expanded their field force to 40,000. This, and attacks on rival nationalists, aroused a French distrust which formal disbandment of the Communist Party little assuaged. For his part, Ho was exercised about France's predilection for keeping Cochin China separate from Annam and Tongking. Fighting flared again in November. Within weeks, the Vietminh had gone over to general insurrection. Their main base areas were in and around the Red River delta.

Well ahead of the military *démarche*, the French lost the political contest. They failed to call forth an alternative nationalist movement tough enough to withstand insurgent intimidation; flexible enough to settle acceptably; and authentic enough to command popular allegiance. Eventually, in 1949, Bao Dai was persuaded to return from a Hong Kong exile to rule a Vietnamese state (Cochin China included) autonomous within the French Union. The masses were already too *dégagé* to notice. Only 1200 people (these mainly expatriates) voted in a Cochin China election held that March.

Militarily, the Vietminh was a three-tier pyramid. At hamlet level were the local militias, usually all but devoid of modern arms. Above them were the regional militias; and above them the field forces. Initially, smuggling through Thailand supplemented stolen weapons. But a regular source of arms and ammunition was afforded in 1949 with the PLA's arrival on the northern border. Two years later, China's railway line to there was completed. **(See 9 (2))**. The reckoning is that, over the next two years, shipments to the Vietminh increased from 50 tons a month to 500. By late 1953, a Vietminh field army of perhaps 125,000 was organized into five light infantry and one heavy division, this last being relatively rich in field and anti-aircraft artillery. There were also 75,000 regional troops and, of course, the hamlet militias.

Legislation precluded French conscripts serving beyond France and Algeria. Therefore the homeland contribution to French Union forces in country was 55,000 regulars. Alongside them stood 20,000 Foreign Legionnaries; 30,000 North Africans (Algerians, Moroccans, Tunisians, Senegalese . . .); 70,000 Vietnamese; montagnard auxiliaries; and 15,000 air and navy personnel. Of the total, one half was assigned to garrison duty. The ratio of security forces to guerrillas looked nothing like enough for the former to win outright, given the geography.

Not that the Vietminh victory came easily. Giap later insisted his army went over to open mobile warfare as early as 1947. However, from December 1950 until succumbing to cancer a year later, General Jean de Lattre de Tassigny became for the Vietminh what Giap gallantly dubbed "an adversary worthy of its steel". In a succession of positional or mobile battles (in one of which his only son died), those under this French Union force commander thwarted a determined effort by Giap to gain the paddy and hamlets of the Red River delta. His death was followed by two more years of French Union decline.

In the Spring of 1953, Giap (perhaps disconcerted by the Korean stalemate) resolved to advance into Laos, his aims being to (a) link with the local Marxian front, the Pathet Lao, (b) get leverage over Laos' opium poppy production and (c) overextend the French. The French high command decided to create a fortified enclave around Dien Bien Phu, a rice-growing and opium-trading hamlet on the Vietnam/Laos border across an old invasion route. It stood in a flat and open valley bottom (19 by 13 kilometres) set midst low well-wooded hills. Airborne regiments (Metropolitan French, Colonial, Legionary and Vietnamese respectively) parachuted in that November.

The ensuing battle for Dien Bien Phu was hard fought. The diplomatic backdrop was that, late in 1953, East and West had begun to discuss informally further (i.e. post-Korea) easing of Far East tension. Early in 1954, a meeting took place in Berlin of the Foreign Ministers of the Four Powers which had occupied Germany in 1945. Vyacheslav Molotov took the opportunity to propose a multilateral conference (with China *inter alia* included) on the proposed subject. Next a British resolution was adopted calling for such a conference in Geneva from 26 April. Representatives of the Vietminh, Vietnam, Laos and Cambodia could join the discussion of Indo-China.

In the field, all hinged on logistics. Dien Bien Phu could be replenished only by air; and 62 French aircraft would be shot down or destroyed on the ground by Vietminh fire. As Giap proudly stressed, his men excelled themselves, lugging into firing positions on the jungly hills 150 mm howitzers, heavy recoilless rifles, *katusha* multiple rocket launchers, and crew-served anti-aircraft guns.[31]

One advantage foot soldiers indigenous to the Far East enjoyed was that rice cooked in water swells to three times its dry bulk. So a man needs but a kilogram of solid rice a day; and Vietminh planners could therefore stipulate that a squaddie bearing 20 kg of rice and vegetables could tramp 400 km through hills before exhausting this supply. Yet a rifle division might still need 50,000 coolies in support. The Vietminh were estimated to have had in 1954 but 1000 lorries.[32]

By January 1954, four Vietminh divisions were investing 9000 French Union troops at Dien Bien Phu. Giap told Stanley Karnow how he originally planned a full-scale attack on the 25th, employing the "human wave" assaults the Chinese had urged on him. But after last minute agonizing, he deferred things till 13 March, leaving himself time to develop a new plan involving extensive tunnelling as a means of remorselessly closing in. The conceptual resemblance to Chiang Kai-shek's 1935 encirclement strategy in Kiangsi is intriguing.

The Vietminh (aided by some Chinese) kept up field artillery and anti-aircraft fire. The French response throughout was of the order of 75 combat aircraft, none well suited to neutralizing rain forest dispositions: "Air power on a more massive scale than was then available could not have changed the outcome of the Indo-China war but it would have saved Dien Bien Phu."[33] As things transpired, the improvised airstrip went finally unserviceable in late March; and the whole garrison surrendered on 7 May, the day before the Geneva conference actually began.

The First Geneva Settlement[34]

Lately Washington had abandoned its previous view that France could win without military intervention by the USA and/or other allies. Secretary Dulles came briefly to favour warning China publicly she would incur aerial reprisals unless aid to the Vietminh ceased. By mid-April, however, he was merely advocating sorties against the Vietminh's logistics around Dien Bien Phu plus, if Geneva collapsed, wider air strikes against resupply. Britain declined to support any immediate military action or to commit itself to anything later. On 21 July, armistice agreements, closely corresponding to an Anglo-American draft, were signed by the French Union commanders and the Vietminh. International Control Commissions proposed by China were to supervise the truce in the respective Indo-Chinese territories. An immediate task for the Vietnam Commission was overviewing the movement of what proved to be 100,000 pro-Vietminh north and 900,000 anti-Vietminh (two-thirds of them Roman Catholics) south, this in relation to a provisional division of Vietnam narrowly along the 17th parallel of latitude. Elections were to be held in Laos and Cambodia in 1955 and throughout Vietnam by July 1956. Military build-ups were not allowed nor, save for two French posts in Laos, were foreign bases. Every delegation, bar Washington and Saigon, orally pledged itself to unifying Vietnam through free elections.

The Vietnamese Communists believed they could have won more than the war-ravaged north if backed more strongly by Moscow and Beijing. At Geneva they had had much contact with the Chinese but little with the Soviets. Afterwards Ho led Hanoi towards a Moscow–Beijing balance. Giap was keen to benefit from Soviet arms technology. Conversely, the ideologue Troung Chinh sought to emulate China's progress to date with agricultural collectivization. But North Vietnam never copied China's "big communes" venture from 1958.

In Laos, the Pathet Lao withdrew to the northernmost provinces pending incorporation in the political process. Among the terms of an agreement signed in November 1957 were that the Pathet Lao should enter the government while providing *inter alia* two battalions to the national army. This accord lapsed because the Pathet Lao did disconcertingly well in the May 1958 elections. Years of bargaining and battling ensued with American and Soviet involvement. By 1962–3 much of the east and north of Laos was under firm Pathet Lao control.

Clearly this afforded an ideal political setting for the "Ho Chi Minh trail", the network of tracks from North Vietnam through Laos that became a conduit of insurgency. Across northern and eastern Laos the vegetation was still very predominantly natural rain forest. Towards the Mekong river line, this partially phased into dry monsoonal woodland and scrub. Despite an adage to the contrary, such jungle and, indeed, such woodland and scrub could never be neutral in a guerrilla war. They hid too much from view. A programme of trail enhancement began in 1959.

Fragile South Vietnam

To be his Premier, Bao Dai appointed Ngo Dinh-diem, a fiercely anti-Communist Catholic from central Vietnam who returned from abroad in 1954. Twenty years before, as a government official, he had apprehended that the Communists would "win by default", win through the shortcomings of their opponents. He now sought somehow to prove himself wrong.

Diem's first move was to depose Bao Dai via a referendum that "like others to follow was a test of authority rather than an exercise in democracy".[35] He always interpreted the art of government as the arbitrary enforcement of personalist rule within a constitutional framework. Now his immediate aim was to stamp Communism out in South Vietnam before the Hanoi regime became assertive enough to inhibit him. He frequently by-passed the National Assembly and the law courts. He set up many detention centres wherein torture and illicit executions were norms. Often repression pre-dated (and maybe precipitated) the subversion it was supposed to thwart. Sometimes, as with the "morality laws" of 1962 and the 1963 crackdown on Buddhist activism, it was plain otiose.

Rice production rose steeply from 1955. But little was done to redistribute in favour of the peasantry the land in the Mekong delta "rice bowl", land still very largely in the hands of a few thousand often absentee landlords. Bitter resentment had been caused by the use of troops to try and collect rent arrears going back perhaps to 1946. Two years' rent usually equalled one year's harvest.

Subject to guidance from a still cautious Hanoi, the Vietcong (the successors within the South of the Vietminh) from 1957 organized 37 armed companies, mostly to operate out of the forests and marshes fringing the west of the delta. Concurrently they began in the hamlets to assassinate elders and public servants. That year 470 were killed. In 1959 the annual total was 1600; and in 1960 it topped 4000. What little hope remained of consensual rural reform was thereby snuffed out.

The Kennedy Contradiction

As his Presidency progressed, John Kennedy grew considerably apprehensive about Indo-China. In 1962, he invited the Senate majority leader, Mike Mansfield (a laconic Catholic liberal) to go to the South to report. His findings were basically that Vietnam could be saved only by the Vietnamese; and that the progressive increase in the American military presence in country (from 700 at the President's inauguration to 12,000 by early 1964) was too open-ended. Kennedy was later to confide to an aide, Kenneth O'Donnell: "I got angry with Mike for disagreeing with our policy so completely; and I got angry with myself because I found myself agreeing with him".

Yet one side of the President's persona was willing enough to get stuck in deeper. The expanding military mission was one manifestation of this. So was the response to a flare-up in Laos in 1963. He sent 3000 troops to Thailand, close by the Laotian border — i.e. the Mekong. So was the support the USA gave to the overthrow of Diem in November 1963, a coup which culminated in his assassination. So, not least, was the relevant passage in the speech John Kennedy would have delivered in Dallas the day he died. Its draft conceded that assistance to nations around the periphery of the Communist world "can be painful, risky and costly as is true in South-East Asia today. But we dare not weary of the task. A successful Communist breakthrough in these areas would necessitate direct military intervention".[36] However, he had been putting the commitment of US troops to Laos or, indeed, Cambodia in the same category as invading North Vietnam or mining Haiphong harbour. In other words, it was subject to a taboo, self-imposed in the hope of keeping the war limited. Assertions that he was resolved to withdraw after the 1964 election are insubstantial.

At the heart of the Vietnam syndrome was the esoteric strategic culture Kennedy and Secretary of Defense McNamara had woven round themselves. Anyone who conjured with such matters in Washington DC in 1962–3 will well hear David Halberstam's relating how a *modus operandi* for counter-insurgency was cobbled together, issues as vexing as those raised by defoliants, napalm and so on being decided veritably *ad hoc*.[37] By much the same token, anyone who spent any time in Saigon in the sixties will appreciate his more general contention that Vietnam was lost not inconsiderably by "the best and the brightest", youthful scholar politicians from Ivy League campuses to whom it was just another containment mission.

Briefed by them, wrote this author several years ago, it might be "all too easy to come back from Vietnam without having been there".[38] This was a harsh rendition but one relating especially to prognosis. The advice still given in April 1966 was that breaking the Communist field forces would take but 12 to 18 months. Next would come several years of localized mopping up. After which peace and democracy should prevail throughout the South.

More generally, the harshness just admitted to related to how slender in war can be the margin between dead right and horribly wrong. Undeniably, the best and the brightest had their strengths, not least unswerving dedication. However, bad misjudgements stemmed from their shortcomings. Here again one saw undue reliance on pleasure–pain action–reaction via punitive deterrence. Also too trite was the interpretation of numerical data drawn, in fact, from skimpy and unsound sources within an alien culture.

An instance may suffice. Within the American mission in-country was a public information service already several hundred strong. One afternoon its head, Barry Zorthian, briefed me a good two hours. My notes show that, towards the end, his advice was that 52 per cent of the population was controlled by the government and 23 by the Vietcong, the rest being "contested". This reckoning invited questions about methodology. How coercively was "control" defined? And must not any calculation be contingent upon the frequent and pronounced mood swings characteristic of situations so fraught? And could snapshots of such situations truly be measurable to one per cent?

This last point was explicitly raised. Zorthian replied that the chief criterion of "control" in a hamlet, village or township was whether the recognized local leaders (elders, teachers, officials ...) resided overnight. Granted this could be positive confirmation of government control in an area where the Vietcong was generally weakish. But anywhere it was strong, the said criterion could surely mislead. An elder, say, sojourning overnight could well be a member of the Vietcong or else bribed or intimidated by them or else plain stupid. Whichever way, he was unreliable.

The Trail Campaign

My view has long remained that a fateful lapse in received wisdom throughout the Second Indo-China War and its subsequent assessment has been a reluctance to appreciate how significant from the outset was Communist infiltration via the Ho Chi Minh trail. To my knowledge, just one book highlights the failure of aerial bombing to interdict this conduit. Its author is a former Marine and Foreign Service officer with Vietnam and Laos experience, 1965–7 and 1969–70. He has told how a sense of hopelessness suddenly overcame him as he looked one time across a boundless jungle canopy about the south of the trail. Neither heavy bombers nor strike fighters were going to be adequately effective from above it.[39]

Those who would discount the part the trail played early on have sometimes been simplistically quantitative. Take a representative guestimate but that 2000 activists moved South thus in 1962,[40] the great majority Southerners who had gone North after the 1954 peace agreement. Would not these returnees have formed a cadre at the heart of the Vietcong? And light though the flow of materials via them may have been, it could well have included field radios, special explosives and other items of high tactical utility. Besides, the discouragement thus visited on those in the South seeking peaceful reform will have been compounded by any infiltration flatly contravening the 1962 Geneva Agreements which covered *inter alia* Laotian neutrality. Needless to say, the same applies in relation to Jeffrey Record's feeling that, as late as 1965, there was not much to interdict along the said corridor.[41] The first North Vietnamese regiment to operate in the South was there by the year's end.

Giap himself intimated in 1990 that a prime reason for this escalation was a concern to block the non-revolutionary transformation of South Vietnam which might have been achievable post-Diem.[42] However, August 1964 had seen a Hanoi-versus-Washington naval clash in the Gulf of Tonking, a clash which gave rise to American air strikes against four North Vietnamese patrol boat bases and an oil depot. The way this incident unfolded owed something to ambiguities in radar and sonar surveillance. One effect soon was to harden Washington opinion, executive and congressional, in favour of on-going selective bombing of the North. Another must have been to accelerate the North Vietnamese build-up in the South. It is to be observed that, shortly after Khruschev's fall from office in October 1964 and then the US Presidential in November, the Kremlin augmented its flow of arms to the North.

Still, this was not to be simply another instance of binary action–reaction. Moscow's weapons supply to Hanoi soon included SAM-2 surface-to-air missiles configured to engage high-altitude strategic bombers. Emplaced just around Hanoi itself and its port Haiphong, these weapons betokened awhile a superpower understanding that the USA would not attack with B-52s those cities or, indeed, anything else in the North while the USSR would not take issue overmuch with a USAF/USN air campaign against tactical targets elsewhere in the North. Nor would the Americans accost Soviet supply ships bound for Haiphong or, from about then, to Sihanoukville in Cambodia. The shared aim was to write China out of the Indo-China calculus. In relation to Hanoi's internal politics, this meant both of them regarding Giap as in the ultimate easier to live with than Truong Chinh.

The probability is that through 1964, most Vietcong weapons derived from the near collapse of the large "strategic hamlets" programme launched in 1962.[43] A promising precedent for it had been afforded by the British during the Malayan Emergency (1948–60) when 500,000 Chinese were resettled in "new villages" protected and monitored by the authorities. However, the rub in South Vietnam was that "strategic hamlets" were too often under-resourced, inexpertly and corruptly managed, liable to involve construction by forced labour, and — not least — too far from ancestral graves. Besides which, Malaya's insurgency had emanated almost exclusively from within a distinguishable half of the population — namely, the Chinese.

An air interdiction campaign against the Ho Chi Minh trail built up through 1965. One week in April 1966 some 27 million leaflets were dropped on Indo-China, a sizeable proportion on the trail. Over the eight years, 1965–73, four million tons of bombs fell somewhere on Indo-China, more than on the Third Reich, 1939–45. Yet such a tonnage could represent but 200 lbs per average day for every square mile the trail might encompass. That could hardly have been decisive without US tactical intelligence being far better than it ever was. All the same, tens of thousands of Communist troops are likely to have been killed thus; and countless more found the bombs and leaflets compounded acutely the stress induced by fatigue, hunger and disease. Also, a high proportion of the motor trucks used were destroyed. Aerial interdiction will, of course, have been least effective during the wet season, mid-June to early October. About 80 per cent of the annual rainfall in eastern Laos (typically 80 inches) came then.

All in all one can understand retiring President Eisenhower's having said "with considerable emotion" at the White House transition meeting in January 1961 that "if we permitted Laos to fall then we would have to write off the whole area". On these grounds alone, one must question Lawrence Freedman's judgement that John Kennedy "left the Cold War in a far less dangerous state than he found it".[44]

The End of South Vietnam

Lyndon Baines Johnson was overwhelmingly re-elected President in 1964 on a ticket of moderation in Vietnam in contrast with the "Extremism in the defence of liberty is no vice" pitch of his Republican opponent, Barry Goldwater. But facing a more rapid deterioration in the situation in the South, "LBJ" authorized in March 1965 the dispatch thence of a Marine battalion, thereby beginning a build-up of ground formations to complement the modulated escalation of an air offensive. American force levels in South Vietnam rose from 23,000 that New Year to a peak of 542,000 in January 1969.

In spite of, or perhaps one should say because of, the incipient American build up, the Communists launched a formidable though inconclusive ground offensive in the summer of 1965. Some months later, their adversaries did likewise. Talking to American paratroops undertaking jungle patrols in one of the "war zones" the Vietcong had designated, one felt quite inspired by the good their common lot was doing for relations between White and Black. Yet one had to feel concerned about morale overall. These lads could not see a defined operational aim. Nor were they happy about tactics, especially how overmuch artillery support was denying their incursions any subtlety. Very similar reservations were being expressed by the British military advisory mission under Sir Robert Thompson.

In 1967, no seismic shifts occurred in-country, which was bad news for Saigon on the argument one used to hear that, when revolutionaries cannot be roundly defeated, they are set to win. Already, too, the battle for "hearts and minds" was extending from the hamlet clusters of the delta more towards the American national and world stage. Opposition to "the war" was melding into the global youth revolt of the late sixties. Additionally, a huge strain was being placed on a turbulent world economy (see 16) as the war undermined the US balance of payments to the tune of 20 billion dollars a year. Viewing the chaotic world of the winter of 1967–8, Giap decided to regain momentum by suddenly playing the urban card. Breaking a truce he had agreed to for Tet (the lunar New Year on 31 January), he had 70,000 People's Army of Vietnam (PAVNM) — North Vietnamese — and Vietcong guerrillas storm into dozens of towns and cities across South Vietnam. They took heavy casualties. They failed to stimulate popular uprisings. Nevertheless, they won the war.

Mass urban violence was, as it will remain, very televisual. In particular, the Communists held the citadel of ancient Hué for a good three weeks against a deter-mined American and South Vietnamese riposte. Also General Westmoreland (US commander-in-chief in Vietnam, 1964–8) was constrained to ask for up to 206 thou-sand additional US troops to turn round what had ostensibly become a new style war. He seemed thus to confirm a widespread impression that he had repeatedly under-rated his opponents' strength — and still more so — their resolve. In March Lyndon Johnson ruled out standing again in November. Since 1966, Bobby Kennedy, now the Democrats' front runner, had been overtly to the Left of him on Vietnam.

Also in March, Clark Clifford replaced Robert McNamara as US Secretary of Defense. Although previously a straight down the line supporter of the Vietnam commitment, he published in *Foreign Affairs* after leaving office the next year an article which smacked of pre-emptive surrender.[45] He well observed that the regional secu-rity situation *vis-à-vis* China and Indonesia had much improved. Witness the visit President Richard Nixon was to pay to Beijing in 1971. However, nothing Clifford said about the situation in South Vietnam would convince anybody that his uncondi-tional advocacy of a full withdrawal of US ground troops before 1971 was other than abdication.

So was there an alternative? GI morale in-country was much lower than in 1966: "Antiwar protests at home had by now spread to the men in the field, many of whom wore peace symbols and refused to go into combat . . . The use of drugs was so wide-spread that, according to an official estimate made in 1971, nearly one third of the troops were addicted to opium or heroin; and marijuana smoking had become routine".[46] The massacre of 300 civilians at Mylai by US infantry in March 1968 was an act of terror directly related to discipline dissolution.

The incoming Nixon administration opted to exit as gracefully as possible. In March 1972, a North Vietnamese offensive into the South (using state of art Soviet tanks) was resisted determinedly by the Army of Vietnam (ARVN) — i.e. the Saigon forces. Meanwhile a temporarily widened air offensive against the North included mining harbours (especially Haiphong) and dislocating air defences (especially around Hanoi). However, that August, the last US ground combat troops left Vietnam; and in January 1973, a Washington–Hanoi peace accord was signed. In March 1975, the PAVNM launched a final armoured offensive. Saigon fell ten weeks later.

The Human Cost

The USA had 47,000 killed in this war, South Vietnam 400,000. North Vietnam lost perhaps a million, many in Laos and the South. A peculiarly ugly derivative of the conflict was the takeover of Cambodia by Pol Pot's Khmer Rouge acting with PAVNM support. Hard upon the fall of Phnom Penh, the capital, in April 1975, a ferocious terror campaign focussed on the city got under way. Three million died in consequence.

History has seen not a few campaigns of terror against big towns and cities *per se*. The Tai Ping rebellion in nineteenth-century China and the Mongol depredations come all too readily to mind. So does Rome's deletion of Carthage, not to mention Corinth the same year — 146 BC. But this Khmer Rouge action must rate among the more flagrant and gratuitous. It cannot reasonably be put down to a reaction against corruption nor related to the USAF's bombing of distributary routes through Cambodia off the Ho Chi Minh trail. Most of the deaths were actually caused by over-work and starvation during "re-education" in the countryside.

What Giap made of it all is not, to my knowledge, recorded. Perhaps it helps explain his withdrawal from the Indo-Chinese and world stage after 1975. Very tough he may have been but he was neither irrational nor dehumanized.

5 The Holy Epicentre

Two contrasting approaches to the Arab–Israel conflict have been adopted since 1945. The one has been to treat it as so special a case as to be best considered apart from the jerky interactions of geopolitics in general. The other has been to regard it as so special as to be in the middle of the geopolitical kaleidoscope, much as medieval cartographers placed Jerusalem in the centre of the world.

My own stand has long been in the latter direction, not so much because one has thought the conflict's instabilities could one day pitch us straight into World War Three as because one has perceived the ultimate possibility of Israeli–Palestinian part-nership as an elixir of planetary peace. The politics of the Holy Land in modern times is further considered in 15 (1). Suffice for now to follow the course of Arab–Israel warfare, 1947–82.

Not least does the sequence exemplify action–reaction. Take the immediate in-country background to Israel's War of Independence in 1948. Critically there was a split within Zionism between the social democratic mainstream led by David Ben-Gurion (1886–1973) and the Revisionists lately led by Vladimir Jabotinsky (1880–1940). An Odessan, the latter was by no means some nativist "Know Nothing". He was multilingual and a gifted orator. Yet, careless of Arab rights, he and his colleagues aspired to create a Jewish state extended to biblical borders beyond the Jordan. Ben-Gurion, a Pole, had come up through social democratic trade unionism.

In 1931, a Revisionist insurgent group, the Irgun Tzvai Leumi, had broken with a Haganah by then established as the main Jewish defence force. The former's worst excess was to be the massacre (with much gratuitous mutilation) of 250 Arab villagers at Deir Yassin in the Judean hills early in 1948. This terror was strongly condemned by social democratic Zionism as well as the world at large. Even so, Deir Yassin long symbolized, for many Palestinians, Israeli ruthlessness and international gaze-averting. Working as a journalist in September 1967, one met on the Jordan's East Bank quite a few of the 90,000 refugees who had fled across the river due to the recent June war. Several told of Israeli troops pressurizing them to "Go to King Hussein". But overwhelmingly, the chief reason given for exiting was the folk memory of Deir Yassin. An American University of Beirut survey concurrently under way was reaching a similar conclusion.

As trustee for the Palestinian mandate, Britain's declared aim in 1945 had been to preserve a communal balance well enough to allow of the emergence of a binational state. From 1936, an Arab insurgency had been the big security problem. But now, in the wake of the Holocaust, Zionists found London's policy intolerable. While over-crowded tramp steamers sought to run the Royal Navy's blockade of the Palestine coast, a Jewish guerrilla campaign inside the country continually grew more intense. Haganah usually sought to restrain the few thousand Revisionists in the Irgun and the one or two hundred members of the totally uncompromising Stern Gang. Jewish guer-rillas killed 150 during 1947. Thirty of them were Jews seen as too possessed by the spirit of compromise. Ninety others were members of the British government secu-rity forces. By then 100,000 British troops were deployed in support of the civil power.

Successive proposals for peace failed to resolve Arab–Zionist differences, partic-ularly as regards partition and allowable rates of Jewish immigration. So in mid-1947 Britain referred the whole question to the UN. In November the General Assembly passed a partition plan, one which allowed for joint control of Jerusalem. Disclaiming further responsibility, Britain announced it would withdraw from Palestine by 15 May 1948.

At the start of the war which then ensued, the Arab armies committed (by Egypt, Jordan, Iraq and Syria) numbered 90,000 men with 200 tanks. Against them stood Haganah's 10,000 riflemen and several tanks plus the Revisionist gunmen they some-times co-operated with locally (e.g. in Jaffa). Except in their failure to wrest the old city of Jerusalem from Jordan's British-led Arab Legion, Haganah was comprehen-sively triumphant. Part of the explanation lies in the respite gained from the first UN truce (11 June to 18 July), a truce which *inter alia* allowed of substantial small arms shipments to Haganah from a now Communist Czechoslovakia. Acting as a proxy for the USSR, Czechoslovakia similarly fed weapons to Egypt from 1955; Nigeria from 1967; and Cyprus in 1971. Causally more basic, however, was the low motivation of most Arab contingents. In 1948, four Iraqi brigades entered Palestine on 15 May and readily reached a point but 10 miles from Israel's middle coastline. There they stopped. The Egyptian performance was, in the main, little better. At the height of the fray, the Engineer Corps was pulled back to build a rest house for King Farouk.

The year 1949 witnessed a series of armistices between individual Arab states and Israel. By the start of the year, over 80 per cent of the 850,000 Palestinian Arabs had fled their homes while 750,000 Jews were within the old mandate boundaries, as against 550,000 in 1945. Then in May 1950, Britain, France and the USA guaran-teed a new Arab-Israel frontier against major violations from either side. Israel was thereby left with an area a quarter as large again as that envisaged in the UN partition plan of 1947. It included all the Negev desert, invaluable for armoured warfare training. It also included the port of Eilat, soon to assume great importance — commercially and politically — as a gateway to the non-Arab developing world. But it also left the Judaean hills in what thenceforward was Jordan's West Bank, over-looking Israel's coastal Plain of Sharon — just the 10 miles across at its narrowest.

Then there was the refugee problem. Historians still fiercely debate in Israel and elsewhere how far the Palestinian exodus was Israeli induced as opposed to being a spontaneous avoidance of a war that those concerned had been assured would soon be victoriously concluded. But Deir Yassin and Jaffa surely highlight a significant degree of coercion. Nor was Arab bitterness assuaged by the migration into Israel of a broadly similar number of Jewish refugees from Iraq and other Arab lands. Their exodus mainly occurred a little later on; it was to the biblically "promised land", not towards the desert. Besides, the context was victory not defeat.

The Palestine War was a climacteric in Arab history. The reaction against this humiliation was manifested in the Syrian army coups of 1949. But it was Egypt, the biggest and the most developed Arab nation state, which saw the most radical upheaval. Apprehension of this caused a pro-loyalist Wafd party government to revive demands that Britain leave the Canal Zone and Egypt be united with Sudan. However, a newly radical nationalism was not diverted quite that easily from the goal of social transformation. One of the first acts of the "young officers" who wrested power from Farouk in July 1952 was to redistribute land from large estates.

They can also be seen today as prime movers in the 1956 Suez crisis, the Arab-Israel conflict most illustrative of how complex may be the problem of whether and how to intervene in a regional crisis situation, intervention which may culminate in regime change. The complexity will always be *sui generis*. But case studies can still be worthwhile.

The Suez War[47]

The USSR's first post-war loan to a non-Communist state was one made in 1954 to Afghanistan. But by 1956 Cairo was focal to Soviet influence building in the region. This targeting involved gaining the goodwill, however contingent, of Gamal Abdel Nasser, the engaging son of a post-office clerk and 1948 war hero who was now undisputed leader of the Egyptian revolution. He continued to repress the Communists and the Moslem Brotherhood. He had negotiated, subject to notional rights of re-entry, Britain's evacuation from the Canal Zone. He had initially sought American defence aid but was told this could not be on a scale compromising to Israel; further, such aid would probably require Egypt's joining West-aligned arrangements for regional security. So in September 1955 a large arms deal was peremptorily signed with Czechoslovakia.

Soon, the question of financing a proposed High Dam on the Nile at Aswan was further exacerbating tensions. Soviet intimations of generous long-term aid were well timed in relation to burgeoning doubts (not least at the World Bank) about Egypt's ability to cope with the financial burden. Germane, too, was the beginning of a reaction across the West against the enthusiasm for big dams that had waxed so strong in the 1930s.[48] Concern about ecological impact was at the heart of the concern. Additionally, strong Congressional lobbying drew strength from an escalating insurgency against Israel from the Gaza strip; the exclusion of Israeli shipping from the Canal; and Egypt's recognition of Communist China.

In mid-July 1956, the USA, duly followed by Britain and the World Bank, cancelled its provisional offers of Aswan aid; the USSR declined to pick up the tab forthwith. The Eisenhower administration had given neither the Egyptians nor, indeed, the British forewarning of its brusque disengagement. Nasser therefore felt constrained to take action which would at least appear to meet his need for extra capital. In a belligerent speech before a huge crowd in Alexandria on the 26th, he announced the nationalization of the Suez Canal company. Some compensation was vaguely promised.

Quite the most disconnected of the lead players in the ensuing saga was Britain's Prime Minister since 1955, Anthony Eden. Once he had seemed possessed of quite singular stature. A close avuncular friend in my youth had served under him as an infantry NCO on the Western Front; and, regardless of their political differences, was adamant about how fine an officer he was to have with you in the trenches. Likewise many would acclaim him as an outstanding Foreign Secretary in Winston Churchill's wartime coalition, 1941–5. Yet well-placed observers, starting with Churchill, never did see in him the kind of extra breadth and depth desirable in the premiership. His limitations were accentuated from 1953 by continual stomach trouble.

In the circumstances now created, he abruptly turned atavistic. He looked forthwith to military intervention, depicting the Egyptian regime as fascist and Nasser as Mussolini. This pitch he was encouraged to sustain by the "Suez group", a faction of Tory MPs led by Julian Amery which had come together in protest against the 1954 decision to withdraw British forces from the Suez "canal zone", all but unconditionally. Amery's father-in-law was Harold Macmillan, then Chancellor of the Exchequer. He assumed a decidedly forked posture towards intervention, quick to support it as a prospect and then to dissociate from it come *dénouement*. Meanwhile, military intelligence waxed too obsessive in what was, after all, the aftermath of its deep involvement in canal zone counter-insurgency, 1952–4.

None the less, there were objective realities to reckon with. Nearly a third of the ships that normally passed through Suez were British; and almost half the company's shares were held by Her Majesty's Government. France was also commercially, involved, and was already incensed by Nasser's aiding the FLN insurgency in Algeria. France and Britain duly led the international opposition to nationalization, contending that the riparian state could not be left in unfettered control of the waterway. Most of Egypt's sterling and franc assets were frozen while Washington was persuaded to join Whitehall and Paris in inviting 24 nations to a London conference. Greece and Egypt declined. But a majority of the conferees supported a proposal tabled by Dulles for international management through a Suez Canal Board. However India, Indonesia, Sri Lanka and the USSR backed Cairo's insistence that, while it might answer to the UN on matters of broad principle, Egypt must be left to operate the waterway as it thought best. On 12 September, Britain and France inaugurated a Users Association to provide pilots, collect dues and so on. The USA acceded but explicitly ruled out ultimate resort to force. The Association never materialized.

Through the summer, Israel had become increasingly concerned. In July, Prime Minister Ben-Gurion had endorsed the act of nationalization; he was to defy Britain again in early October with a counter-insurgent strike against Jordan. Moreover, he remained reluctant to get locked into an anti-Arab tripartite alliance. On the other hand, relatively massive retaliation was seen as the only way of curbing mounting trans-border insurgency. Yet Egypt's rapid and on-going acquisition from Czechoslovakia of modern Soviet-design weapons was overshadowing the prospects for this and, indeed, for more basic defence. Israel's air arm as yet lacked the strength and the depth of air space to protect its cities and support its armies adequately. Some kind of pre-emption was urgently called for.

In other respects, too, the time constraints on intervention were tight. Early in August an *ad hoc* Anglo-French command had started to prepare for a large-scale amphibious landing. This was anything but easy. A good deal of improvisation was involved. Not till mid-September was the wherewithal at all ready. But the run-up to the US Presidential election, to be held on 6 November, was a time to avoid if one did not want to be accused of very sharp practice. Yet one also had to reckon with the approach of winter, given the character of the amphibious assault envisaged. One well-informed opinion afterwards was that, as events actually transpired, a freshening north-easterly breeze would have made the landings hazardous eight hours later and impossible if attempted the following day.[49] This was because wreckage, blockships and suspected mines had already made the harbour at Port Said almost unusable. Narrow and exposed beaches had to be resorted to instead. Besides which, there were general reasons why any war needed to be brought to a satisfactory conclusion as quickly as possible. One was that Western Europe's oil reserves were equivalent to but three or four weeks normal consumption.

Tripartite Collusion

The French took upon themselves the difficult task of joining up Israeli and British strategies. In September they began regular staff discussions with the Israelis. Substantially, the only help the Israelis really sought was protection against air attack. Meanwhile, the British government sought grounds for intervention more persuasive than nationalization *per se* but not involving too overt an accord with Israel.

Intrigue culminated in a secret conference held at Sèvres in France on 23 October, two days after a pro-Nasser swing in Jordanian elections had heralded the formation of a joint command with Egypt and Syria. Christian Pineau and Selwyn Lloyd, the French and British foreign ministers, attended, as did David Ben-Gurion. Pineau later confirmed that the key moves the next ten days were co-ordinated there.

Israel's covert mobilization began on the 25th. Although tactical surprise was achieved against the Egyptians on the 29th, the relevant moves had been noticed by the Americans who had duly started evacuating nationals from Alexandria on the 28th. The same day, there were trilateral diplomatic exchanges between the Americans and the Anglo-French, with the latter giving nothing away. Meanwhile, the French task force was putting to sea from Toulon and so on. France was also lending Israel more direct support in several ways. Sixty French air force Mystères arrived there on the 28th. When war broke out the next day, they joined with Israeli planes in attacking Egyptian ground forces with cannon, rockets and napalm. There was also a bombardment of Rafa by French warships.

The British remained more concerned to conceal collusion. There is little reason to contest the claim that British units in Malta had no prior notice of the Anglo-French 12-hour ultimatum sent to Egypt and Israel on the 29th, telling each to stand back at least ten miles from the canal and allow the joint task force to occupy the buffer zone thus created: a truce plan which fitted the specifics of geography but was conspicuously one-sided.

The Anglo-French forces committed totalled 90,000 men, half from each nation. Their order of battle included the equivalent of roughly two divisions of ground troops; several hundred warplanes; and six carriers plus other vessels. Some RAF planes were based on Malta but most were crammed into three airfields in Cyprus. These remained very inadequate despite reconstruction that summer. This inadequacy restricted the scale of the parachute landing. Since Cyprus was deficient in port facilities, too, the British half of the invasion fleet had to proceed at a steady 10 knots from Malta while the French came in from points west. As each steamed east, ships, submerged submarines and aircraft from the US 6th Fleet subjected them to some hassle.

Allied air attacks began 90 minutes after the rejection of the ultimatum by Egypt. Initially the emphasis was on the crippling of Egyptian air power and preventing block-ships being sunk in the canal. By 2 November, the first aim had been attained and the second frustrated. Next, the main object was to isolate Port Said; and a subsidiary one, readily accomplished, the immobilization of Radio Cairo. A significant part in the counter airfield operations was played by the RAF's Valiant V-bombers — planes more basically earmarked for strategic nuclear encounter. On 4 November, their sorties finally ceased for lack of suitable targets. The following sunrise the British and French paratroopers dropped about Port Said. Next day, seaborne landings began, a day ahead of those American elections.

The Egyptian dispositions in the Sinai derived from the assumption that they would only have to fight on the one front. Several brigades were dug in about the metalled coast road in the Rafa–Gaza sector. Though their immediate flank was covered by the fortifications of Abu Agweiba, the line more to southward was thinly held. The concept was that the air force plus mobile army reserves from the Delta would curtail any enveloping movement. Under the circumstances now obtaining, however, the emphasis was more on withdrawal from the Sinai than on reinforcement.

Permeating the specific tactical implications of the Anglo-French intervention was its psychological impact. Egyptian resistance was often determined. Against lesser odds, it might have been more universal and sustained. In particular, the Israeli offensive might have been thwarted. After all, their 5000 vehicles were so ill suited to the desert surfaces that, during a mere five days' fighting, 2000 broke down beyond immediate repair.[50]

Nineteen hours after the first seaborne forces had landed, a UN ceasefire came into effect. By that time, interventionary forces were 25 miles south of Port Said. So the French high command were expecting Suez to be reached the next day though Britain's War Office felt another four were needed. That would have meant the whole operation taking 12 days, the original planning estimate. However, Eden later stressed it could have been longer.[51] Indeed, it is a moot point whether effective control was gainable at all. Egyptian soldiers coming out of the Sinai were merging with the 400,000-strong population of the Canal Zone, many of whom had themselves received arms from the government over the last few days. In 1954, some 80,000 British troops had been tied down on anti-guerrilla operations in this very area. In other words, the intervention could soon have turned non-viable regardless of reaction in the wider world.

Nevertheless, external pressure was most immediately compelling. About 15 per cent of Britain's gold and currency reserves left London during the operation; and it had therefore to seek a $1.5 billion loan from the International Monetary Fund to uphold the pound, a loan the Americans would have to underwrite. In regard to this and other aspects, the USA used its superpower status to obtain an unconditional restoration of the territorial *status quo ante*. In the UN, this pitch enjoyed solid support, except from Australia and New Zealand. What went unexplained was how Israel might otherwise have coped with Egypt's present of arms.

The canal was soon functioning smoothly enough again, thanks considerably to Soviet pilots. The Soviets agreed to finance Aswan as well. But silt retention was to prove for that project a major and chronic problem.[52] Meanwhile, Israel laid more emphasis on making her cadre-conscript army tank-rich, while pressing on to build an air arm of exceptional strength and quality. Come 1967 it would have 320 warplanes, all bar 48 Skyhawks of French design. Also, in 1964, France had contracted to supply Israel with the Dimona nuclear reactor, this without anti-proliferation strings. As for France itself, "Suez" can be seen as one of several factors leading to De Gaulle's foreclosure of the Fourth Republic in 1958. Soon France and Germany would be drawing closer, within the "Europe of the Six".

As for Britain, "Suez" led directly to Harold Macmillan's taking over as Premier. More generally, this crisis has often been identified as the point at which Britain was forced to acknowledge that the Empire was finished. Whether it stands out thus compared with Singapore, 1942; India, 1947; sterling, 1949 and 1967; or Hong Kong, 1999 could be questioned.

Washington sought urgently to fill the geopolitical vacuum it sensed was left by "Suez". Hence, January 1957 saw the enunciation of the Eisenhower Doctrine: that the USA would help, on request, any nation or group of nations in the Middle East to "resist overt armed aggression from any country controlled by international Communism". It explicitly remained a benchmark for about two years. It did so most conspicuously when, in July 1958, President Chamoun of the Lebanon, a Maronite, sought support in his struggle against Sunni Moslem radicalism backed from Syria. The USA dispatched a strong marine force to the Lebanon. Collaterally, Britain sent paratroops to Jordan. More extensive contingency preparations were also in hand.

For several weeks during that summer, this author, as a young naval lieutenant then serving with the Mediterranean fleet in a cruiser, was one of a small team detailed to the RN Fleet Operations Centre in Valetta to help distribute incoming cable traffic. What struck one was the degree of unambivalent harmony that prevailed between the British and American navies in theatre, this not two years post-Suez.

The June War

A systemic weakness in the international political process is not knowing when to go hard for the lasting resolution of conflict situations. While war is raging, this goal seems unattainable. If peace prevails, the disposition in both government and the media is to push it down the agenda in favour of more urgent or exotic concerns. Such two-forked evasion was very characteristic of, for instance, the Cyprus question from 1960 through 1975. So has it been of the Arab–Israel conflict.

An especially golden opportunity *vis-à-vis* the latter may have been lost in 1962 to 1964. During the 1958 crisis mentioned above, a union of Egypt and Syria had been enacted. But this had collapsed in 1961, being too much at odds with History and Geography. Soon after which, the resolution of the Cuba missile crisis generated awhile an atmosphere worldwide conducive to real accommodations. Furthermore, neither the leadership proffered to the Arab world by President Nasser and King Hussein nor that to the Israelis by Prime Minister Eshkol and Foreign Minister Eban was ultimately sustained by a fear of lasting peace. Yet Washington and the international community in general let this transient opportunity pass unheeded. Late in 1963, Israel detailed a unilateral plan to divert upper Jordan water. In 1964, Khrushchev fell from power in Moscow; the Soviet fleet returned to the Mediterranean after a six-year absence; the Franco-Israeli reactor deal was struck; and the formation of the Palestine Liberation Organization (PLO) gave insurgency more cohesion and thrust. Shortly, the sixties would turn turbulent around the world.

In November 1966, Syria and the United Arab Republic (alias Egypt) signed a defence pact, its significance lying in its committing the two countries to head the struggle against Zionism. Come January, this author was in Cairo as a reporter. He was advised by quite high authority that war with Israel was inevitable; and when it came, the Arabs would win whether or not the Americans and British joined in. However, the Arab world in general and Egypt in particular would not be ready for several years. One immediate reason in the latter case was the recent launch of a nationwide programme of family planning. Another was Egypt's still being saddled with her Yemen commitment.

A "republican" coup had been essayed in the Yemen in 1962. But the pre-existing "royalist" regime had been fighting back with Saudi support in what for centuries had been rated a very enclosed quasi-hinterland of the old *Arabia Felix*. (See 2 (2)). Nasser displayed characteristic impetuosity as regards rallying to a progressivist Arab cause, however ill-founded. He had dispatched 70,000 of his best troops to back the Yemeni republicans; he still had 50,000 there in 1967. It does seem, too, that since 1965 they had been using mustard gas in the field, something President Nasser would not likely have countenanced had their position been tolerably secure.[53]

Reservists included, the Egyptian army in 1967 numbered 240,000. So the relevance of the Yemen commitment is partly that a good third must have been in that deployment or directly supporting it. Regardless of this, however, that impetuous streak was soon to surface anew. Cairo did nothing when, on 7 April, Israel shot down six Syrian MiGs. But then, early in May, "Russian friends" intimated to an Egyptian parliamentary delegation visiting Moscow that the Israelis were making ready to attack Syria, currently the front-runner in insurgency support. On 13 May, Syria's Foreign Minister said Israel was planning aggression. On 15 May, Israel held an Independence Day military parade in Jerusalem. Under the 1949 armistice terms, armour was excluded from such an event in that locale. However, Egyptian intelligence is thought to have advised the Syrians that the absence therefrom of heavy weapons was because "14 Israeli brigades had been concentrated within a few miles of the border".[54] On 18 May, Egypt formally requested the withdrawal from the Gaza and Sinai borders of the United Nations Emergency Force peacekeepers in place since 1957. On the 23rd, Egypt (having garrisoned Sharm-el-Sheikh and deployed a third of her navy in the Red Sea) closed the Gulf of Aqaba to Israeli shipping.

War duly began, on 5 June, with pre-emption against 16 Arab front-line airfields by the enhanced Israeli air force, typically approaching at 500 mph below 150 feet. Some 300 planes were thus destroyed (nearly all while still on the ground) and many runways shattered. The dispatch of the Egyptian, Syrian and Jordanian armies was thereafter but a matter of time. No matter that historic Jerusalem (where air power was circumscribed) was hard contested. The perfection of this pre-emptive blow owed much to "precise timing, hard training, accurate striking power, understanding of their own limitations and proper intelligence, including an accurate psychological assessment of their opponents".[55]

Although this one stratagem made the war's outcome a foregone conclusion, many other features of the campaign could be mentioned. Several might now be. One is whether King Hussein could and should have responded positively to Levi Eshkol's plea on 5 June to stay out of the conflict. His negative reaction has been basic to all that has happened since. Yet perhaps the prior question for the moment should be whether it was wise or needful militarily to hand command of the Jordan Arab Army over beforehand to a middling Egyptian general as per the accord struck with President

Nasser when they had met on 20 May. A more enterprising commander might have gone over to a light infantry offensive in Jerusalem, even allowing that the Jewish troops in the Holy City were suffused with added inspiration. As the Israelis said, "The Arab Legion always fights well".

Still, the comparison of most concern was that with Egypt, the biblical superpower. Israelis (not least General Ariel Sharon) tended to speak quite highly of the competence and motivation Egypt's poor bloody infanteers often displayed. But they had undisguised contempt for the slothful way they were commanded: "The officers are too fat and the men are too thin". Why was the 15 miles of the nodal Mitla Pass not held? Why was Israel's naval arm (with but 3000 regulars) able, without air support, to sink two Egyptian submarines and several fast patrol boats while losing no vessels itself?

Much wider and deeper questions were posed by the outcome of the June War, not the least of them being how Israel could remain an essentially Jewish democracy if it was going to consolidate through settlements and in other ways its hold on the Golan, West Bank and Gaza. One answer too often proffered on the Revisionist Right of Israeli politics was "population transfer" — alias ethnic cleansing. That would have generated an even more searing identity crisis for Zionism at home and abroad.

To Yom Kippur

Israel's air arm appeared hardly less devastating in response to Egypt's "war of attrition" along the Suez canal from March 1969. At little cost to itself, it inflicted much damage and dislocation on the Egyptian army there, not least on its air defence umbrella. Then in January 1970, deeper penetration was essayed, some targets in the outskirts of Cairo being bombed in the hope of destabilizing President Nasser.[56] His response was to visit Moscow to procure more adaptive surface-to-air weapons — notably, SAM-3 crew-served missiles and ZSU-23–4 multiple machine-guns on tracks. The upshot was a ceasefire that August. Given how fast and low they overflew the populous Nile valley, Israeli pilots had been unable to match the unerring selection of strictly military targets they had almost always achieved previously.

Much as the American Civil War had looked back to 1815 and forward towards 1914, so the struggle of Yom Kippur (alias Ramadan) evolved from what can be characterized as an essentially mechanical past to a more electronic future — 1940 versus 2000. The number of tanks committed all round was 5000, over half as many again as in 1967 and comparable to the numbers deployed at the epochal Battle of Kursk in 1943. On the other hand, the Israelis came to use extensively the Maverick, an airborne Precision Guided Munition of American design. Its electro-optical guidance reportedly ensured a 95 per cent success rate against Egyptian tanks on open landscapes near the canal.

For obvious reasons, the Arab belligerent states (now just Egypt and Syria) were anxious to land this time a mortifying first blow. The chosen date was 6 October 1973. Good tidal conditions obtained in the canal. Moreover, the 6th was half-way through the Ramadan fast as well as being the anniversary of the Prophet's victory at the Battle of Badr in 626. It was also, of course, Yom Kippur, the Jewish religious day. Nor should operations be delayed beyond October since rain and snow would effect the higher ground.

Faced with accumulating evidence of Arab preparations for war, Israel began again to consider aerial pre-emption. By 5 October, her air reservists had been mobilized. Early on the 6th, Chief of Staff David Elazar urged a strike be launched at 1300. Two years later, a USAF study concluded that it could have destroyed within a few hours up to 90 per cent of the Arab SAM sites for the loss by Israel of fewer than ten planes.[57] But overall, the Israelis felt unable to read the signs with enough assurance to convince not just themselves but the Americans that an onslaught was in the offing.

One factor in this may have been Arab regimes' gradually assuming a better mien post-1967. Certainly their public pronouncements became less divorced from tactical or strategic reality. Probably, too, the confidentiality of their in-house business was better maintained. In 1968 one had been told by a State Department official serving in the Mediterranean that the Israelis had sometimes passed to them credible accounts of Arab cabinet meetings. Such could hardly have applied to this 1973 situation.

Decisively important, however, was the Egyptians' amazingly bold deception plan. In addition to spreading their assault forces all along the Canal, they employed a clutch of more specific tricks: positioning tanks and artillery pieces as though for defensive action; hiding bridging equipment in crates and pits; using underground cables to transmit sensitive messages; obscuring the general forward movement by unit rotation; and so on. As General el Shazli was to put it, "The last three days were especially difficult . . . we did not expect the enemy to be taken in as easily as he was".[58]

The combined Arab offensive was launched at 1400 on the 6th. Thenceforward the course of the campaign was determined considerably by geographic peculiarities. Take the Golan. On a casual overviewing, this short-grassed plateau looks almost formless, topographically speaking. Yet the variations within it of geology and thence landscape proved to be of tactical importance. Thus the Israeli forward defence rested on 14 *telal*: small hills of volcanic origin, each of which had been capped by a complex of barbed wire, concrete, metal sheeting and sandbags with bunkers below. Though subject to heavy bombardment from land and air, a number of these strongpoints survived the entire war.

Then again, the Israelis eventually turned the contouring of that sector to advantage in Close Air Support. The first afternoon, they suffered mortifying losses from Surface-to-Air Missiles (SAMs) as well as from mobile anti-aircraft guns. As dusk approached, however, pairs of Skyhawks "skimmed in a low northward curve over Jordanian territory, hugging the contours until they rocketed up and over the Golan plateau to take the Syrian armour in the flank, then curved away west of Mount Hermon — hopefully without passing over the deadly SAM sites".[59] That response was most effective once Syrian armour had crossed onto ground sloping more towards Galilee. Within days, the Syrian army was falling back in some disarray.

Meanwhile, the Egyptians had confronted the Bar Lev line, some 16 fortified complexes positioned along the canal's east bank at 8 to 13 kilometre intervals. On the 6 October, they crossed the canal in dozens of places; and within two days had brought ten heavy duty bridges into service. At that stage the collateral air offensive largely took the form of "shallow interdiction" attacks on airstrips and master radars along with camps and local headquarters. A deeper attack had been one on the 6th against the airfield and communications centre at Bir Gifgata.

Israeli land forces launched 23 local counterattacks the first three days but had their confidence "badly shaken by failure and heavy losses."[60] However, the Egyptians found it impracticable to move much of their SAM screen east of the canal. Obversely, armoured forces advancing beyond the cover it proffered soon proved extremely vulnerable. Therefore the posture they had adopted (linear deployment in little depth in front of a waterway) was a classically insecure one. **(See 10)**. Once the Israelis had accepted the loss of the Bar Lev line (save, in fact, for one complex), they were well placed to use open mobile warfare to break up the Egyptian front. On the 16th they forced their way onto the west bank of the canal. The next day they bridged it firmly, using tailor-made prefabrication. Meanwhile, much "state of art" ordnance was reaching Israel from the USA.

By the 20th, the situation on the canal's west bank was deteriorating alarmingly for Egypt's 3rd Army, in effect her right flank. That day the Saudis stopped oil exports to the USA. On the 24th, Israel, having surrounded the 3rd Army, bowed to pressure from Washington for a genuine ceasefire.

A matter of dispute afterwards was how far the dismay and confusion thus engendered in the 3rd Army reflected Egypt's war-waging potential overall. It was argued authoritatively that its army had quickly re-established its SAM network with a switch line. Accordingly, any attempt by the Israel Defence Force (IDF) to "carry open warfare into Egypt would have led it to disaster".[61] What this view may too heavily discount, however, is the value to Israel of the Emergent Technologies (ET), especially the air-to-surface missiles, being rushed across the Atlantic. To cite President Sadat's reported message to President Assad on 20 October: "To put it bluntly, I cannot fight the United States or accept the responsibility before history of destroying our armed forces a second time."[62]

Postscript

Militarily, Israel had won again. On the other hand, the Arab (and especially the Egyptian) forces had much more to be proud of on this occasion. Besides which, Arab oil seemed at long last to be a real influence geopolitically, in that this crisis became an opportunity to correct (nay, overcorrect) the underpricing of oil and natural gas so evident in the sixties and fifties. But the Saudis and other conservative oil producers were not going to use the oil weapon to try and regain Jerusalem or other ulterior aims. They would thereby risk the collapse of the West and Soviet ascendancy after all. In short, they were not afraid of losing an oil war. They were afraid of winning it.

Some, noting the Arabs' improved performance on the battlefield, concluded that after 1980 they would be overhauling the Israelis militarily. However, arms races are rarely that linear. During Israel's 1982 invasion of the Lebanon, her aircraft shot down 80 Syrian warplanes for the loss of but one or two of their own. As in previous Middle East conflicts as well as those in Korea and above the Straits of Taiwan, fighter aircraft produced in the West outfought those produced in the USSR. The most crucial reason for this was better electronics — a dimension that assumed ever more importance.

Notes

1   Athanasios Lykogiannis, "The Early Post-War Greek Economy: from Liberation to the Truman Doctrine", *Journal of European Economic History* 23, 2 (Fall 1994): 345–64.
2   Robert Thompson, "When Greek Meets Greek" in Robert Thompson (ed.), *War in Peace* (London: Orbis, 1981), pp. 16–20.
3   Peveril Meigs, "Some Geographical Factors in the Peloponnesian War", *Geographical Review* LI, 3 (July 1961): 370–80.

4 Neville Brown, "Cyprus: a Study in Unresolved Conflict", *The World Today* 23, 11 (November 1967): 396–405.

5 Alex Danchev and Daniel Todman (eds.), *The War Diaries of Field Marshal Lord Alanbrooke* (London: Weidenfeld and Nicolson, 2001), pp. 458–9.

6 David Watt in Michael Sissons and Philip French (eds.), *The Age of Austerity* (London: Hodder and Stoughton, 1963, Chapter 5).

7 Malcolm Mackintosh, *Juggernaut* (London: Secker and Warburg, 1967), Chapter 12.

8 George H. Kerr, *Formosa Betrayed* (London: Eyre and Spottiswoode, 1966), Chapters XII and XIV.

9 Kufi-Sheng Chang, "The Changing Railroad Pattern in Mainland China", *Geographical Review* 11, 4 (October 1961): 534–48.

10 For a fuller treatment of the post-1945 politics see Donald Watt, Frank Spencer and Neville Brown, *A History of the World in the Twentieth Century* (New York: William Morrow, 1968) Part Three, Chapter XI(i).

11 David Rees, *Korea, The Limited War* (London: Macmillan, 1964).

12 Michael Hickey, *The Korean War* (London: John Murray, 1999), p. 19.

13 Brian Catchpole, *The Korean War* (London: Constable and Robinson, 2000), p. 36.

14 Hickey, *The Korean War*, pp. 85–7.

15 Catchpole, *The Korean War*, pp. 157–60.

16 James T. Stewart, *Air Power: The Decisive Force in Korea* (Princeton: Van Nostrand, 1957), Preface.

17 "The Attack on the Irrigation Dams in North Korea", *Air University Quarterly Review* VI, 4 (Winter 1953–4): 40–61.

18 Rees, *Korea, The Limited War*, Chapter 20.

19 M. S. Knock, *Encyclopedia of US Air Force Aircraft and Missile Systems* (Washington DC: Office of Air Force History, 1978), Vol. 1, p. 63.

20 R. M. Bueschel, *Communist Chinese Air Power* (New York: Praeger, 1968), p. 64.

21 W. W. Momyer, *Air Power in Three Wars* (Washington DC: Lavalle, 1978), p. 115.

22 Maurice Allward, *F-86 Sabre* (London: Ian Allan, 1978), p. 46

23 Colonel-General G. F. Krivosheev (ed.), *Soviet Casualties and Combat Losses in the 20th Century* (London: Greenhill Books, 1997), p. 90.

24 Catchpole, *The Korean War*, Appendix 2.

25 Richard Harris, "Indo-China and the French", *History Today* 2, 2 (February 1955): 84–94.

26 See the author's *EC*, Chapter 14.

27 Stanley Karnow, *Vietnam* (New York: Viking Press, 1983), p. 106.

28 Christopher Partridge (ed.), *Encyclopedia of New Religions* (Oxford: Lion, 2004), pp. 234–6.

29 Neville Brown, "Britain and South-East Asia", *New Statesman* (3 June 1966): 801–3.

30 Robert Thompson in Robert Thompson (ed.), *War in Peace* (London: Orbis, 1981), p. 89.

31 Stanley Karnow, "An Interview with General Giap" in Walter Capps (ed.), *The Vietnam Reader* (Routledge: New York, 1991), Chapter 12.

32 George Tanham, *Communist Revolutionary Warfare; the Vietminh in Indo-China* (New York: Praeger, 1961), Chapter III.

33 Bernard Fall, *Hell in a Very Small Place* (London: Pall Mall Press, 1967), p. 455.

34 This section draws upon D. C. Watt, Frank Spencer, and Neville Brown, *A History of the World in the Twentieth Century* (London: Hodder and Stoughton, 1966), Part 3, Chapter XVI (1).

35 Karnow, *Vietnam*, p. 223.

36 William F. Kaufman, *The McNamara Strategy* (New York; Harper and Row, 1964), pp. 316–17.

37 David Halberstam, *The Best and the Brightest* (New York: Random House, 1972), p. 316 and pp. 210–11.

38 Quoted from the author's *Global Instability and Strategic Crisis* (London: Routledge, 2004), p. 35.

39 Richard L. Stevens, *The Trail* (Hamden: Garland, 1993). These comments owe much to discussion with Dr Stevens.

40 Lawrence Freedman, *Kennedy's Wars* (New York: Oxford University Press, 2000), pp. 325 and 353.

41 Jeffrey Record, *The Wrong War. Why We Lost in Vietnam* (Annapolis: The Naval Institute, 1998), p. 177.

42 Karnow in Capps (ed.), *The Vietnam Reader*, Chapter 12.

43 Julian Thompson, *The Lifeblood of War* (London: Brasseys, 1991), pp. 191–2.

44 Freedman, *Kennedy's Wars*, p. 419.

45 Clark Clifford, "A Vietnam Reappraisal", *Foreign Affairs* 47, 4 (July 1969): 601–22.

46 Karnow, *Vietnam*, p. 23

47 For a fuller account of the operational aspects, see the author's *Strategic Mobility* (London: Chatto and Windus for the Institute for Strategic Studies (now the IISS), 1963), Chapter 4 (1).

48 Peter H. Gleick, *The World's Water* (Covello: Island Press, 1998), Chapter 3.

49 Bernard Fergusson, *The Watery Maze* (London: Collins, 1961), Chapter XVI.

50 Edgar O'Balance, *The Sinai Campaign* (London: Faber and Faber, 1959), Chapter III.

51 Anthony Eden, *Full Circle* (London: Cassell, 1960), p. 254.

52 Clive Ponting, *A Green History of the World* (London: Penguin, 1992), pp. 86–7.

53 Doubts have been expressed as to whether the Egyptians did turn to mustard gas in the Yemen. Visiting Aden and the Radfan as a journalist, 1966–7, one heard convincing reports they had.

54 Peter Young, *The Israeli Campaign* (London: William Kimber, 1967), p. 77.

55 Robert Higham, *Air Power, A Concise History* (London: MacDonald, 1972), p. 229.

56 A. Schlaim and R. Tranter, "Decision Process, Choice and Consequence", *World Politics* 30, 4 (July 1978): 483.

57 S. R. Rosen and M. Indyk, "The Temptation to Pre-Empt in a Fifth Arab–Israeli War", *Orbis* 20, 2 (Summer 1976): 265–85.

58 *Insight on the Middle East War* London: André Deutsch (1975), p. 110.

59 H. P. Willmott in Robert Thompson (ed.), *War and Peace* (London: Orbis, 1981), p. 233.

60 Anthony Farrar-Hockley, *The Arab-Israel War. October 1973, Background and Events*, Adelphi Paper 111 (London: International Institute for Strategic Studies, 1974), p. 30.

61 *Ibid.*

62 Quoted in Mohammed Heikal, *The Road to Ramadan* (London: Collins, 1975), pp. 238–9.

10 | The Iron Curtain

The geopolitics of the Cold War is loosely analogous with World War One as Britain's Easterners would have had it. Hold firm on the main front in Europe while exercising leverage in the Mediterranean and Near East. However, until 1918 that front was structurally very immobile thanks to high force densities and to the two sides remaining closely equivalent. Contrariwise, confrontation along the Iron Curtain from 1946 onwards was characterized by asymmetries which could have much complicated the control of a warlike crisis. A good year to focus on with a view to gleaning lessons could be 1986, the first year to register Moscow's making ready to wind down the Cold War.

Discord had lately arisen out of the USSR's embarking on the modernization of her echelon of Intermediate- and Medium-Range Ballistic Missiles (IRBMs and MRBMs[1]) close inside her western borders. Since 1977 Moscow had been installing there the SS-20, a road-mobile IRBM able to dispatch, across a good 3000 miles, three warheads of perhaps 150 kilotons apiece. In response, NATO Council had in December 1979 confirmed plans to deploy on European soil 464 BGM-109A Ground-Launched Cruise Missiles plus 102 Pershing II ballistic missiles. The latter's nuclear warload might comprise five individual warheads of maybe 75 kilotons apiece. Come 1986, the Pershings had been deployed only in Germany but the GLCMs in Britain, France and Italy as well.

The Berlin Enclave

By dint of geography, however, West Berlin posed a singularly difficult security problem throughout the Cold War. It therefore calls for prior consideration now. The official advice one was always receiving, at least from the sixties onwards, was that the Soviets well knew that the West would go to all-out war to protect the two million West Berliners. The subsidies the city regularly received were held to betoken this resolve.

Unfortunately, such averration glossed over the stark reality that Moscow could make West Berlin feel insecure in manifold ways far too subtle to elicit a Counterforce thermonuclear response. Thus in June 1948 the Western occupation powers, along with the Low Countries, came out in support of German currency reform. On the

18th, three days after the Soviets had closed "for repairs" the Helmstedt–Berlin access road, a new *deutschmark* currency was introduced into the western zones of Germany, the pre-existing *reichsmark* remaining the sole tender throughout Berlin. Then on June 22nd, the Soviets announced reform of the east zone currency. Next day, the West declared itself willing that both currencies should circulate in West Berlin, a solution bound to lead to a collapse of the East's new *deutschmark*. That night, the Soviets curtailed for "technical" reasons West Berlin's electricity supplies and closed every access route from the West.

The alliance response was the flying in of relief supplies. Three airfields were available for off-loading; and soon a plane was landing at one of them every third minute when weather allowed. The cargo carried reached 6000 tons a day, usually over half of it coal. Duly only several percent of West Berliners availed themselves of an offer to register for food in the Soviet sector. This "Berlin Airlift" ended in May 1949 with the Soviets agreeing to the West's new currency being the sole legal tender in its sectors. Granted, inordinate usage was by then inflicting acute wear and tear on the recipient runways. But Moscow could no longer push its luck, particularly since its first nuclear test was still some months away.

The next major crisis was the erection of the "Berlin Wall". By August 1961, almost three million East Germans had fled to the West via the city's sector boundaries. Currently the daily efflux topped a thousand a day. In desperation, the East German regime illicitly enclosed West Berlin in a concrete and steel circumvallation able to thwart all but the most daring. During the resulting tension, the USA took certain military precautions. France remained studiedly nonchalant.

The breaking up of the Berlin Wall in 1989–90 heralded Germany's reunification. In the interim, it had occasionally seemed possible West Berlin would figure in Limited World War. When, during the 1962 Cuba crisis, the Warsaw Pact did bring its forces near Berlin to alert status, Bonn, Washington, London and Paris worked together on contingent countermeasures.[2]

Then again, the psychological pressure on West Berlin during the early months of 1965 and again of 1968 (mainly the increased air activity) coincided with "tipping points" in the Vietnam conflict although the reasons proffered were couched strictly in local terms. Suffice to add that an important contribution throughout to stability in the Germanies was the near absence of intrusions by either side into the other's air space.

WMD

During the Kennedy–McNamara years (1961–3), the number of US nuclear warheads in the European theatre had more than doubled to 6000, their aggregate explosive yield then exceeding 400 megatons: well over 100 times the chemical explosive energy released on the Third Reich by allied bombers. By 1986, this warhead total had fallen by 1000 or so. But several hundred British and French tactical warheads were also available.

A sizeable proportion of the American warheads had been allocated to NATO allies under "dual key" (i.e. binational) provisions for precluding their unauthorized use. And of the grand total, the great majority will have been placed in the Federal Republic.

Regarding warhead types, the lowest on the yield spectrum were 1750 used in certain demolition charges or air-to-air missiles. A detonation would correspond to one or two hundred tons of TNT. At the top end were some of the 1100 warheads earmarked for NATO strike aircraft. These could individually extend into the low megaton range. Nearly two-thirds of the remainder was in heavy artillery shells and the rest for surface-to-surface missiles.

About the Soviet nuclear panoply, there was within NATO much uncertainty. Nuclear artillery shells with yields around a kiloton were believed to be deployed down to divisional level in Ground Forces Soviet Germany (GFSG). Still it was very likely that a Soviet division on a war footing held significantly fewer nuclear warheads than might a NATO one. A similar comparison could be drawn on the air side.

Not that one could just ignore those 400 IRBMs and MRBMs (plus some 300 strategic bombers) inside the USSR's western border. A further complication was that, since 1984, SS-23 missiles had been deployed in East Germany. These could travel 500 miles.

Turning from nuclear warheads to chemical, one enters a sphere in which the Pact superiority was understood to be pronounced, potentially if not in situ. Some guestimated the USSR's total holdings as high as 300,000 tons. Contrariwise, American production had essentially been halted 15 years previously. Only these two Superpowers plus France had offensive capabilities. The challenge thus presented was complicated by how poison gases can range from non-injurious disablers to means of truly mass murder.

The Wider Arms Balance

The armies of NATO had 27 divisions or division equivalents in Federal Germany and the Low Countries. Twelve were German, five American, four Low Countries, three British and three French. There was, too, a Canadian brigade.

On the Pact side of Central Europe (East Germany, Czechoslovakia and Poland) stood Ground Forces Soviet Germany (GFSG). To its 19 or so divisions could be added two more covering lines of communication through Poland plus the five-division Soviet garrison established in Czechoslovakia in 1968, following the Warsaw Pact invasion. Indigenously there were six East German, ten Czech and 15 Polish.

Thus the Pact had a divisional order of battle nearly twice as strong. So how did that preponderance relate to a much less acute one apropos soldiers with the colours: 1,000,000 versus NATO's 800,000? The answer lies, in part, in the average manning level of Pact divisions being barely half their wartime establishment. But even at war strength, a Pact division would represent but half the divisional slice a western division would connote.

However, Pact members continued to derive dividends, in terms of economy and flexibility, from greater standardization of organization and equipment. They were sustained, too, by relatively long conscript service: 18 to 24 months as against the eight to 15 operative by then within NATO. The USA, Britain and Canada relied entirely on volunteers. Armies comprised of long-service volunteers may, of course, be better trained and more cohesive. But they generate thinner echelons of recallable reserves.

Then there is air power. Attention has been drawn above to the much more prominent role it had assumed compared even with 1945. Some concern was duly felt that NATO's warplanes numbered 1700 as against the Pact's 3100. However, into such a comparison could be fed attributes pointing to the conclusion that the aggregate sortie rates the Pact might sustain over time would fall well below NATO's. These attributes included electronic aids; the scale of ground support; flight simulation training; and the calibre of flying training. Granted, a 1976 analysis at the Arms Control and Disarmament Agency (ACDA) in Washington put the Pact's sortie rate at 20 per cent above NATO's.[3] Perhaps, however, this should be read in relation to other American admonitions that NATO air staffs paid insufficient attention to surge capability, particularly in regard to pre-emptive attack by whichever side. A 1981 study of pre-emptive NATO action had suggested that a Pershing caliber missile bearing multiple small high-explosive warheads could pose a big threat to airfields. The dispatch of 100 by NATO might cut the Pact sortie rate by a third.[4] Presumably its SS-23s could be very similarly potent.

Strategic Reinforcement

Though good size airfields were more abundant in Central Europe than in other theatres, their immobilization through hostile action could likewise have been a limiting factor *vis-à-vis* strategic reinforcement. A further risk for NATO was engagement of transatlantic flights and sailings by the considerable Soviet naval presence on Atlantic stations.

Otherwise the USAF counted on deploying across the ocean within ten days 1000 warplanes plus requisite support. Also the USA with Canada could hopefully have sent by air and sea six or seven ground divisions. Meantime, several divisions should have arrived from France and the equivalent of two from Britain.

Even so, the basic military balance within central Europe and the theatre as a whole could have tilted more against NATO during several weeks of competitive reinforcement, even presuming adequate warning time and political expedition. Thus each and every division in the East German, Czech and Polish armies could have been raised to battle strength. Also Ground Forces Soviet Germany (GFSG) could have been boosted by an estimated 40 divisions, this being the number garrisoned (at varying levels of readiness) in the European USSR exclusive of Caucasia.

Moreover, a good 5000 tactical warplanes were operational somewhere inside Soviet borders. Half were within the air force proper; perhaps half of these were contingently earmarked to move into Central Europe. The remainder were within PVO-Strany, the independent air command tasked with the comprehensive defence of the homeland against aerial threats. (See 8). There had recently been some convergence, regarding command structures, between PVO-Strany and the Soviet air force proper. Whether that meant the former might supplement the latter in theatre conflict was hard to say.

Soviet Doctrine

Throughout the Cold War, those with an especially strong aversion to modern war protested it was otiose to plan for limited wars against the Soviet bloc. We were advised that, all else apart, the Soviet leaders would, if attacked, "almost certainly retaliate with all the forces at their disposal", to cite a 1982 working party report to the Anglican Synod.[5] Nor can one deny that Soviet thinking owed much to the continentalist perspectives of Karl von Clausewitz (1780–1831), the Prussian staff officer seen as very much a pioneer of modern strategic thought. A prime conviction of his was that military violence was too rough, clumsy and befogged to lend itself to subtle modulation. He hailed from the same Hegelian ambience as Marx. Lenin made notes on his writings, occasionally referring to them. Likewise Leo Tolstoy (1828–1910) had dismissed the idea of assured control. He even persuaded himself that it was an invalid corollary of what was to him the erroneous philosophy of free will. This confusion is abundantly evident in the relevant pages of *War and Peace* (1862–9).

So what might this have meant operationally, particularly in Central Europe? How much Clausewitzian emphasis was put on offensive or counter-offensive action? The answer seems to be a lot. A major study published in Moscow in 1970 averred that "the defence is a temporary type of combat action. It is organized and conducted in those cases where it is necessary to gain time and create favourable conditions for the launching of an offensive . . . "[6] A further averration was that, though both sides would have to reckon with heavy losses and widespread zones of contamination, general recourse to nuclear warheads would on balance advantage the attack, not least because they could smash fixed defences without the preliminary concentration of masses of artillery.[7]

Mercifully, interpreting Soviet declared doctrine in the light of the configuration of Pact armies led to a less wild interpretation. This could be that, though Moscow and its allies remained scornful of the American-inspired notions about subtly graduated escalation, they did see nuclear firepower as separable into several steps. Any held in limbo could still have an overhanging influence.

Besides which, there was interest in waging war below the nuclear threshold, at least for quite an interval. Confirmation came from manoeuvres. In the Pact exercises October Storm (1965) and Vitava (1966), escalation above the threshold occurred only after several days, the initiative being taken by the "NATO invaders". Then in Exercise Dnepr in the autumn of 1967, the Soviets had conducted purely non-nuclear operations across an area of the southern USSR comparable in size and shape to Western Europe. However, during Exercise Dvina in 1970, their armed forces went nuclear extensively.

Speaking at Tula in January 1977, Leonid Brezhnev prepared the way for a nuclear "no first use" posture. In 1982 this duly became Pact doctrine. Then in 1985 the Deputy Chief of the Soviet General Staff, M. A. Gareev, acknowledged that the "massed employment" of nuclears could entail "catastrophic consequences for both sides". He noted that NATO had lately emphasized "the development of highly accurate guided weapons which in terms of effectiveness are close to low-power nuclear weapons. Under these conditions . . . there will be greater opportunity for conducting a comparatively long war employing conventional weapons and primarily new types of high precision weapons."[8]

Forward Defence by NATO

In a *volte face* in September 1950, three months after the *blitzkrieg* against South Korea, NATO had asked the German Federal Republic to contribute to alliance rearmament. This was agreed though only within the context of a strategy designed to forestall any cession of German territory — in other words, forward defence. Bonn did not want the German lands to become yet again the "cockpit of Europe". It was mindful, too, of the fact that, even within 30 miles of the inter-German border, lay such towns and cities as Lübeck, Hamburg, Göttingen, Kassel, Schweinfurt and Bayreuth.

Then again, the southern flank of that border with East Germany formed a concave arc, the most obtrusive part of which was over against the Fulda Gap, a col leading to the Main–Rhine confluence not a 100 miles away. Hostile forces at the Rhine would split NATO's armies. Pertinent here is a maxim enunciated well over two millennia ago by Sun Tzu, the Chinese scholar and general. It was that a defending army should never allow itself to be with its back to a river.[9] May one note that the Roman army's defence posture for the said river line included (*c.* AD 260 to 410) a number of forts on the east bank individually guarding key crossing points. But its main forces were to the west.[10]

Indentations apart, the said border was 400 miles long; its continuation as the Bohemian frontier spanned another hundred. Additionally, a Pact *coup de main* against neutral Austria could readily have opened up 200 miles more. Accordingly the MC-70 plan NATO formulated in 1957 called for 30 full divisions to be located in West Germany and the Netherlands to enable a tolerably forward coverage to be maintained. This target level related more to geographical and other realities then had the 96–division called by NATO Ministers meeting in Lisbon in May 1952. It was also more mindful of regional geography than any Soviet or Warsaw Pact doctrine one knows of. However, the alliance was still several divisions short when, in 1962, it resolved to implement MC-70 in view of the regional and world situation. The 30-division target never was reached.

At least the landscape was not inimical to defensive war. Bavaria mainly consists of a plateau (1500 to 3000 feet high) incised by deepish valleys, some well oriented to act as defence lines. Moreover, the Federal Republic's exposed borders tended to follow sharp natural divides – the Thuringian hills, the Bohemian rimland, the Tyrol. Then again, roughly a third of its total land area was wooded; another third urbanized or otherwise "enclosed".[11] An alternative criterion concerns the density of individual towns and villages. It was said that a typical brigade sector in Germany might embrace 85 villages and townships, separated from their neighbours averagely by four kilometres.[12]

Yet how landscape character would have affected an actual war with its attendant anguish and confusion is hard to say. What, for example, of the prevalence of motorized refugees? In any case, NATO was virtually denied the advantages accruing for the Defence by (a) divisions typically having to screen a sector of as much as 25 miles would be obliged to rely heavily at brigade level on counter-attacks and (b) Bonn's refusal to countenance fixed defences near the inter-German border (see below). A strategy of counter-attacks can draw added resilience from certain kinds of prepared positions.

In any case, forward defence was *a priori* unstable. What if Pact forces concentrated for offensive action had broken through NATO's cordon, perhaps after selecting a sector not held by a strong division? Might not they have either enveloped NATO divisions close by or else rampaged through the rear areas disrupting command and control, logistics and air support. Early on in World War Two, an essentially political requirement to defend the entire homeland by stretching one's forces all along the threatened border had facilitated the defeat by blitzkrieg of four armies in succession: the Polish in 1939; the French in 1940; the Greek and Russian in 1941. Visiting Paris in May 1940, as the allied front was crumbling, Churchill urged the French to commit their strategic reserve. He was told none existed. Would not much the same have applied above divisional level to NATO Germany in 1986?

At all events, many commentators were persuaded, on military operational grounds alone, that NATO Germany could not hold out beyond several days on a non-nuclear basis. In any case, could financial collapse have been averted more than a day or two? The great bourses of Europe (Basle, Zurich, Frankfurt, London . . .) stood desperately vulnerable to distraction or disruption, even by local hostilities which were ongoing but well short of full scale. Collaterally, wholesale debt renunciation was a hazard, given that the total volume of international indebtedness was 700 billion dollars, 80 billion of it owing by the Soviet bloc. Here was a potentially catastrophic contradiction between military and fiscal geography. Due account of the former was taken by NATO. But the latter was consistently ignored

The chief architect of NATO Europe's tactical nuclear posture was Robert McNamara, acting in his capacity as US Secretary of Defense (1960–6). It is therefore disturbingly paradoxical that in 1983 he could insist that "nuclear weapons serve no military purpose whatsoever. They are totally useless — except only to deter one's opponents from using them." The paradox is rendered the more acute by his having gone on to insist that "This is my view today. It was my view in the early 1960's. At that time, in long private conversations with successive Presidents — Kennedy and Johnson — I recommended that they never initiate under any circumstances, the use of nuclear weapons. I believe they accepted my recommendations."[13] He still presents "no first use" as his abiding belief, never mind how he actually proceeded in office. The ambiguity thus created within NATO could have been much more dangerous in time of crisis than a firm stance one way or the other.

Nuclear Doctrine Evolves

By New Year 1954, substantial numbers of battlefield nuclear weapons (artillery shells; aerial bombs; short-range missiles; nuclear mines . . .) had been deployed by the USA in Europe. Soon some argued influentially that "tactical nuclears" would enable the Defence actually to prevail. They invoked, in a context he can hardly have approved of, Liddell Hart's precept, debatable in any case, about more firepower tending to favour the Defence since it enables a division, say, to cover a wider front. According to another source, influential at the time, this would typically mean more than 20 miles if it is nuclear firepower one talks of.[14] Some suggested, too, that NATO armies were further advantaged in the battlefield nuclear context by an ethos conducive to the devolution of command to small mobile units. In his neo-classical study of 1958, Henry Kissinger said that pattern is what the said situation would demand.[15] For some years, indeed, the US army experimented with a "pentomic" divisional structure — i.e. one with a five regiment structure adapted to all-round defence in depth. In 1960 it placed a provisional order for 6000 Davy Crockett mortars able to fire across perhaps two miles shells weighing 150 to 375 lbs and yielding a quarter of a kiloton.

During 1962, some 200 Davy Crocketts were acquired by the US 7th Army in southern Germany. Already, however, their days were numbered. Opinion was turning against them as part of a wider reaction against any idea that battlefield nuclears were Defence-friendly in some meaningful sense, even when both sides had them.

Similarly questionable though never overtly questioned was the doctrine of the "pause". The Kennedy administration assumed office in 1961 committed to keeping

NATO's options in Germany wide open by beefing up its military presence in every way. One alternative hopefully presented thereby would be to use on the battlefield enough nuclear firepower to interrupt an enemy advance and betoken a wider intent. Come that November, Laurie Norstad, then NATO's Supreme Allied Commander Europe, encapsulated the notion thus. Should an attack come, the immediate aim would be to break its continuity and so create breathing space. To this end, the nuclear capability organic to major formations "could, if the situation demands, be joined, promptly and effectively, with the conventional effort to force a pause . . . ".[16]

"Nuclear first use" interpreted in this way was open to two objections *a priori*. Most obvious was the heavy reliance for escalation deterrence on manned warplanes, poised vulnerably around even more vulnerable runways. Not a lot more subtle was the way an operational requirement for pause reinforcement was interwoven with the desire to send a cautionary signal direct to Moscow. The battlefield imperative might be determined on the basis of, let us say, a report by a reconnaissance pilot about traffic over a particular bridge.

These contradictions were made the sharper by the authorization process for nuclear release. The President of the United States would have to consent. So would the government of the country owning the weaponry assigned; and that of the territory (most likely the Germanies) where the firing would take place from. Going up then down the chain of command to request and receive permission could take in excess of a full 24-hour day.

My own view as expressed these past 40 years has been that, had a nuclear warning shot ever seemed needful, it might better have been delivered against a target outside the main battle area. It was a stratagem which phased into limited strategic war. (See 11). Applied to NATO, it would have required contingent preparations — doctrinal, institutional and material. And one cannot pretend it could ever have been risk free, At best less risky.

A Radical Alternative

Some on the anti-nuclear Left in West Germany argued instead for a strong infantry-rich cordon along the Federal Republic's eastern border, a cordon from behind which counter-attacks might come. Reference was made to a plan formulated in the early fifties by Bogislav von Bonin, a staff colonel in the Amt Blank, a precursor of Bonn's Ministry of Defence. He had proposed that 200,000 German light infanteers be deployed within a forward defence belt 80 km deep which would be well laced with minefields and contain hundreds of camouflaged bunkers. Though accorded much press coverage awhile, the Bonin plan was never implemented. Two political objections were that (a) it seemed to underscore too firmly the division of Germany and (b) manning the cordon just by German troops was deemed a kind of reverse exceptionalism.[17] Then there was the military question as to whether due provision had been made for counter-attacks to recoup territorial losses and, more generally, to punish the transgressor.

The Flanks of Europe, North and South

The military situation on the flanks of NATO Europe was discussed to an extent by this author in 1972.[18] Come 1986, things had changed little. Still it was the northern flank that caused most concern. This was for strategic but also for operational reasons. Wintertime in Finnmark is dark and stormy though not as unsuitable for overland deployment as the marshy summers.

Compounding the difficulties of tactical engagement were those of strategic reinforcement. The allied intention in warlike crisis would be to buttress Norway's mobilization with British and Canadian units plus two NATO multinational groupings: Allied Command Europe (ACE) Mobile Force — comprising land and air elements — and the naval Standing Force Atlantic. However, the Soviets were well placed to interdict the main road to the North or, of course, the sea and air approaches.

Evidently a cardinal point about overseas reinforcement was demonstration of intent — meaning, in the ultimate, a readiness to use tactical nuclears to shield Finnmark and so on. As early as 1957, a NATO Strike Fleet on exercise had simulated nuclear attacks on Norwegian targets. Some ships sailed beyond 70°N although, in those auroral latitudes, magnetic storms compromised communications.[19]

Neither Norway nor Denmark accepted nuclear warheads on their soil in peacetime. Yet either could be a useful prize. Denmark's exposure was accented by plans contingently to mine the Kattegat to inhibit the open sea deployment of the USSR's Baltic Fleet. Norway's was by the utility of Tromso and Finnmark for monitoring its Northern Fleet.[20]

A further complication was American cogitation as to whether the Greenland–Iceland–United Kingdom (GIUK) line should receive anti-submarine minefields. This posture, which excluded Norway, of course, had been adopted in two World Wars with very modest results.[21] But now it could specifically be argued that the ridged submarine contours precluded deep-diving nuclear submarines transitting over a 1000 feet below to avoid contact.

The Royal Navy was always more skeptical than were its American colleagues. Their divergence was sharpened by Washington's geopolitically-inspired backing of Revkjavik during the Anglo-Icelandic "cod war" which was resolved in Iceland's favour in 1977.[22] Having been approached by Britain's Jim Callaghan to put pressure on Reykjavik, Henry Kissinger mischievously quoted Bismarck, "How great is the tyranny to which small nations can subject great".[23]

Soon, however, the American naval emphasis had switched towards an assertive "forward strategy" — one of presenting at high latitudes. It was apparently calculated to compromise (a) the defensive "bastions" into which the USSR's Fleet Ballistic Missile submarines on patrol might group, and (b) their transit routes to and from Kola. It could therefore be seen as complementary to the philosophy of encompassment inherent in the bolder renderings of SDI. No matter that SDI was considerably unpopular in Nordic Europe, especially Greenland and Denmark. Nor that compromising a Superpower's nuclear deterrent might not be sensible.

Arms Control

Many asymmetries and contradictions characterized the question of European Security. Only a few have here been mentioned. There was, in fact, a school of thought (notably on the German Right) which contended that a rather loose-ended state of affairs sustained deterrence because it meant that any transgression would draw its perpetrator onto a deadly escalator of "atomic uncertainty".[24] Franz Josef Strauss, the Bavarian leader of the Christian Social Union and from 1956 Chancellor Adenauer's

Minister of Defence, argued that "pause" and "nuclear threshold" should not be seen as concepts invariably applicable.[25] Undoubtedly Berlin weighed in his thinking.

Those Europeans who found such declamation just too alarming felt drawn towards either neutralism or arms control. The former drew the widest constituency. But the latter appealed more to strategists, nuclear free zones and "open skies" being much favoured.

In 1966 NATO endorsed a study it had launched under Pierre Harmel, then Belgium's Foreign Minister. Following up seminal academic work (notably by the late Hedley Bull of Australia), this Harmel report said that the quests for defence preparedness and for arms control/phased disarmament should be seen as twin aspects of the same theme — "not contradictory but complementary".

In 1963, several of us staffers at the Institute for Strategic Studies worked on a contract from the Arms Control and Disarmament Agency (ACDA) in Washington, a progeny of the Kennedy administration. We were mandated to look afresh at the interwar distinction between "offensive" and "defensive" weapons — the former essentially comprising warplanes, tanks and maybe field artillery. The aim was to further arms reductions in Europe.

Recently, it had become customary in professional circles to talk of a "divisional slice": the number of troops within a division plus a due fraction of the corps and army echelons behind it. A standard American divisional slice in Germany was then put at 43,000 and a Soviet at 20,000.[26] Now we proposed a rider, the "battalion slice". Our calculations were that a tank or artillery slice was 3200 and an infantry one 2000. We inferred that reciprocal cuts in heavy offensive weapons could make it possible to generate more light divisions with the same manpower levels. Therefore divisional frontages in Germany could be shortened, itself a contribution to stability. Furthermore, any fighting that did break out would occasion lower casualty rates. Then not least, a simpler division would be easier to maintain on a cadre basis in peacetime.[27]

Our report was warmly received by ACDA. We felt content ourselves. Today one might not have. Nor would this be just because the offensive versus defensive distinction is always problematic. It would be more that so radical a reorganization of armies and air forces would have left them a long while inchoate and therefore unreliable. Too often one has to say that arms control could only be feasible once it is superfluous.

Sobering Lessons

What then of a clean sweep alternative? Through the middle 1980s, the politics of neutralism was constrained at governmental level in the two European "swing" states, France and Italy. In Paris, socialist President François Mitterand was putting a positive slant on the Gaullist precept of co-operation with the Atlantic Alliance. Meanwhile in Rome, socialist Premier Bettino Craxi stood ready to accept 112 of the cruise missiles proposed for INF modification.

At grass-roots level, however, Leftist neutrality had currently waxed strong. It had done so possessed of a decidedly millenarian mien, epitomized by the slogan "better red than dead". In fact, there were by 1986 abundant indications from the real world that meekly accepting Soviet Communism would not ensure lasting peace. The Cold War years had seen much armed conflict between Communist regimes: Yugoslavia and her neighbours; the USSR and Hungary; the USSR and China; the Warsaw Pact and Czechoslovakia; Vietnam and China; and Vietnam and Kampuchea. First red then dead?

The indicated aim of this short review of the Iron Curtain viewed as a subject for international politico-military planning was to look for lessons of lasting import. Those gleaned are surely salutary for all of us persuaded of the need for closer international collaboration in a variety of fields if we are to weather a manifold world crisis. For the fact was that diplomatic accommodation left many pertinent questions unanswered or even not asked. Yet the alternatives radical critics advanced might all too consistently have left things worse not better. And even if the posture adopted had been better founded, there was still no guarantee all would have been right "on the day". There would have been too much devil in the detail for that.

Notes

1 MRBMs are rockets seen as strategic although their ranges are below 1500 miles.
2 See the author's *Strategic Mobility* (London: Chatto and Windus for Institute for Strategic Studies, 1963), p. 102.
3 R. L. Fischer, Adelphi Paper 127, *Defending the Central Front: The Balance of Forces*, London, International Institute for Strategic Studies, 1976, p. 33.
4 *Flight International*, vol. 120, no. 3770, 8 August 1981, p. 434.
5 Bishop John Baker of Salisbury *et al.*, *The Church and the Bomb* (London: Hodder and Stoughton, 1982), p. 45.
6 A. A. Sidorenko, *The Offensive* (Moscow: Translated in Washington DC by the USAF), 1970, p. 51.
7 Sidorenko, p. 66.
8 Christoph Bluth, *New Thinking in Soviet Military Policy* (London: Royal Institute of International Affairs, 1990), p. 15.
9 Sun Tzu (Lionel Giles, trans.), *The Art of War* (Taipeh: Civilian Publishing Service, 1953), Chapter 8.
10 Murray Eiland, "Maritime Mainz and Rome's Defence of the Rhine", *Minerva*, vol. 15 no. 2 (March–April 2004) : 42–3.
11 David Gates, "Area Defence Concepts: the West German debate", *Survival* XXIX, 4 (July–August 1987): 303–17.
12 Paul Bracken, "Urban Sprawl and NATO Defence", *Survival* XVIII, 6 (November/December 1976): 254–60.
13 Robert McNamara, "The Military Role of Nuclear Weapons: Perceptions and Misperceptions", *Foreign Affairs*, 62, i (Fall 1983): 59–80.
14 T. C. Matakis and S. I. Goldberg, *Nuclear Tactics* (Harrisburg: Stackpole, 1958), p. 151.
15 Henry Kissinger, *Nuclear Weapons and Foreign Policy* (New York: Harper, 1957), 174–83.
16 Quoted in *The Guardian*, 24 November 1961.
17 Colonel Hans D Lemke in Brian Holden Reid and Michael Dewar (ed.), *Military Strategy in a Changing Europe* (London: Brassey's, 1991), Chapter 9.
18 *European Security, 1972–80* (London: Royal United Services Institute, 1972), Chapter 7.
19 Mats Berdal, *British Naval policy and Norwegian Security* (Oslo: Institutt for Forsvarsstudier, 1992), pp. 37–8.
20 Olav Riste, *The Norwegian Intelligence Service, 1945–70* (London: Frank Cass, 1999), Chapter 4.
21 J. S. Cowie, *Mines, Mine-Layers and Minelaying* (Oxford: Oxford University Press, 1949), Chapter 7.
22 For the resource background, see Bruce Mitchel, "Politics, Fish and International Resource Management: the British-Icelandic Cod War", *Geographical Review* 66, 2 (April 1976): 127–30.

23 Roy Hattersley, *Who Goes Home* (London: Little Brown, 1995), p. 143.

24 Adelbert Weinstein in *Frankfurter Allegemeine Zeitung*, 30 December 1963.

25 Alastair Buchan and Philip Windsor, *Arms and Stability in Europe* (London: Chatto and Windus for Institute of Strategic Studies, 1963), p. 81.

26 Brown, *Strategic Mobility*, Chapter 8 (II).

27 See the author in Evan Luard (ed.), *First Steps to Disarmament* (London: Thames and Hudson, 1965), Chapter 9.

11 | Intercontinental Deterrence

My career as the second Research Associate at the Institute for Strategic Studies (now the IISS) ran from autumn 1962 to late 1964. Mostly one was contributing to the *Military Balance*, the annual compilation of the orders of battle of states around the world. It was the most generally remarked of several literary innovations which soon made the Institute *primus inter pares* as a fount of analysis. Within several years, it would be an accepted data base in arms control talks.

One problem to remedy was the opacity of the Heartland, an attribute reaching way back. How strong were Attila or Genghiz Khan? When or where would they strike? Grappling with such enigmas *à la mode moderne* one had privileged if constrained access to certain informed channels. A more basic reality was that overhead reconnaissance (airborne or Spaceborne) had lately been rendering more visible a variety of assets, civil and military. Among the latter were missile emplacements and bomber bases.

In the early fifties, Britain's Royal Air Force had conducted several probes into Soviet airspace, it having been agreed with Washington that, should any mishap occur, the diplomatic repercussions would be less severe than had these missions been by the USAF.[1] Then from 1956 to May 1960 (when the programme was aborted by the Soviets' shooting down Gary Powers), Lockheed U-2 reconnaissance planes ranged over the USSR relying for their survival on ultra-high-altitudes.

That August the CIA's Discovery satellite returned the first orbital pictures of Soviet territory; within a year these transits became routine. Perforce the Soviets tracked them from the outset, and awhile were exercised. However, their aggravation eased somewhat during 1963 as they themselves began orbital reconnaissance, most advantageously over China.

Meantime, Washington remained anxious not to appear triumphalist about prising open the Heartland thus. President Eisenhower had allowed the U-2s to proceed because of his military instinct to know "the other side of the hill". However, he warned of a need "to be wise and careful" so as "to give the Soviets every chance to move in peaceful directions". Granted, open discussion was soon taking place about the new fruits of overhead intelligence. Even so, the first presidential acknowledgement of the programme *per se* is thought to have been a remark Lyndon Johnson casually made in March 1967.[2]

Bombers or Missiles?

The ensuing procurement debates needfully assumed the form of a dialectic between the strategic missile and the strategic bomber. A geodetic limitation on an "intercontinental" (i.e. inter-hemispherical) missile as late as 1955, say, would have been continuing uncertainty about the diameter of the Earth to the nearest few hundred metres in whichever direction. By 1960, this obstacle to precision was being eliminated. Indeed, it could then be surmised that, given aircrew fallibility under extreme stress, missile standard accuracy should soon be the greater. Besides, a land-based strategic missile would always be cheaper to install and much cheaper to maintain than its bomber counterparts. Emplaced missiles would also be easier to protect versus blast and other effects than would parked bombers; and, of course, much harder to intercept in flight.

On the other hand, bombers carried a bigger warload. They might also make repeated sorties; and, most would say, be more readily divertible to non-nuclear missions. Nor was the weather going to impede strategic bombing the way it so often did in World War Two. For one thing, preferred cruising altitudes would always lie well within, in higher latitudes for sure, the almost weather-free stratosphere. Nor would fog or cloud obscure the Heartland's surface the way it so often did the German in 1943.

A Bomber Gap?[3]

In 1955, the USAF brought into service the B-52, its first intercontinental bomber of post-war design. By this time the Soviet air force was receiving the Tupolev-20, designated by NATO the Bear. It, too, could make an intercontinental round trip. Since the USSR's strategic bombing effort against the Third Reich had been sporadic, it was adjudged to have made big strides in terms of concept and application. Whence arose a "bomber gap" legend which induced Congress to vote extra air force appropriations in 1956.

Yet the U-2 evidence would soon indicate that Moscow's top priority had been not strategic air attack but strategic air defence. Authoritative assessments were that, through the fifties, the USSR spent three times as much on this complementary aim as the USA did. Moreover, within the realm of strategic air attack, the prime Soviet concern was to create an air fleet equipped to strike at Western Europe, not one designed to hit North America. By 1961 the Soviets had, instead of the 1000 intercontinental bombers predicted by US intelligence in 1956, a mere 200 — just over half of them Bears and the rest a smaller type. Yet they also had in service 1000 Tupolev 16 Badgers. These medium range strategic bombers were predominantly located not far inside her main western border.

A Missile Gap?

Hard on the heels of the bomber gap fallacy came that of a "missile gap". In 1955, American radar located in Turkey detected a Soviet long-range rocket being test-fired. Then in 1957, the launchings of Sputniks 1 and 2, the world's first artificial satellites, connoted the USSR's possessing rockets of sufficient strength and accuracy for intercontinental engagement.

Electoral fear of an impending "missile gap" sufficed to clinch the desperately narrow victory Senator John Kennedy gained over Vice President Richard Nixon in the 1960 Presidential. Nor was it just in the USA that the Sputniks were read as a sign. "The East Wind," said Mao Tse-tung, was "prevailing over the West Wind." Correspondingly, the *People's Daily*, mouthpiece of Beijing's Maoist establishment, advised its readers that the ascendancy of the "anti-imperialist bloc" had "reached unprecedented heights", thus effecting a "major turning point" in world affairs.

A mystified China's impatience with continued Soviet caution sharply increased during the Taiwan Straits and Lebanon summer crises of 1958; and, indeed, signified the commencement of the great Sino-Soviet split. One could further surmise that the quiescence of Communist Eastern Europe between the invasion of Hungary in 1956 and of Czechoslovakia in 1968 owed something to the Sputniks seeming to betoken a broader superiority. The "missile gap" myth was so widely believed that it acquired awhile some reality.

As to what actually happened, the Soviet planners seemed once again to be preoccupied with Western Europe as either a potential hostage or a source of "revanchist" aggression. Between 1961 and 1963, an echelon of 700 fixed and mobile Medium and Intermediate Range Ballistic Missiles was built up inside the USSR's main western border.[4] Yet at the time of the 1962 Cuba crisis, the USSR had but 75 ICBMs deployed whereas the USA had three times as many plus *c.* 30 Jupiter IRBMs in Turkey, trainable on much Soviet territory. Between late 1959 and late 1961, American predictions of Soviet ICBM acquisitions by 1963 shrank by 96 per cent.[5]

The American–Soviet disparity in strategic orders of battle currently looked at least as great at sea. Nearly 100 Polaris A-1 Submarine-Launched Ballistic Missiles (SLBMs) with a range of 1200 miles were already installed in six submarines. Nuclear-driven, these vessels combined high underwater speed with agility and quietness, this by virtue of their revolutionary tear-drop hull profiles. They could patrol under the Arctic ice cap for weeks on end. The concealment thus afforded would be much accentuated by how multiple reverberations off the roughened ice-water interface attenuated sound waves an order of magnitude more than simple salt water absorption might.[6] To keep radio contact or fire missiles, the submarines could access near the surface the leads of open water quite often encountered even in those days to quite high latitudes.

The USSR had launched the world's first nuclear-propelled merchant ship, the icebreaker *Lenin*, in 1957; and four years later had started to commission nuclear-powered submarines, having 25 in service by 1964. However, barely half these boats carried nuclear missiles. Those which did bore three of 100 or 400 miles range. Moreover, these weapons could only be fired after elevation from having lain flush with the weather-deck, the boat in question having come to semi-surface.

The Advent of Minuteman

Missiles fired from submarines have to have solid not liquid fuels. However, the first generation ICBMs on each side were fuelled by liquids. This was because the even or smoothly changing burning rates which much facilitated accurate guidance were initially more easily secured through a system of pumps and valves. Moreover, in the earlier models (which included all ICBMs in Soviet service in 1962) the liquids employed were "unstorable" – i.e. could not long be held outside a refrigerated environment. Missile tanks had therefore to be filled immediately prior to firing which, given the ancillary equipment needed for fuel storage and control, made providing emplacements with ferro-concrete shielding costly. Furthermore the ICBMs in question would have taken a good hour to tank up whereas half an hour was all an ICBM needed to reach the central USSR from the central USA or *vice versa*.

In the Spring of 1962, the USAF began deploying the Minuteman, a solid-fuelled ICBM; and soon proceeded to install two every three days. Accordingly, at the Ann Arbor campus of the University of Michigan in June 1962, Secretary of Defense Robert McNamara spoke thus: "if, in spite of all our efforts, nuclear war should occur, our best hope lies in conducting a centrally controlled campaign against all of the enemy's vital nuclear capabilities".[7]

This averration was primarily to refute Gaullist suggestions that the American guarantee to Europe had weakened because strategic nuclear parity already prevailed between the superpowers. It related as well to McNamara's dislike, also expressed at Ann Arbor, of Britain and France respectively developing independent nuclear deterrents. Inevitably, however, this enunciation of a Counterforce strategy was also heard in Moscow. The Soviet attempt that autumn to establish nuclear missiles in Cuba should surely be seen, first and foremost, as an attempt to offset a critical margin of nuclear inferiority. The whole episode was a case study of how, in our shrinking and often transparent world, it is hard to send a diplomatic signal to one player without its being heard and judged by others. No less currently applies in the realm of high finance.

Mutual Assured Destruction

Though the Cuban foray was abortive, several developments in train promised soon to make the superpower balance stable in this domain. During 1963, the USSR brought into service an ICBM which (like the Titan II then complementing the Minuteman force) was fuelled by "storable liquids" – i.e. fuels which could be held in rockets for extended intervals. ICBMs with tanks thus filled could be fired well within the time between detection of an incoming salvo and its impact. Such "launch on warning" could entail a risk of catastrophic error. But neither side ever renounced it formally.

The Victory Parade held in Moscow in May 1965 served to show the world an ICBM similar in configuration to the Minuteman. Judging from Minuteman experience then, the sliced cost of installing one in a "silo" — i.e. hardened emplacement, might be a quarter of that for a liquid fuelled counterpart. Systems so compact, cheap and readily installed or made mobile (as well as easy to fake or camouflage) would not lend themselves well to monitoring within an arms control framework. A good measure of mutual candour and trust would be called for. Still, their acquisition by each side held out good hope of this aspect of the arms race winding down of its own accord.

The general assumption in public debate was that a Minuteman silo could withstand shock waves not exceeding 300 pounds per square inch (psi). A megaton warhead delivered with a standard error of half a mile would have only a one in three chance of destroying an emplacement hardened to that extent. Whether any ICBMs were as yet that accurate was doubtful. Also to reckon with, however, was a lip of crater debris with a radius, in limestone say, close to half a mile. So the silo might be buried instead.

Back of postcard calculation can take the argument further. Let us postulate that each side would likely find an "unacceptable" damage level one involving the destruction of sixty large cities in a retaliatory Countervalue second strike. Then, working from the hardening guestimates above, its protagonist would have simultaneously to deliver about 100 warheads of megaton yield against an ICBM field of 90 in order to exclude such a response. Suppose instead the side under attack had 300 ICBMs deployed, then a Counterforce salvo of some 900 would be needed to ensure or almost ensure this presumptively adequate result.

Out of such rough and ready reasoning came by 1964 acknowledgements that an era of Mutual Assured Destruction (MAD) was dawning between the two Superpowers. The gist was to be neatly expressed in 1970 by a ranking quasi-governmental think-tank in Stockholm: "The balance of terror is not delicate: quite substantial changes in the numbers of warheads on either side would leave both sides with a second-strike capability."[8] Bilateral Strategic Arms Limitation Talks duly began in 1969 and culminated in the SALT-1 agreement of 1972. National means of verification were its lynchpin; these (meaning Space reconnaissance) depended on co-operative interplay. Both party's ICBMs stood uncamouflaged in static emplacements. Each could soon have acquired road-or-rail-mobile models to disperse, disguise or conceal. But so what?

Stability at Sea

Many of us concluded forthwith that this MAD regime was unshakeable. Nevertheless, certain questions did come up operationally. One theme in the literature was the putative vulnerability of SLBM fleets to radical breakthroughs in Anti-Submarine Warfare. In which connection, may one just add this to what has been said above about under-ice patrols. May 1943 was a decisive month for the Allies in their "Battle of the Atlantic" against Germany's U-boats, *untersee botts*. In April, Dr Goebbels had been claiming, with some justification, that they "had England by the throat". During May, however, the extension over mid-ocean of long-range aerial reconnaissance from shore bases weighed in against the submarines as did free-ranging "hunter killer" groups of warships. After this, the U-boat fleet was withdrawn for prolonged recuperation, never again to be more than a nuisance.

Even so, the rate of attrition during May of the U-boat fleet on station had been barely one per cent a day. A pre-emptive Counterforce strike against what would be a well dispersed SLBM force would have had to register close to 100 per cent success in under half an hour. Stated thus, the stipulation was too far fetched to warrant further discussion apropos superpower SLBMs or, indeed, the smaller SLBM deterrents then being created by Britain and France.

Nevertheless, more or less all of us with an interest in such matters were much in error, it now seems to me, in one important respect. We disregarded the dependence of SLBM vessels on a very few home ports and maybe Very Low Frequency (VLF) wireless transmitters. The ready destruction of such facilities could have been a deadly serious matter especially if a warlike crisis unfolded over weeks or even months, not an implausible circumstance in the MAD era. Thus in the event of such a showdown, communication by VLF might have provided a crucial back up. The basic physics is that five to 15 kilocycles per second is the part of the spectrum within which electromagnetic waves travel through sea water tolerably well. To be a little more explicit, especially powerful transmitters can communicate across oceanic distances with submarines trailing aerials to within perhaps ten metres of the surface. By 1964, the United States Navy (USN) was operating two such transmitters on the USA's western and eastern seaboards respectively.

In 1965, the Pentagon finally withdrew fleet carriers from its contingency plans for a strategic nuclear war with the Soviets. This decision at last snuffed out the notion, beloved of certain lobbyists in the early nuclear years, that the supercarrier was a good alternative to the long-range bomber even against a land-locked and ice-locked Heartland. Intrinsically, this claim was always shaky. And now that the strategic bomber was itself giving way considerably to the ICBM and the SLBM, it was losing all relevance.

Soviet C3

A collateral consideration was how the Soviets might manage a strategic thermonuclear conflict. Eventually, Washington specialists would feel themselves to have a goodish understanding of the material infrastructure of the Soviet strategic Command, Control and Communications (C3) network. There was much duplication and diversification; and a marked emphasis on bunkers and deep shelters. Six VLF transmitters might communicate with SLBMs on station. Mobile command posts (ground, sea and air) were available, not least for the very "top brass". Special communications satellites could be launched contingently.[9] Elaborate precautions were in place against unauthorised recourse to nuclear weaponry.[10]

Much of this provision will not have been made come 1970; and whatever was will not have been fully registered in the West. Evident, too, is an on-going disagreement about how much of Moscow's leadership cadre should be left intact with a view to truce negotiation. In particular, Whitehall's preoccupation with the destruction of Moscow was not universally endorsed. Over and beyond which, criteria were, and probably still are, lacking for judging anybody's C3 in the round.

Civil Defence and Recuperation

What attracted more attention in the public domain, however, were Soviet civil defence preparations, real or imagined. A RAND study inferred in 1961 that the USSR was still proceeding with the construction of large deep shelters for the long-duration use of the population at large.[11] [12] This inference was received sceptically. Some warned that articles about shelter construction were no proof much action was being taken.

What can generally be said, indeed, is that the surest sign of Soviet officialdom's positive interest in a military subject was its disappearance as a topic in the open literature. Take the threat posed to radio reception by the ElectroMagnetic Pulse (EMP) caused by nuclear explosions in Near Space. In 1986, against the background of the worldwide debate on President Reagan's Strategic Defense Initiative, this suddenly emerged as a live issue in Moscow as elsewhere. Yet a scan of Soviet technical journals completed in 1985 had come up with but three articles, all decidedly dated, which addressed it at any length.[13]

On balance, the signs were that institutionalized enthusiasm for civil defence peaked in the USSR rather ahead of its doing so (1961–2) in the USA.[14] From 1958,

Civil Defence was deleted from the list of authorities required to approve building plans. Then, come 1962, Marshal Rodion Malinowsky (by then Minister of Defence) dismissed the shelter concept as "but a coffin or a grave prepared in advance". Something which will have been borne in upon him and colleagues is how limited is the role of shelters in protecting economic and cultural resources. After all, it was Nikita Khruschev who warned us all that, in the aftermath of thermonuclear war, "the survivors will envy the dead". He was saying, in effect, that econometric calculation could never encompass the holistic horror engendered.

Weighing in from a contrary perspective, Herman Kahn, the founder of the Hudson Institute, insisted that econometric analyses were still worth doing. In 1960, he published the results of an in-house study conducted in association with RAND. A key theme was how quickly the American nation, 200 million strong, could regain pre-war living standards from differing levels of fatality. Ten million dead was reckoned to allow of full recuperation within five years. Forty million would within 20. He further opined that, although misery would increase, this "would not preclude normal and happy lives for a majority of survivors".[15]

Yet surely econometrics could never gauge the likelihood that terrible mortality, savage uglification and a gripping fear of radioactive after effects would turn a crazed people via social breakdown to manic iconoclasm, political and religious. Take the First World War. The real victors were not the Allies at Versailles, they were the political religionists — communist and fascist. Their *modus operandi* was to be the systematic denial of plain truth and sweet reason.

Threats Overlooked

Two particularly morbid operational hazards, nuclear winter and biowarfare, hardly figured in debate through the sixties and seventies. In 1965, Professor Robert Ayres completed at the Hudson Institute a three-volume study portending a deep global freeze hard upon a major nuclear war.[16] On the Soviet side, this subject was raised by Mikhail Budyko, a ranking climatologist at the State Hydrological Institute in Leningrad, who was gaining good repute in the West. Again the theme did not catch on.

It took a double feature in *Science* in December 1983 by R. G. Turco *et al.* to kick start active debate.[17] It appeared at a time of renewed altercation about military nuclear modernization, in Europe and *vis-à-vis* the Strategic Defense Initiative. Earlier that year, 500 scientists from 15 nations conferring in Washington had expressed concern lest a major nuclear exchange brought on months of bitter cold and gloom through soot and dust being projected into the atmosphere. Donald Kennedy, President of Stanford, apprehended the biological trauma would be more profound than anything the last 65 million years. He was alluding to the natural cataclysm we see as marking the divide between the Cretaceous and Tertiary eras. This "KT" event eliminated *c.* 80 per cent of all living species. It accelerated sharply the demise of the dinosaurs.

In 1980, it had been proposed that KT was caused by a unitary asteroid six miles across, impacting with a force of maybe 200,000 megatons.[18] Iridium traces confirmed this explanation, pointing to the impact being located in the Yucatan. A critical comparison is that groundbursting just a one megaton bomb will cause dust particles and vapours to rise a good 70,000 feet, well into the rainless stratosphere the world over. In the absence of rain, the materials thus entrained cannot just be washed out in a week or two. Instead, the finer particles will descend into the troposphere (i.e. the rain swept lower atmosphere) over the course of several years.

The amount of dust projected skywards in a large-scale nuclear exchange would be of much the same order as that from a major volcanic eruption. Meanwhile, an even bigger "nuclear winter" problem, early on for sure, would be the accumulation (largely in the high troposphere) of soot from countless fires. In 1988, Budyko and colleagues produced a comprehensive review of how the debate had progressed in the USSR and, more particularly, in the West. Though helpful in various respects, this compilation focussed disproportionately and prejudicially on a 1984 article in *Nature* by the late Edward Teller.[19]

Teller was by 1988 a *bête noire* among progressives in the USA and beyond for having testified with studied convolution against Robert Oppenheimer's security clearance being extended to work on the hydrogen bomb[20] and for having fathered lately the Strategic Defense Initiative (see below). Now this Soviet team portrayed him as having *inter alia* concluded (*sic*) that the 1983–4 "calculations overestimated the expected temperature decrease by ten times".[21] What he actually did was admit that even a "decrease of 5° or 6° between northern latitudes 30° and 70° during summer, rather than the predicted 50° or 60°, could still lead to widespread failure of harvests and famine". Certainly, "the possibility of nuclear winter has not been excluded". In any case, the ozone would thin by a third for a year or two, threatening many eco-systems. Teller did call for more research and better modelling especially in regard to irregular soot dispersion. That was surely fair enough.

So even in the Gorbachev era even someone of Budyko's stature felt constrained periodically to genuflect to political correctness, Soviet style. This instance stands sadly in contrast with the determined stand he was by then taking against the disposition of the Muscovite establishment to discount the role played in climate change by variations in atmospheric carbon dioxide.[22]

In 1990, Turco's team were to present in *Science* the results of more robust modelling and analysis. The main contributor to nuclear winter would be soot from urban fires caused by Countervalue targetting. This smoke would be protected considerably against rainfall wash out by convective lofting as temperatures on the upper surfaces of the smoke clouds rose by perhaps 100°C. Average land cooling below these clouds could amount to 10° to 20°C; and in intercontinental interiors might be by 40°C in summer. Mean rainfall over land between 30° and 70° N would diminish by 75 per cent. The ozone layer over the northern hemisphere would weaken 40 to 50 per cent. Only after several weeks would any recovery be discernible.[23]

What was thus anticipated was a sudden descent into a global freeze twice as pronounced as during the last ice age. Were the dip down but a third, say, of what was predicted, the outcome worldwide could still aggravate terribly the post-nuclear trauma. By 1990, of course, many were expressing confidence that Moscow-Washington tensions could no longer climax thus. They would have felt less sure about Washington and Beijing.

The other big subject virtually ignored in the early MAD debates was biowarfare. Speaking in 1938 as Commissar for Defence, Marshal Kliment Voroshilov indicated the USSR was working on this option. Then in 1945 the Soviets picked up a goodly proportion of Nazi Germany's advanced weapons scientists. Yet in 1965 a senior staffer at the Hudson Institute told me that biowarfare never figured in Washington perspectives on strategic deterrence. Nor did he see why it should.

MAD between the superpowers never became as well defined as not a few of us students of such matters understood. Moreover, many of our lay contemporaries perceived it as emerging earlier than the 1967/9 time frame suggested by those standard assumptions about numbers and attributes of missiles and targets. Many saw the 1962 Cuba crisis as confirmation, the Ann Arbor speech having passed them by. On all sides, there was a tendency to talk with more assurance than was warranted, given the unprecedented nature of the interplay between psychology, technology and geography.

Limited World War?

So what about the effect this novel prospect might have on the incidence of local wars? How conducive would it be to linkage being forged between widely separated conflicts either to help resolve them or else to compound the geopolitical pressures? At a conference in Philadelphia in 1969, called just ahead of the commencement of Strategic Arms Limitation Talks (SALT), the Europeans present were struck by how earnestly our American colleagues were looking to SALT to pave the way for conflict resolution in Indo-China and the Middle East.

Not that everyone was persuaded that the advent of MAD would likely prove benign. Around 1960, a more prevalent view had been that America's loss of strategic nuclear superiority would encourage the Soviets to probe more daringly. General Maxwell Taylor wrote in 1959 that, in the years ahead, the Soviets, blessed with what he, like others, then foresaw as their actual superiority in strategic missilry, "may be expected to press harder than ever before, counting on submissiveness arising from our consciousness of weakness".[24] At the same time, Sir Basil Liddell Hart was warning that "nuclear nullity", as he termed it, "inherently favours and fosters a renewal of non-nuclear aggressive activity".[25] The impression thus made on him by these two scholarly and not illiberal students of strategy was a mainspring of President Kennedy's drive to build up American and allied non-nuclear capabilities. General Taylor served in his "New Frontier" administration as presidential adviser and then Chairman of the Joint Chiefs of Staff.

Limited Strategic War?

This was conceived of as a core response to the MAD situation. Klaus Knorr of Princeton defined the concept thus. It mainly "involves the use of strategic or long-range weapons" and is "deliberately and voluntarily limited in the total amount of damage threatened, planned and done as well as in the kinds of targets attacked". Though a majority of strikes would probably be intended to reduce the adversary's military capability, "the primary purpose is to act on his will . . . a contest of resolve".[26]

Doctrinal formulation proved vexatious. Kahn looked for a comprehensive blueprint for steady and universal application. In 1965 he claimed to have discovered one, cast as a "ladder of escalation" of 44 rungs.[27] Unfortunately, however, this linear perspective made it impossible to subsume the contingent variables (locale, area, duration, intensity, timing . . .) which were bound to determine how escalatory or otherwise a given riposte might be. Nor could anyone properly allow for those mood swings in time of crisis that could so affect reactions.

The formulation by Klaus Knorr cited above was more rounded and cursive. But still it failed to get to grips with critical dilemmas. How might one predict an adversary's reaction to a specific gambit? Will he merely be deterred or might he turn more positive? How can any humiliation inflicted be suitably modulated? How, might one add, can doctrine for Limited Strategic War relate to that for Limited World War?

The difficulties inherent in trying to forge any such operational concept were always liable to translate into an intractable problem for Alliance cohesion. Writing early in 1962, the incumbent first Director of the ISS, Alastair Buchan, called for changes in Alliance governance which he claimed could enable Washington to present its allies with "a clear and continuous picture of its strategic policy . . . the general considerations by which it would fight a war as well as deter one . . . ". He suggested this would *inter alia* remove the "main motive" of the Europeans "for wishing to acquire operational control of nuclear weapons".[28] In this rendition there was much to question. Suffice now to submit that the clarity and continuity sought could only be pipe-dreams, given the nature of the subject matter.

Take, too, the maritime "forward strategy" enunciated by the United States Navy *c.* 1977 and endorsed by NATO in 1981. The following year John Lehmann, US Secretary of the Navy (1980–87), described the Soviets' massive infrastructure, military and naval, on the Kola peninsula as "the most valuable piece of real estate on the Earth." He further averred that the way to keep Soviet naval forces well above the Greenland–Iceland–United Kingdom (GIUK) line was "to be up there . . . forcing them onto the defensive . . . to protect their assets".[29] Did that not make explicit a direct threat, at least to the protective "bastions" the USSR's SLBM force was organized into at sea? But how much would that quite representative interpretation tell anybody about actual behaviour in a real emergency?

Ballistic Missile Defence

Pentagon studies completed by 1964 indicated that Anti-Ballistic Missile (ABM) rocket batteries located around 30 American cities might hold down to several tens of millions the fatalities to be expected in the event of a "small to moderate" Countervalue rocket salvo launched the next decade or so. Yet on virtually any reckoning, the capital outlay would be several times as much as Moscow or, indeed, Beijing, might expend to make ready to overwhelm the said panoply. Moreover, such comparisons became ever more unfavourable the further the proposed coverage was extended. Rather more than 30 million Americans lived in those 30 largest cities. Only as many again did in the next 300.

None the less, in the autumn of 1967, Secretary of Defense Robert McNamara at long last authorized, wearily and reluctantly, a limited ABM deployment. This Sentinel programme would proffer a "thin screen" defence of American cities through the mid-1970s, one able to ward off a strategic attack of the kind Maoist China might by then present. In 1969, President Richard Nixon outlined his Safeguard modification of Sentinel. Priority would be given to a two-layer defence of ICBM silos though an eventual reversion to city defence versus Beijing was not ruled out. The one "layer" would involve interception at high altitude and the other engagement close in.

The China that Sentinel had primarily been configured to protect against was embarked upon the viciously irrational Cultural Revolution (1966–9). Nevertheless, the programme's inception stemmed more directly from a political need to match somehow a move the Soviets had begun in 1966 — namely, the deployment around Moscow of an ABM rocket code-named by NATO *Galosh*.

This Soviet departure, too, bespoke concern about Chinese pre-emption or escalation. The reasons for saying as much were manifold. What could be divined of the system's technical attributes favoured such an interpretation. So did its deployment being confined to date to the national capital/command centre. So, of course, did the close concurrence with the onset of the Cultural Revolution. So did the commencement of a big build-up of Soviet ground and tactical air forces about the long Chinese border, the evident aim being to dampen Maoist ardour. So, not least, was the way this move bucked what had been a trend in Soviet military thinking away from strategic ABM/BMD.[30] Relevant, too, was Beijing's having exploded a nuclear fission device in 1964; and Moscow's perhaps having an inkling of the thermonuclear test to come in 1967.

SALT and MIRVs

The 1972 SALT 1 agreement comprised two parts: an interim agreement curbing the procurement of strategic offensive missiles; and a treaty curtailing the deployment of ABMs. To be more specific, the latter allowed each superpower to possess two ABM screens, each to comprise not more than 100 static launchers. One screen could be to defend the national capital while the other could cover an ICBM field. Come 1974, a codicil required the signatories to settle for one or the other of these assignments.

The next year the Pentagon cancelled Safeguard, having done enough to indicate some generic feasibility, at least against single-warhead capsules.[31] It embarked on no other ABM programme *pro tem*. Nor was the USSR extending *Galosh* outside the Moscow area.

One complication would soon be the advent of Multiple Independently Targetable Re-Entry Vehicles (MIRVs). The principle underlying them is that the warload capsule of an offensive missile ejects soon after rocket burn-out a number of nuclear/thermonuclear warheads along with decoys — each warhead then being steered by vernier mini-rockets programmed to feed in late corrections. In other words, the warheads became more accurate as well as more numerous.

It proved to be another instance of the superpowers achieving nearly concurrently a most exotic step forward in military technology. In 1968, the USA test fired two MIRV-ed systems: the Minuteman 3 ICBM and Poseidon, a new SLBM. Around that time, too, MIRVs were released in firings across the Pacific of Soviet SS-9 ICBMs. Three warheads of several megatons apiece were ejected from each SS-9 capsule or "bus". Their descent to target was in patterns adjudged to indicate an ability to strike at three Minuteman sites with standard accuracies so high that usually all would be wrecked outright or else engulfed in crater debris.

The USA had MIRVed all its ICBMs and SLBMs by 1975. By then, too, the USSR was MIRV-ing its ICBMs; and by the turn of that decade would be following suit with SLBMs. Stability would thereby be firmly restored albeit at higher warhead levels. Unfortunately, this progression was bound to engender cynicism among those emergent nations expected to subscribe to the nuclear Non-Proliferation Treaty of 1968 in exchange for the superpowers' stabilizing their inventories. More basic, however, was the question of wide-area Ballistic Missile Defence (BMD). Would not multiple warheads always make this goal unapproachable except perhaps by intercepting ascending missiles prior to burn-out — intercepting them, that is to say, in their boost phase by means of battle stations in orbit? From then on, support for such orbital battle stations steadily became stronger, more orchestrated and more vocal on the Republican Right in the USA.

Edward Teller had an audience with President Ronald Reagan on 14 September 1982. Hence the latter's announcement on 23 March 1983 of the Strategic Defense Initiative (SDI), what the sardonic soon dubbed "Star Wars".[32] Along with everybody else, Dr Teller stressed throughout that foolproof protection against strategic missile attack was not on offer because, in the face of so novel and fast- breaking a contingency, one could never anticipate what would be the optimum command and control structure.

Notes

1 Robert Jackson, *High Cold War* (London: Patrack Stephens, 1958), Chapter 5.

2 John Lewis Gaddis, *The Long Peace* (New York: Oxford University Press, 1987), Ch. 7.

3 This section and the next draw considerably on the author's "Towards the Superpower Deadlock", *World Today* XXII, 9 (September 1966): 366–74.

4 A Medium-Range Ballistic Missile (MRBM) is defined as having a range between 600 and 1,500 nautical miles while an Intermediate-Range Ballistic Missile (IRBM) covers 1,500 to 3,000 nm.

5 P. M. S. Blackett, "Steps Towards Disarmament", *Scientific American* 206, 4 (April 1962): 45–63.

6 *Technical Report* 8089 (Newport: Naval Underwater Systems Centre, 1987), p. 1.

7 *Survival* 4, 5 (September–October 1962): 195.

8 *SIPRI Yearbook of World Armaments and Disarmament 1969/70* (Stockholm: Almqvist and Wiksell for Stockholm International Peace Research Institute, 1970), pp. xviii.

9 Bruce G. Blair, *The Logic of Accidental Nuclear War* (Washington DC: Brookings Institution, 1993), Chapter 5.

10 *Ibid.*, pp. 63–4.

11 RAND Corporation was and is a think-tank funded by the USAF. Its overall pitch has not been illiberal.

12 Leon Gouré, *Civil Defense in the Soviet Union* (Berkeley and Los Angeles: University of California Press, 1962), Chapter 10.

13 David G. Chizum, *Soviet Radioelectronic Combat* (Boulder: Westview, 1985), p. 48.

14 Arthur L. Waskow and Stanley L. Newman, *American in Hiding* (New York: Ballantine, 1962), Chapter 12.

15 Herman Kahn, *On Thermonuclear War* (Princeton: Princeton University Press, 1960, p. 21.

16 Cited in Leonard Bertin, Background Paper No. 3, *Nuclear Winter* (Toronto: Canadian Institute for International Peace and Security, March 1986), p. 1.

17 R. P. Turco *et al.*, "Nuclear Winter: global consequences of multiple nuclear explosions" and Paul R. Erlich *et al.*, "Long-term biological consequences of nuclear war" in *Science* 222, 4630 (23 December 1983: 1283–92 and 1293–6.

18 L. E. Alvarez *et al.*, "Extraterrestrial cause for the Cretaceous-Tertiary extinction", *Science* 208 (1980): 1095–108.

19 Edward Teller "Widespread after-effects of nuclear war", *Nature* 310 (23 August 1984): 623–4.

20 Rudolph Peierls *Atomic Histories* (New York: Springer Verlag, 1997), pp. 283–4.

21 M. I. Budyko, G. S. Golitsyn, Y. A. Izrael (V. O. Vanuta, trans.), *Global Climate Catastrophes* (New York: Springer-Verlag, 1988), p. 83.

22 M. I. Budyko, "Climate Conditions of the Future", *Conference on Climate and Water* (2 Vols.), (Helsinki: Valtion Pianatuskeskus, 1989), Vol. 1, pp. 9–25.

23 R. P. Turco *et al.*, "Climate and Smoke: an appraisal of Nuclear Winter", *Science* 247, 4939 (12 January 1990): 166–76.

24 Maxwell D. Taylor, *The Uncertain Trumpet* (London: Stevens, 1959), p. 136.

25 B. H. Liddell Hart, *Deterrent Defence* (London: Stevens, 1960), p. 43.

26 Klaus Knorr in Klaus Knorr and Thornton Read (eds.), *Limited Strategic War* (London: Pall Mall, 1962), pp. 3–4.

27 Herman Kahn, *On Escalation: Metaphors and Scenarios* (New York: Praeger, 1965), pp. 186, 190, 217–20.

28 Alastair Buchan, "The Reform of NATO", *Foreign Affairs* 40, 2 (January 1962): 165–82.

29 Michael Getler, "Lehman sees Norwegian Sea as a key to Soviet naval strategy", *The Washington Post*, 29 December 1982.

30 Christoph Bluth, *The Collapse of Soviet Military Power* (Aldershot: Dartmouth, 1995), p. 179.

31 Paul Stockton, *Stratgegic Stability between the Super-Powers*, Adelphi Paper 213 (London: International Institute for Strategic Studies, Winter 1986), p. 13.

32 William J. Broad, *Teller's War* (New York: Simon and Schuster, 1992), pp. 102–19.

The Western Question

12 | The Highest Frontier

Currently the only countries with Space satellite launch facilities are the USA, Russia, China, India, Japan, France and Israel. However, some 50 countries now have at least one satellite in orbit. President Bush announced in January 2004 a new *Vision for Space Exploration* which has refocused NASA's objectives towards human missions to the Moon and Mars. The European Space Agency's *Aurora* programme embraces very similar aims.

In September 2007, Japan dispatched a satellite to orbit the Moon. In October, China did likewise, and plans to land a man on the Moon by 2024. In October 2008, India dispatched a lunar orbital. Given recent growth rates, the global Space market could be 15 times its current size by 2025. Thus far only "a small fraction of the technological potential of satellites has been realised".[1] When and how most has will bear quite heavily on the Western Question.

For one thing, Space usage may soon loom as large in public consciousness as aviation did after 1918. Already the fictional imagery can range anywhere between utopian serenity and savage bellicosity. Either way, the media coverage has often been blatantly escapist. Buck Rogers made his debut in 1932, as the world slump passed its nadir. *Star Trek* dates from 1969 when South Vietnam was being abandoned. The 1999 revival of *Star Wars* took place against a backdrop of anxiety about the disorderliness of the "new world order".

Early on, the printed word then radio had conveyed something of the sweetness of cosmic melancholy. From 1877 and 1906 respectively, two ranking field astronomers, Giovanni Schiaparelli of Italy (1835–1910) and Percival Lowell (1855–1916), founder of the Flagstaff observatory in Arizona, allowed themselves to be the exploiters of, though soon the exploited by, the notion of a criss-crossing of the surface of Mars by what Schiaparelli called "canali", artificial water courses. They proposed these betokened an advanced civilization. Alas, it was a hypothesis dependent on reading far too much into such striations as Schiaparelli may have discerned. Yet despite intense professional protestation, it was influential for decades. Meanwhile, in 1938, two million listeners panicked over the "breaking news" of a Martian invasion in the Orson Welles dramatization on CBS of some H. G. Wells.

In due course, a succession of astral extravaganzas were published as supposedly confirmed historical truth. In 1950 came Immanuel Velikovsky's *Worlds in Collision*. It would have had us believe that *c*. 1500 BC a comet issued from Jupiter, made several passes by the Earth and then settled in orbit as the planet Venus. Among the depositions from those passes were said to be oilfields, locusts, hail and manna. In reality, the nucleus of a comet would be between a millionth and a 100 billionth the mass of such a planet as Venus.

UFOs and Astrology

Other astral fantasies have welled up more spontaneously at popular level. Imaginings of Unidentified Flying Objects (UFOs) — "flying saucers" — may no longer attract the attention they once did. But throughout the Cold War they showed a marked propensity to proliferate during times of heightened tensions. In 1986/7, as the USSR slid into terminal crisis, UFOs aroused much interest in the Soviet media.[2] Carl Jung read into UFOs a mandala, a symbol of protection which readily presents itself in various cultures in times of stress.[3]

Astrology is an activity that has shown a remarkable ability to survive endless pronouncements of demise. The first edition of the *Encyclopedia Britannica*, published in 1771, dismissed it as a discipline which long ago became "a just subject of contempt and ridicule". Yet come the turbulent turn of that century, astrological almanacs were selling well. In the case of Moore's, this meant up to half a million copies a year.[4]

Lately the astrology business has experienced an episodic resurgence. In India and parts of East Asia, the subject is strong; it ticks over steadily in Britain and elsewhere. As often as not, its acolytes seem able to extol its nostra, amiably imbibing the while the findings of modern astrophysics. Then again, the Falun Gong movement in China (officially estimated to be peaking at 30 million in 1996) attached to itself a ludicrously arbitrary and intricate pseudo-cosmology.[5]

Left Critiques

Early on in the "Soviet experiment", an ambivalence towards Space research was manifest. Evidently, the dark night skies were a good backdrop for the martial puritanism that suffused the regime. Space could also be seen as a relatively short cut to technological parity with the West — a way to break out of the country's geographical gridlock. On all counts, it was deemed to reinforce the revolutionary legitimacy of a "dictatorship of the proletariat" committed to ushering in "scientific socialism", the word "scientific" here connoting at once supremely "rational" and "science based". Besides, it enabled atheists to reach out to Heaven.

None the less, in the early twenties Stalin and Trotsky each rounded on a burgeoning cult of Cosmism. It was forcefully portrayed by the latter as outright escapism. Then by the sixties, Moscow was evidently exercised about the exposure of its once closed polity to broadcasts and surveillance from on high. On the other hand, Sputnik could fairly be seen as a way to break out with a flourish. But then in 1983 came the Strategic Defense Initiative. The non-weaponization of Space is one of several aims Moscow under Vladimir Putin came to share through triadic diplomacy with Beijing and New Delhi.

The broad Left across the West has likewise been ambivalent about Space in general. By 1968, a Californian *avant garde* sculptor, Theodore Roszak, was well to the fore in a counter-culture campaign against "big science", a campaign galvanized by "Vietnam". In 1971, he was to be found scorning "too much limelighted posturing by the astronauts . . . and the research teams . . . the boyish modesty, the understatement, the winsome embarrassment at all the applause".[6] Yet in 1979, with "Vietnam" over and done with, he could appreciate natural science as a whole as a "vast many-sided" adventure. He cited with approval the two most celebrated British astronomers of the inter-war period, Sir James Jeans and Sir Arthur Eddington. He did so partly by virtue of their neo-Idealist disposition to see Mind as more fundamental than Matter.[7]

The least congruent perspective has to be that proffered by Bertrand Russell. The genesis of his blanket rejection of Space travel can be traced back to 1924 which is when he recalled in all seriousness how, in classical legend, "Icarus, having been taught to fly by his father, Daedalus, was destroyed by his rashness. I fear that the same fate may overtake the populations whom modern men of science have taught to fly". Ever after, he insisted that (a) the Space programme lacked scientific impartiality, (b) Space exploration could spread human foolishness, (c) we should address terrestrial problems first, and (d) any gain in human understanding was problematic.[8] The climax was his reaction to the Apollo 11 lunar landing. He saw it as precursory to voyages to other stars "by savage pilgrims of hate . . . grown old during the long journey". In contrast, the wise Immanuel Kant "never travelled more than 10 miles (*sic*) from Konigsberg", his native city.[9] Granted, the case against manned flight into Deep Space is strong. But Russell's diatribe against manned aerospace in its entirety was altogether too peevish, particularly as coming from a Logical Positivist and inveterate jet-setter. Also he cavalierly begged the question as to what attitude Kant, a keen and contributive student of astronomy, would have adopted to manned Spaceflight.

Aerospace Travel

Air travel emerged in 2007 as a major bone of contention *vis-à-vis* global warming. Suggestions that for exhaust gases to be released on high (i.e. at 8 to 15 km) was worse gram for gram than if near the surface are related to the formation of contrails and ozone.[10]

Yet if objections can thus be raised to routinized air travel, then the more readily may they be against Space tourism: a possibility various investors currently express interest in, the first Space tourist having flown in 2001. In the frame at present is a Spaceplane able to hold several passengers in orbit for several minutes before gliding back to Earth. The charge to each Spaceplane passenger would be around £100,000, much of this to cover fuel costs. However, the emission of greenhouse gases (carbon dioxide? methane? water?) will be negated above 80 kilometres by the tendency for molecules to dissociate under the influence of intense short-wave radiation.

Space-based Manufacture

A quarter of a century ago, a possibility ardently foreseen was that of micro-manufacturing in Space.[11] With the virtual exclusion of gravity as a factor in movement and association, scientists could focus on more subtle interactions: diffusion, solubility, capillary action and electrical attraction or repulsion. Benefits could accrue in terms of quality control, low cost and, indeed, certain chemical syntheses impossible on Earth. The large molecules characteristic of organic substances may form more readily outside gravity. So may large crystals.

Although interest initially waxed strong in the USA and Europe, one of the fullest accounts of the opportunities came from Moscow.[12] Among them was the construction of optical fibres, making them the more efficient as conduits of photonic energy. In Washington, however, the Office of Technology Assessment, the erstwhile Congressional think-tank, warned that the "true value of the microgravity sciences" would not be calculable for years. For some time now, in fact, little has been heard of this approach despite the burgeoning revolution in nanotechnology, the study of structures and processes on the scale of billionths of a metre. Presumably one reason has been the effect over time of weightlessness on human participants.

Remedies for Greenhouse

For 30 years and more, coteries of specialists have enthused about easing global warming through the placing of installations in Space, these not manned continuously but subject to regular servicings. It has been accepted throughout that economies of scale are such that any structures will necessarily be very large. Moreover, thorough political and legal process will be required to address all the economic, ecological, security and managerial questions posed.

An array of solar cells locked together in a geostationary orbit may receive virtually 100 per cent of the heat energy incident from the Sun, there being no interference from clouds, weather, atmospheric gases or whatever. The solar energy intensity in such a Near Space location is 32,600 watt hours per square metre per day whereas the average daily inflow at the Earth's surface is given as 8,600 in sunny Phoenix, Arizona, and in cloudier Boston 3500. Microwave transmission to a ground station of energy gathered in orbit may be rather over 60 per cent efficient.

However, technical doubts are expressed on several counts — an orbital mass of several tons for every megawatt generated; heavy assembly costs; and the possible effect of microwave overspill on radio transmissions and, indeed, life forms.[13,14] Meanwhile a raft of technical objections have been raised to an alternative concept of reflecting a proportion of the Sun's rays back into Space by means of a diaphanous mirror parked in orbit.

Openness

The pressures worldwide towards more open societies have in no small measure been generated by a trend to more open skies. This has meant, above all, orbital satellites achieving ever more potency as harbingers of ubiquitous communications and surveillance. Telstar, which entered service in 1962, was the first satellite to relay television "in real time".

Of course, openness can effect transformation deeper than the routinized political process. It can involve a renaissance of national culture. As Prime Minister of India, Indira Gandhi used to express the hope that Indian television would underpin nationwide unity by strengthening the identities of the big regions as opposed to more local affinities. She observed how on the north-east border "the vociferous demand of elder tribal people that their customs should be left undisturbed found support from noted anthropologists". But against this had to be set "the protest of the younger elements that . . . they were being preserved as museum pieces".[15]

That consideration may initially weigh in strongly *vis-à-vis* a national television service. But how about exposure in due course to scores of channels transmitting within an utterly cosmopolitan milieu? By no means only the more authoritarian regimes (China, Iran, Saudi Arabia . . .) have waxed apprehensive on this score as satellite dishes have become smaller and therefore harder to regulate. In 1986, Papua New Guinea deferred a decision about television since "We do not want to become Americans or Australians"; and because too many soap operas would exacerbate further the high incidence of violent crime, alias "rascalism". More generally, many people have dreaded the creation of "cultural deserts" as traditional mores are eroded in favour of a glitzy consumerism with general dumbing down.

In July 1941, the Americans began to establish a substantial military presence in Iceland. In 1966, the Icelandic National Broadcasting Service started television broadcasts "with extremely meagre means". About the same time, the United States forces inaugurated a better financed channel. However, this soon engendered acute controversy which duly became entrained with the then widespread opposition to the foreign military presence. For a while, the Americans were limited to cable transmissions just within their Keflavik base. But by the turn of the century there were available nationally three Icelandic channels and many foreign channels; Reykjavik enjoyed some fame in the pop world. Meanwhile the strategic revolution was transforming the external scene but also impacting on Iceland's shifting politics domestically. So no one any longer even claimed to care whether the television channel at Keflavik was "seen by Icelandic eyes". But was this professed indifference based on new-found cultural confidence or weary resignation? Most likely, a measure of both.

Military Surveillance

With integrated multisensing, comparisons of image definition as between different surveillance regimes can be complicated. What mixes of technologies and wavebands are involved? What light levels? What field of view? How many transits? These days "definition" may be taken to mean the widths of the smallest features discernible. "One metre" is often a "sweet spot" for, let us say, distinguishing the individual vehicles in a road convoy. Commercial reconnaissance is now on that threshold. With American military optical surveillance, five centimetres may sometimes be achieved.[16] Ten to 20 centimetres has been presumed *vis-à-vis* the Chinese programme.[17] Concurrently the locating of one's own mobile assets has become very precise, save sometimes as yet within urban areas.

Yet even with such capabilities (plus much electronic eavesdropping, especially on short-wave radio or radar), surveillance from Near Space of an adversary still has limitations akin to those affecting other modes of reconnaissance. Looking through smoke and cloud is very difficult and into solid structures still virtually impossible. Observation in darkness is ruled out except by radar. Most particularly, satellites are very unlikely to monitor properly, if at all, work on biological weapons. Nor may they probe the innermost workings of the human mind. Nor may they successfully counter deception by thorough camouflage or the deployment of dummies.

World War Two and the Cold War afford various instances, even on open landscapes, of recourse to dummies in order to vitiate low level aerial reconnaissance, a mode usually harder to hoodwink than the orbital might be. Before the fateful Battle of Alamein in 1942, the British liberally faked vehicles, pipelines and supply dumps, a ruse which made it harder for Rommel to determine assuredly the main thrust of their incoming attack.[18] Similarly, before the Israelis attacked in the Sinai in 1967, they moved a brigade to just opposite El Kuntilla. Then dummy tanks, masked by studiedly inadequate camouflage netting, were fed in to comprise two phantom brigades in support.

Strategic Surveillance

Still, the advantage that may accrue from orbital or, if appropriate, aerial observation may be greatest at continental level. An instance to date, positive enough in itself, was the downward revision from 1956 (thanks to the U-2 overflights and then to orbital reconnaissance) of the Red Army's order of battle. It was down to 140 divisions after mobilization instead of anything up to 450.[19] But how had the latter number ever come to be thought sustainable? Could such a mass ever have been deployable? And had the strategic geography of the USSR ever been quite that opaque?

During the fifties, the Soviet army became far more mechanized. That transformation would have been extremely hard to effect without its divisional strength being duly constrained. Besides which, a mechanized division requires, when embattled, at least one two-lane axial road to keep supplies flowing; and several such roads to redeploy in good order. Instead, in all directions, the Red Army's strategic mobility was limited by the physical and economic landscape, especially the lack of good roads. In other words, Moscow's imperial bounds never had the smooth uniformity of the billiard table seemingly implied by Mackinder's mental map of the "heartland".

Over and beyond which, the internal stability of the USSR still depended crucially on a culture of immobility. No deep insight should have been needed to divine connections, in terms of official attitudes, between near serfhood restrictions on rural domestic travel, the dismal state of so many roads, rigid occupational zoning of residential areas, a lack of decent maps, and countless obstacles to access to officialdom. Granted, the reinforcement of the Moscow front the Red Army effected late in 1941 was very crucial. But the number of divisions involved was only about a dozen.[20] So all in all, any western analyst making a guestimate without benefit of Space reconnaissance ought to have put the divisional order of battle much closer to 140 than 450. May one stress, too, that reconnaissance from Space ought to have allowed of much more measured official assessments in 2002–3 of Iraq's status *vis-à-vis* Weapons of Mass Destruction.

A further rider is that capabilities for orbital military reconnaissance will tend to proliferate, thereby giving more scope regionally for pre-emptive strikes. The general presumption has to be that this will be destabilizing, though whether on balance it actually is will depend on the specifics, geographical and political, of a given situation.

In arguments about Space surveillance, wounded pride figures in what might otherwise be deemed irrational hostility. From Brazil there have been protestations against ecological data being processed outside the country. India and Pakistan have objected to monitoring from Space being allowed for in any Comprehensive Test Ban (CTB) on nuclear warheads.[21]

The question of a contingent refusal to share military Space imagery has also come up recurrently. Israel's access to American images has often been constrained, notably after her attack on Iraq's Osirak reactor in 1981 and during the 1991–2 Gulf War. In 1994, responding to a congressional order to cease support for the UN arms embargo against Bosnia, President Clinton cut his NATO allies off from satellite intelligence about ship movements in the Adriatic. Rome, Madrid and Athens have since sought to obviate the risk of the like happening again by sharing financially in the French reconnaissance programme.

Ballistic Missile Defence (BMD)

Acute technical problems are always posed when one makes ready to intercept from the Earth's surface either ballistic or the aerodynamic "cruise" missiles, this within the context of localized conflict. None the less, no philosophic questions come up save for eternal ones concerning war in general. On the other hand, specific issues of principle can be raised should one seek to alter the global balance by making ready to intercept a barrage of strategic missiles.

This especially applies if the plan is to engage them with orbital satellites. The prime operational reason for wanting to do this is to allow of the interception of a hostile rocket towards the end of its boost phase — that is to say, just before it could have released a spread of warheads and decoys. (See 11). Experience suggests that, once a state has developed strategic rockets, it should be able within a few years to install in them decoys and simple multiple warheads: a combination liable to pose a long-range threat to large cities. Admittedly Chevaline, the British programme for MIRV development, cost a billion pounds. But the cost might have come right down had independent targeting for each RV released not been sought.

These next 20 years or so, only the USA could hope to introduce the orbital interception of ascending rockets. This would be with laser beams, the two alternatives considered early on (particle beams and Electro-Magnetic Launchers) having been effectively discarded since 1990. Washington has not committed itself to make this departure. Nor is it committed never to. Meantime a sophisticated debate is on-going within the American military and scientific communities as to the merits. Within the United States Air Force, in particular, it is given added spice in that some reduction in the role of the manned warplane could revive questions about an independent air arm's *raison d'être*. Space-based Lasers (SBLs) would very likely be a USAF responsibility.

My current perspective on the above was outlined sufficient for present purposes in 2004.[22] The case made can rest, except that it fails to recognize (a) just how diverse are the other purposes to which lasers generated in Space might be put and how these could complicate arms control provisions and (b) the incredible flux intensities pulsed lasers are now achieving through pulse compression.[23] The higher the flux intensity an SBL can achieve, the less acute will be the problem of target lock on.

Suffice for now though to indicate two respects in which orbiting SBLs could exercise physically an oppressive influence on the world scene. The one is the "dwell time" over a given target area of orbital platforms revolving round a rotating Earth. An SBL constellation that was always to have four platforms overhead to Iran would *ipso facto* have to have 24 overhead to China. And so on.

The other is the ability of laser beams on certain wavelengths to pass through the Earth's atmosphere without severe absorption or other interference, an attribute which opens up the possibility of targeting anywhere on the Earth's surface. This prospect was discussed in the eighties in connection with the Free Electron Laser, a non-chemical device with therefore tunable wavelengths. Access can, of course, work both ways. China pointedly turns ground-based radar skyward, and even Libya jammed thus a satellite telephone system for some months in 2006.[24]

Since 1975 the USAF has been developing what is now known as the AL-1A Airborne Laser, an aircraft originally due to enter service in 2008. Its continuous wave Chemical Oxygen Iodine Laser (COIL) can likely sweep the Earth's surface because its wavelength (1315 nanometres) lies just within the sharp boundary of a pronounced window of low atmospheric absorption on the electromagnetic spectrum. It is the one through which floods in the sunlight our eyes are sensitized to. They process what we know as "visible light" between 400 and 800 nm but are vulnerable to overly intense radiation across the said window's span. An AL-1A is to attempt the late boost phase interception of a strategic missile over the Pacific this summer.

The Mythic and the Mundane

That "death rays" have long been a stable of science fiction is expressive of how such imagery evokes, from deep within our psyche, graphic folklore: lightning shafts; the basilisk's glance; the meteoritic "firedrake" of Germanic tradition; the "evil eye" set deep in Sicilian and in MesoAmerican folklore . . . Even the prosaic Anglo-Saxon Chronicle tells how, among the "foreboding omens" that "wretchedly terrified" the people of Northumbria in the year 793 were "lightning storms and fiery dragons . . . flying in the sky". However phantasmogoric latter-day resurfacings of such concerns may seem to ultra-rational strategic analysts, they would do well to reflect upon their impact.[25] Perceptions embedded within our ancestral unconscious guide our mundane preferences.

SBLs able to incinerate targets on the Earth's surface would underscore heavily the continuing superiority of advanced Western nations across almost the whole ambit of military science. In 1986, Herbert York, (formerly Presidential chief scientist to Dwight Eisenhower) warned that such an approach would be "to take our shield from in front of our body and shove it up against our opponent's nose even in peacetime". He was then alluding to strategic relations with the USSR. But the metaphor evidently has more general application.[26]

We may be deluding ourselves if we think that the disadvantaged peoples of this planet would abjectly acquiesce in such a state of affairs. On the contrary, they may become more prepared to resort, whether via statehoods or insurgency cells, to the most lethal option remaining, biowarfare — spoken of by Dr York on another occasion as "the most dangerous thing on the horizon".[27] This could transpire against the background of a syndrome of malaise encompassing resource limitation, ecological distortion and social alienation.

The ABM Treaty

In December 2001, President Bush withdrew the USA from the Anti-Ballistic Missile Treaty. He did so in the face of progressivist opposition especially in the USA itself. The fact remained there was ever-widening discordance between this 1972 formulation and what was actually happening, with White House and Kremlin involvement or connivance.

It was not just that Article V committed both signatories "not to develop, test or deploy ABM systems or components which are sea-based, air based, Space-based or mobile land-based". Other articles or appended statements were likewise mocked by changing realities. Thus the 1972 preamble stated that the treaty's prime aim was to curb the "strategic" arms race. Since when, the difference between "strategic" and "tactical" had been blurred by technical dexterity and geopolitical change. A neat distinction between "anti-missile" and "anti-aircraft" was fading too. That between ABM and Anti-Satellite (ASAT) had never been convincing in any case.

Since 1996 at the latest, it had looked improbable that one could "preserve far into the next century the 1972 accord".[28] Such was the case, regardless of any nuances of "broad" interpretation or "clarification of its terms", to cite the two nostra beloved of certain liberals. The blunt truth was that the text dealt with a weapons genre that had lost its identity. One might as well have tried to forge an arms pact around the light bomber or the battle cruiser or the siege gun. No less obvious an anachronism was that the said treaty was just as between two nation states. Nevertheless, President Bush was surely at fault in that he renounced it without propounding an alternative mode of strategic arms restraint. But neither did his critics come up with anything germane.

No Fourth Domain

The fact that an orbital satellite is intrinsically so different from any other vehicle invites *a priori* the thought that Near Space is a conflict arena which stands apart from the land, maritime and aerial. However, this is not really a helpful perspective. A submarine is just as different from all other platforms. But almost nobody has ever said there is a separate domain of war below the sea's surface.

Today there is, most evidently in the civil sphere, a telecommunications network which is almost seamlessly integrated as between land, sea and the skies above. Clearly this militates against separatist interpretations. Still, the bottom line must be whether attempts to conceptualize thus the military Space environment do anything to further our understanding thereof. Are analogies with command of the sea instructive? Does it help, for instance, to regard Near Space as akin to Alfred Mahan's "indivisible world ocean"?[29] Or does it resemble the sea dimension Julian Corbett would have had us exploit more circumspectly and contingently?[30]

What of those five shifting points between the Moon and the Earth identified by the French mathematician and astronomer, Joseph LaGrange (1736–1831) as being always, in principle, gravity-free? Everett C. Dolman of Berry College, a veteran Space buff, has described the military and commercial value of at least two of these LaGrange Libration Points as "highly speculative but imaginatively immense". Angels on pinheads?

So what then of Dolman's makeover of Mackinder's tripartite division of the world into the Heartland, the Inner Crescent and the Outer Crescent? My own prior assumption had been that he would distinguish between the Earth, the Atmosphere and Near Space. Instead Everett Dolman proposed a quite different quadripartite interpretation: the *Earth*, its atmosphere included; *Earth Space* from the lowest viable orbit to just beyond the geostationary altitude; *Lunar Space*, from a geostationary position to just beyond the lunar orbit; and *Solar Space* for everything in the solar system outside the lunar orbit.[31] If it is that problematic what Mackinder-style format may be most appropriate to conflict in Space, one may fairly conclude that any of the alternatives would constrict and distort our understanding of reality.

But is not a clinching argument in favour of Near Space being regarded as subsidiary to the three terrestrial domains simply this? On-going warfare in Space would be bound to extend to the Earth's surface but violent war on the surface need not extend into Space. However, cyberwar might. When the Galaxy IV satellite lost its bearings in 1998, scores of millions of people around the world experienced information blockage.

Debris

So should we not be seeking a softer perspective on Near Space, one sufficiently free from geopolitical contrivance as to be truly planetary? One particular but persuasive reason for preferring this is the accumulation in Near Space of orbital debris. This built up a lot during the late seventies on account of firings of US Delta launch rockets plus Anti-Satellite (ASAT) interceptions by the Soviets. From 1981, the Deltas were set to burn out lower down; and, in 1982, Moscow suspended ASAT tests. Nevertheless, the number of observable objects continued to rise from 5000 in 1981 to 8000 in 1995. Now the figure has topped 10,000, only 600 being active satellites. China's ill-considered ASAT test of January 2007, a successful interception 530 miles out, initially added 2800 significant fragments.

The number of pieces more than one centimetre wide residing in Space well exceeds 100,000. Co-ordination on this subject between national Space agencies has been developing. The best hope is that debris amounts may be stabilised.[32] For in the rain-free environment above the tropopause (the well-defined boundary between the lower atmosphere and the stratosphere), the tinier particles descend to Earth very, very slowly. While they persist, they pose a threat of abrasive collision even though, for the time being at least, this is eased by the debris being fairly well layered, much of it being between 500 and 1000 km (and, more especially, near 900 km).

One proposal has been that lethal laser beams could break the particles up, making them averagely smaller. Yet this would accentuate an opacity problem in the night sky. It could also increase the frequency of erosive collisions while decreasing the risk of catastrophic ones. What is difficult to accept, in any case, is the notion that the energy beams could so retard a piece of debris as to make it descend below 200 km where "drag increases sufficiently to terminate the object within a few hours".[33] May one add that a general ASAT war that did not leave a sizeable fraction of the resultant debris orbiting is implausible. In the context, indeed, the deliberate creation of fields of erosive debris is all too easy to imagine.

The Heavens Above

In 1983, Gro Harlem Brundtland (sometime Swedish Prime Minister) was invited by the UN Secretary General to chair a World Commission on Environment and Development. Its report designated "Outer Space" as one of three "global commons", the other two being the Oceans and Antarctica. Acting on this principle could become crucial, this most urgently in respect of "Outer Space", alias "Near Space". The Counter-Culture precept that "The Heavens were made for wonder not for war" can be a point of departure, provided it be well understood that the taboo *sine qua non* is the weaponization of Space, not the military use of that milieu for early warning, surveillance and crisis communication.

Many regimes still feel unhappy about their economic and environmental travails being exposed, this by outsiders. Some, indeed, will especially dislike having revealed dubiety within their international statistical returns, not to mention their operation of "gulag archipelagos" and so on.

Nevertheless, there are big practical gains for humankind at large from rapid assemblages of global data (synoptic and hard) about such matters as floods and drought; deforestation; shifting land use; rainfall over the oceans; the density of fish stocks and the phytoplankton that sustain them . . . For Geographers this can allow of a shift of emphasis away from building a data base and towards interpretation and thence policy prescriptions. Maps that can be computer-produced in hours rather than months will be invaluable on occasion in peace as well as often in war.

Notes

1 *Strategic Survey, 2007* (London: International Institute for Strategic Studies, 2006), pp. 69–74.
2 Martin Walker, *The Guardian* (4 February 1987).
3 C. G. Jung (R. F. C. Hull, trans.), *Flying Saucers* (London: Routledge and Kegan Paul, 1959), pp. 18–19.
4 Peter Whitfield, *Astrology. A History* (London: British Library, 2001), Chapter 5.
5 Maria Hsia Chang, *Falun Gong. The End of Days* (New Haven: Yale University Press, 2004), Chapter 3.
6 Theodore Roszak, "Autopsy on Science", *New Scientist* 57, 830 (11 March 1971): 536–8.
7 Theodore Roszak, *Person/Planet* (London: Victor Gollancz, 1979), p. 50.
8 Chad Trainer, "Earth to Russell", *Philosophy Now* 40 (March/April 2003): 20–2.
9 Bertrand Russell, "Why Men Should Keep Away From The Moon", *The Times*, 15 July 1969.
10 "Green Sky Thinking", *New Scientist* 193, 2592 (24 February 2007): 33–8.
11 Peter Marsh, *The Space Business* (Harmondsworth: Penguin, 1985), Chapter 7.
12 V. C. Avduyevsky (ed.), *Manufacturing in Space: Processing Problems and Advances* (Moscow: 1985), pp. 155–6.
13 Clarke Covington, "Solar Power and the Future in Space", *Journal of the Royal United Services Institute for Defence Studies* 124, 1 (March 1979): 12–21.
14 Peter E. Glazer, "Evolution of the Solar Power Satellite concept: the utilization of energy from Space", *Space Solar Power Review* 4 (1983): 11–21.
15 Indira Gandhi, *People and Politics* (London: Hodder and Stoughton, 1982), p. 63.
16 Jeffrey T. Richelson, "The Whole World is Watching", *Bulletin of the Atomic Scientists* 62, 1 (January/February 2006): 26–35.
17 Joan Johnson-Freeze, "China's Manned Space Program, *Naval War College Review* LVI, 2 (Summer 2003): 31–71.
18 Fred Majdalany, *The Battle of Alamein* (London: Weidenfeld and Nicolson, 1965), pp. 73–4.
19 B. H. Liddell Hart (ed.), *The Soviet Army* (London: Weidenfeld and Nicolson, 1956), Chapter 21.
20 Inferred from Malcolm Mackintosh, *Juggernaut* (London: Secker and Warburg, 1967), pp. 164–6.
21 Tara Patel, "Nuclear treaty flounders as Asia steps up demands", *New Scientist* 149, 2020 (9 March 1996): 6.
22 *GISC*, Part III.
23 Mark E. Rogers, *Lasers in Space. Technological Options for Enhancing US Military*

Capabilities (Maxwell Air Force Base, Center for Strategy and Technology at the Air War College, Occasional Paper 2, November 1997).

24 "Disharmony in the Spheres", *The Economist* 386, 8563 (19 January 2008): 23–5.
25 *EC*, p. 158.
26 Herbert York, "Does Strategic Defense breed Offense?", *1986 Lamont Lecture* (Cambridge, MA: Harvard University Press, 1967).
27 Quoted in George Dyson, *Project Orion* (London: Allen Lane, 2002), pp 276–7.
28 Neville Brown, *The Fundamental Issues Study within the British BMD Review* (Oxford: Mansfield College, 1998), p. 71.
29 "Maritime Strategy in Space?", *Naval Review* 91, 4 (November 2003: 204–7.
30 John J. Klein, "Corbett in Orbit", *Naval War College Review* LVII, 1 (Spring 2004): 60–74.
31 Everett C. Dolman in Colin S. Gray and Geoffrey Sloan (eds.), *Geopolitics, Geography and Strategy* (London: Frank Cass, 1999), Chapter 5.
32 Leonard David, "The Clutter Above Us", *Bulletin of the Atomic Scientists* 61, 4 (July/August 2005): 32–7.
33 Jonathan Campbell, "Using Lasers to Remove Orbital Debris", *Aerospace Power Journal* XIV, 4 (Winter 2000): 90–2.

13 | The Strategic Revolution

1 Fragile Globalism

The outreach effected by the West Europeans was extraordinarily rapid and ubiquitous. But the globalization on-going today has stronger impetus and a deeper import. Quite the most dramatic global growth is that in data flow. But that seems more or less impossible to gauge comprehensively. So let us look at the international trade in merchandise which grows at *c.* 15 per cent per year. Roughly 80 per cent of it involves the sea lanes; 90 percent of that will travel by container. In 2005, fifty million containers were landed worldwide. Incidentally, only several per cent were subject to anything like proper inspection in the ports receiving them.[1]

The Credit Crunch

You could well say the world economy was synchronized closely throughout the twentieth century. Even so, it is remarkable how worldwide was the property price boom which very generally peaked in 2005; with the USA and many countries experiencing downturn since. However, neither rapportage nor analysis effected anything like a consensus about whether steadily rising house prices stimulate consumer spending and therefore economic confidence. What few seemed to doubt was that a sharp fall consequent upon too heady a rise could engender a wider deflation. But there was little disposition pre-2008 to see in this the seeds of a banking crisis.

It is pertinent to ask whether a Ricardian/Malthusian principle is at work here, the pressure of people on finite amounts of land causing erratic though often fast inflating real estate prices. An answer can be that house sales at least are considerably a function of institutional factors (e.g. savings to incomes ratios) which do not relate too obviously to anything David Ricardo or Thomas Malthus said. In fact, one study showed house prices measured as a multiple of personal income to lie quite regularly through the nineties between 12 and 18 across the Atlantic region.

Japan had become a great exception. There the said multiple was 67 in 1989 and still 27 a decade later, a reckoning closely related to the high population densities across *kanto heiya* — the coastal plain of Honshu. One consequence was to be that the collapse of an overheated property market caused Japan's recovery from the world recession of 1990–2 to be very drawn out. In 2007–8, however, Japanese property prices have been among the most stable.

In Britain population pressure has lately been mounting strongly across the South-East of England and incipiently in the South-West. This tends to be reflected in a "nimby" ("not in my backyard") reaction to development proposals. A 2007 study found that 35 per cent of Britons were opposed to new building in their neighbourhood as compared with 6 per cent of Americans. That October, the IMF adjudged the British housing market to be 40 per cent overvalued.

Even in the far flung United States, however, the lead factor in the threat of recession looming by November 2007 was an implosion of the housing market, something experts had said could never happen across the nation. Most worrisome was a drying up of new residential building plus a "glut of unsold homes".[2] The limited number of prime sites in leafy suburbia had become overvalued. In both the USA and Britain the banking sector had for some years been allowed, nay encouraged, to extend credit too freely, especially in respect of housing mortgages.

Immigration

Meanwhile immigration is a subject in ascension, certainly in Western Europe. It is no longer too "politically incorrect" to caution that, above some lowish threshold, it could unduly weaken national identity, strain public services and overload the environment.[3] Undue influx could easily compromise the good progress made this last quarter of a century in inter-community relations in most democracies, this despite a concurrent rise in individual self concern.

Badly neglected nowadays is the other side of this coin — the loss of talent by the countries of origin, given that immigrants are often well-endowed with skills. In the sixties and seventies, concern was frequently expressed in social democrat circles in Britain about a "brain drain" from the developing world. Let us recall that in 1914 the explorer Fridtjof Nansen (a Nobel Peace Laureate to be) wrote that, in his native Norway, "where we have had far too much emigration for many years, we are not slow to see the harmful effects of this on the nation".[4]

Metropolitan Nodality

Any review of population geography is bound to address the status of the capital of any state under consideration. Nodality, natural and reinforced, is the crux. Rome acquired historically a widely set nodality, being the centrepiece of an Italian peninsula which splits the Mediterranean ("the sea between lands") into two quite equivalent basins. Then again, Paris and London each stand on a major river which bisects a basin landscape some 300 kilometres across, the common geological origin being Tertiary folding.

Once made evident, being pivotal builds on itself, external economies of scale being realized more and more. In London's case, her resurgence since 1986 as the world's greatest financial centre, this hard upon deregulation, will have been an effect and therefore a further cause. The disposition of young graduates to gravitate to London to be in touch with peers will have been contributive. So, in the final analysis, will proximity to the centre of Earth's land hemisphere with all this connotes for time zoning. None the less, other financial centres will rise as data flows ever more freely the world over.

Intra-national Regionalism

In Britain, as elsewhere, there has been since 1945 a determined effort to spread the national culture through the regions whence, after all, much of it originated. Lately, too, goodly measures of parliamentary self government have been accorded Wales, Scotland and — once again — Northern Ireland. Even so, the great British tradition of a local press has weakened considerably. Newspaper reporting even of local and regional news has been mediated more than ever by London perspectives, priorities and decision taking. Most revealingly, the proportion of the House of Lords member-ship domiciled in London or the South-East has lately risen past 40 per cent.[5] By that measure, this second chamber proffers no check on metropolitan domination, the "Washington outsider" corrective.

Much the same will apply in other spheres concerned with public affairs, art and recreation. A fairly recent government survey shows the South of England, and espe-cially the South East, still to have the advantage over the English North in regard to not a few of, at any rate, the more quantifiable criteria of the quality of life. However, London itself was actually the worst region in the whole country for the incidence of armed robbery as well as of children in low-income households.[6]

The connotations of the hegemony of the South-East can be seen in the studied way the English have applied the term "Establishment" to politics and the like. To Matthew Arnold (1822–88), sometime Professor of Poetry at Oxford, it was a community of singularly high-minded cognescenti essentially located in Oxford and London. Its propensity for keeping at bay Nonconformist "provinciality" was, in the opinion of Arnold, an arch-conservative and philosophic nihilist, our one great hope.[7]

From the middle nineteen fifties, this term was accented afresh, again from within the conservative literati. This time round, the connotations were negative. They were of an intellectually incestuous clique of the "great and the good", stratospherically detached from the broad streams of national sentiment. But again, too, the perception was of overlording from the lower Thames valley, the axis of what Mackinder had called the "metropolitan" quadrant of England.

In federal democracies (e.g. the USA, Germany, Canada and Australia), the national press tends to be headquartered very considerably outside the capital. Another way in which the overweaning strength of a metropolis may be reined back is through the growth (spontaneous and/or officially promoted) of a second city: Alexandria with Cairo; Tel Aviv with Jerusalem; Johannesburg with Cape Town; Mumbai with New Delhi; St Petersburg with Moscow; Shanghai with Beijing; Sydney with Canberra; Osaka with Edo/Tokyo; Montreal with Ottawa; New York with Washington; Glasgow with Edinburgh . . . In 1850, Manchester was widely seen as the counterpoise to London. Today a more compact English scene looks less bifocal.

The Spanish Experience

Madrid is the ultimate example of a city centrally placed within a country strikingly symmetrical in regard to its borders, sea and land. It is currently devolving considerable political power to 17 newly autonomous regions, a consummation sought by political activists, though less by others.

The rationale for this departure is so variegated as to amount to mutual cancellation. Catalonia derives legal and linguistic singularity from having been the Spanish March created by Charlemagne in the ninth century. Andalusia draws on an agrarian radicalism rooted in how the *reconquista* frontier came to be characterized by large ranches — *latifundia*. In the 1931 election, for instance, there was a close correspondence between the *latifundia* area and that of socialist voting.[8] The border of Andalusia also fitted well the template.

Since time immemorial, Basque society has been delineated by a quite distinct language; a high degree of agricultural self-sufficiency; a pride in self-government; and adherence to a highly puritanical Catholicism. Yet even in this epic example of community identity, the binding forces weaken progressively. Not half the people in the four Basque provinces today speak Basque. Furthermore, even the true Basques have been split as between the militants and the moderates; the former have divided between those who have endorsed insurgency and the rest; and the former again between the radical nationalists and the social revolutionaries. Meanwhile, in the three departments of France customarily regarded as Basque, assimilation is well under way.

From late 1936 to early 1939, Madrid experienced military nodality as a forward citadel of Republican resistance to Francisco Franco's fascist forces in the Spanish Civil War. Yet so did it as the seat of a government also struggling to impose centralized war management on the Anarchist deviation in Catalonia. Madrid's seizure by Franco in March 1939 effectively ended hostilities. Shortly beforehand, Franco could boast that he had columns advancing on the city from the four cardinal points of the compass. Within its bounds, he had a "fifth column". This term soon became a byword for cliques plotting internal violence in cahoots with foreign fascism.

Wartime Nodality

One of the grimmest aspects of war is how capitals and other large cities can become all the more harshly iconic on account of their ability to "absorb punishment", to draw fire. The two cities which became under siege especially expressive of Soviet resolve prevailing over Nazi were Leningrad (1941–4) and Stalingrad (1942–3). Paris assumed special significance during the French revolutions of 1789 and 1848 and then during the Franco-Prussian war of 1870–1. Its surrender to German forces in January 1871 after a four-month siege all but sealed France's defeat. Next a leftist Commune of Paris, inspired by the 1792 precedent, formed up to reject a peace settlement and adopt a broad revolutionary agenda. That May, the Commune was brutally crushed by government forces, 17,000 people being executed in the ensuing purge. A legacy was a deepening of the Left versus Right divide within the French body politic, a divide which proved disastrous in 1940.

The geography of its site and of its regional setting have bestowed a military nodality on London since Roman times. From 1300 almost to 2000, a major concern of English/British policy was that the continental coast opposite the Thames estuary must always be denied to unfriendly hands — Spanish, Dutch, French, German, Soviet . . . Come 1914–18, a novel threat from that general direction was posed by mines laid by U-boats negotiating the estuarine shallows on the strong spring tides. Anti-submarine nets and defensive minefields helped keep at least the Black Deep channel open to shipping more or less continually. But overseas trade through the Port of London was still badly affected.[9]

Come World War Two, the Luftwaffe soon launched a mining offensive against London and other seaports. The instruments thereof were magnetic mines laid on the sea bed. Only brilliantly swift "degaussing" innovation saved the situation for Britain. It involved cancelling out the magnetic field of a ship with an electric cable strapped round its hull.

During the Battle of Britain, London's ability to absorb punishment became a crucial factor. By early September, continual bombing of the forward airfields had brought RAF Fighter Command to the brink. Then RAF Bomber Command launched its raid on Berlin. (See 6 (I)). Duly, the Führer turned his attention to London, bombing the city by night or day for — in this first instance — a good ten weeks.

Against the impassioned advice of senior military, Hitler insisted that the V-1 cruise missile (or "flying bomb") offensive against England from France and the Low Countries starting June 1944 be directed exclusively against London rather than the embarkation ports for the Second Front. To which one may add that the 40 per cent interception of the 16,000 V-1s launched owed a lot to the cover afforded London by anti-aircraft emplacements along the North Downs.[10]

Cities Too Soon?

Large urban conglomerations are peculiarly prone to sudden descent into instability when a situation develops in which every negative trend (poverty, undue inward migration, disease, filth, crime . . .) is reinforcing acutely every other one. China must be fairly close to this state of affairs in some of her burgeoning cities. It is reckoned currently to have 16 of the world's 20 most polluted.[11] Its mammoth apartment blocks have about them a cheerlessness sufficient to make Britain's incongruent departures in that direction in the fifties look downright cosy. Much of its historic architecture has been demolished for redevelopment. China looks ill-prepared to cope with a disposal crisis for rubbish already amounting to 400 kg per urban dweller per year.[12] Meanwhile a UN study has estimated that, across China, 400,000 are dying each year from air pollution. As 2008 dawned, Beijing was struggling to bring its levels suffi-ciently under control in time for the Olympics. Tight restraints on intra-city transport enabled it to do this.

Among the metropoli lately cited as being comprehensively "feral" (literally, untamed) are Johannesburg (population 2,900,000) and Mexico City (population 20+ million). In the former, many key buildings in the Central Business District have been abandoned. The police admit to not controlling wide areas. At a conservatively estimated 170 per 100,000 per year, the murder rate is horrific. River pollution is dreadful. So is the incidence of AIDS.

Meanwhile, Mexico City's "air is so polluted that it is routinely rated . . . unfit to breathe". Over 15 million assaults are recorded every year. Both the police force and the judiciary are seriously corrupt[13], a reality that no doubt reflects the situation nation-wide. The population total just cited compares with one million in 1940.

New York Turns Round

Nevertheless, it has proved possible to turn city decline around, especially when a motivated individual uses their mayoral or similar authority in propitious circumstances. The rebuilding of devastated cities after World War Two was certainly encouraging in this respect. But there have also been recovery sagas in what one can term peacetime situations proper.

New York is a renowned example. Between 1945 and 1990, the murder rate multiplied sixfold. Correspondingly, from 1970 to 1990 the recorded incidence of assaults and thefts climbed steeply, largely because of a burgeoning drug culture but also as a collateral to financial corruption in higher places.[14] Many New York public schools had become bog standard with much violence and terrible malaise.

Then from 1993, Mayor Rudolph Giuliani built to effect on a backlash against all this. The most renowned aspect of his approach was the "broken windows" philosophy of cracking down on minor offences to encourage a culture of lawfulness. His successor from 2002, Michael Bloomberg, was to carry things forward. As of 2007 recorded crime is down 25 per cent since 2001 and 75 per cent since 1992. Especially effective has been Operation Impact, the short-term flooding with police officers of a troubled district. The parallel with the military's 2007 "surge" stratagem in Baghdad is obvious.

In December 2006, Mayor Bloomberg introduced a 25-year plan intended to extend considerably the work already in hand. His objectives include a massive increase in affordable housing; the cleanest air of any American city; a 30 per cent reduction by 2050 in greenhouse emissions; 90 per cent of the river and bay frontages made suitable for recreation; and the sustaining of big improvements in infrastructure, not least school buildings. Then in November 2007, he launched a new education strategy intended to consolidate impressive gains lately made in public school test-scores and graduation levels. Its main thrust is making schools more accountable and subject to a "stick or carrot" regime. High performing establishments will get more money. Low performing ones will be penalized and ultimately closed.

Such transformations will often be harder to achieve in developing world cities, not least because of high rates percentagewise of net immigration from rural areas. However, Michael Bloomberg tells how "a new urban global community is emerging in which cities are collaborating". New York itself is, he says, learning much from Hong Kong, Shanghai, New Delhi and Bogota about rapid transit systems. Containing global warming is likewise becoming a common urban theme. One just hopes the Kiberas of this world can benefit from such sharing of ideas. (See 14 (2)).

Internetted Conflation

Half a century ago, an American sociologist coined the term "rurbanization" to signify a melding of lifestyles between town and country. Since when, the process has progressed considerably further the world over. However, it has by no means completely suppressed the customary dialectic between the two contrasting domains. Both patterns can be read into internet impact as interpreted by Manuel Castells at Oxford in the 2000 Clarendon Lectures on Management.[15]

He refuted assertions that the Internet Age meant "the end of geography" (p. 207) though it does recognize the need to redefine distance as "new territorial configurations emerge from the simultaneous processes of spatial concentration, decentralization and connection" (p. 207). After all, it involves "an unprecedented combination of . . . co-ordinated decision-making and decentralised execution, of individualised expression and global horizontal communication" (p. 2). But "overall, there is a strong correlation between metropolitan dominance and early adoption of Internet use" (p. 212). Indeed, the Internet is "the technological medium that allows metropolitan concentration and global networking to proceed simultaneously" (p. 225).

However, the Internet revolution is characterized, thus far at least, by a pronounced "digital divide" between those who have access to the Internet and those who do not. The former were the more likely to be urban , young and quite well off. Africa has been extremely backward in this regard. Nevertheless, Castells did see the Internet as "a fundamental instrument" for development across what he curiously still knew by that dated Khruschevian term, "Third World" (p. 5). Yet even if so, one has still to ask whether the countries thus benefiting will therefore be more likely to develop along democratic as opposed to autocratic lines.

Thus far, the signs are mixed. The accelerated weakening of traditional culture can have a two-edged effect. Highly ambiguous, too, in terms of democratic connotations is the fact that of the one-and-a-half billion Internet users currently extant, by far the biggest share is the Chinese. (See 15 (2)). Elsewhere, however, democratic mass movements have already been facilitated, notably in the Ukraine and the Philippines. Nevertheless, one has encountered the military view that "cyber-mobilization" has made it harder for the peacebuilding to prevail in the circumstances prevailing in Iraq.[16]

Then there is the question of how menacing to social cohesion and rationality may be the readiness with which the Internet enables people virtually to create for themselves an alternative identity — an avatur. A leading British commentator on current affairs has articulated anxieties many of us feel. A young father struggling "to make ends meet becomes a powerful financier with his own tropical island". The "misfit with no friends becomes a hunky sex god". We could be heading for a dystopian future in which vast numbers of people who cannot "quite cope with the demands of the real world escape into an artificial one and become increasingly detached from their responsibilities".[17] Not a few might be susceptible to the blandishments of fanatic cells of one kind or another.

An aspect of the cybernetics revolution of still more direct interest to students of international security is "cyberwar" waged as an alternative to lethal engagement or else to complement it. The most remarked example to date has been the offensive against Estonia (apparently by Russia) from April 2007, hard upon the authorities' removing from the centre of Talinn a monument honouring Soviet soldiers of the Second World War. Heavy cyber reliance had made this Baltic state vulnerable. "E-stonia" said some.

The attacks came in waves over several weeks, the computers deployed having protocol addresses the world over. The targets were mainly governmental but included websites for newspapers, internet banks and so on. This time the Estonians managed to keep all sites going by agile manipulation of access intervals.[18] But their adversaries were probably making a less than maximum effort, very cost effectively.

Background Factors

Many criteria have been advanced for determining what the Classical Athenians came to call "the efficiency" of a citizen's life style.[19] How far did it achieve whatever might be deemed sound social and philosophic aims? These days there is a strong disposition to quantify observations or predictions concerning this kind of thing, a disposition which leads on ineluctably to cartographical representation. Among the less tangible attributes thus addressed are economic freedom, economic competitiveness, environmental change and corruption. See, too, the most intricate map of "tranquillity" differentials produced by the Campaign to Protect Rural England in 2006.[20]

In reality, big problems can arise when such compilations, cartographic or otherwise, are used for purposes of international comparison. These mainly concern evidence and definition, and can lead to absurd conclusions. Witness tables of recorded crime which ostensibly show peaceable Finland and New Zealand to be exceptionally delinquent![21]

Murder rates are usually based on reasonably solid fact. However, their impact across society can be too piecemeal to admit of using them to signify social cohesion or breakdown. One should bear in mind, too, that levels of violent crime can, to quite an extent, be a function of what proportion of a population are males aged between 15 and 30. Also, reporting rates for delinquency can vary markedly.

Corruption can be even harder to quantify or even define. None the less in 1996, against a background of rising popular concern more or less worldwide, the World Bank abruptly ended its studied indifference to the topic. Since when, it has been collecting a plethora of data.

After ten years, alas, it felt unable to report real progress. Most countries had changed little in this regard. Some, notably Indonesia, had progressed. Others — including China, South Korea, Vietnam, the Philippines and Malaysia — were deemed to have deteriorated.[22] We needs become better able, as regards information and methodology, to make international comparisons and ascertain trends.[23]

Meantime, redundancy linked to IT and globalization has thus far been coming through not as geographical pockets of high unemployment but in terms of individuals losing their jobs, maybe to someone in the same firm but thousands of miles away. In Europe, the most characteristic result of displacement is taken to be joblessness. In the USA, it is retrained individuals accepting jobs at lower rates. This contrast is not unconnected with a greater American preparedness still to identify certain positives as attaching to economic recession.[24] A legacy of the moving frontier?

A more universal tendency has been for salary gaps between the better qualified workers and the rest to widen, partly because the former now find it much easier to relocate abroad. A further trend potentially productive of social stress is for labour as a whole to lose out against what classically have been identified as the other factors of production — land, capital and enterprise. It loses out against land for the Ricardian reason that land remains immobile with increments hard to come by. Ironically, it loses out against capital because that is now so mobile. If you are the entrepreneur "you don't have to move manufacturing, you just have to be able to threaten to do so".[25] And, of course, enterprise as such is translatable in the time it takes to make a phone call.

Income Divergence Internationally

Economic relations between the developed and the developing world always command attention. A key issue of late has been whether the latter is at last closing with the former as regards annual income per head. In 2006, the world's Gross Domestic Product (GDP) rose 4 to 5 per cent. In the main developed regions (the USA, the EU, Japan) the mean rate percentagewise was three or just over. But developing and transitional economies rose averagely 7 per cent.[26]

However, such comparison requires careful interpretation. Most obviously, those proportional increments have been measured in relation to widely varying GDP baselines. Nor do the above calculations take any account of differential population growth. What should also be borne in mind is that, in the developing world, many and perhaps most workers are part or full-time members of an "informal economy", much of which is not too dreadfully illicit but all of which is effectively beyond governmental ken.[27] Besides all of which, economic calculations made for the developing world very consistently look less favourable if China is excluded.

Still, it should be possible to maintain this next decade the progress lately made with lifting people out of primary poverty, defined as below two dollars per head per day (1995 US dollars). According to the World Bank, the percentage dropped from 24 to 20 through the nineties. Unfortunately, it is in the cities of the developing world that one finds the biggest gaps between expectations aroused by promotional drives and what the local facilities can proffer.

Alienation

Adam Smith (1723–90) was one of several leaders of Scotland's vigorous Enlightenment destined to exercise a lasting influence worldwide. He did this by dint of being accepted as the paramount prophet of industrial capitalism. Yet in his *magnum opus* of 1770, *The Wealth of Nations*, he is starkly admonitory about the numbing mental effects of the division of labour — the principle guiding the new factory system: "The man whose whole life is spent in performing a few simple operations of which the effects, too, are perhaps always the same or very nearly the same, has no occasion to exert his understanding. He naturally loses, therefore, the habit of such exertion; and generally becomes as stupid and ignorant as it is possible for a human creature to become."

Explicitly following in Smith's footsteps, Karl Marx (1818–83) also averred that factory labour was causing, to an unconscionable degree, the alienation of working people from their labour and the products thereof. At the same time, they were becoming detached from local communities and also from Nature. As the Industrial Revolution progressed, many other commentators argued similarly, though often placing special emphasis on the erosion of Nature and/or Community rather than on factory routines.

In those days, contributions which focussed primarily on Nature routinely disparaged Science as a blinkered pursuit leading us whither it neither knew nor cared. Witness the acrid comment Samuel Coleridge famously made in 1801: "I believe the souls of 500 Sir Isaac Newtons would go to the making up of a Shakespeare or a Milton."

Happiness or Fulfilment?

Over time, a large volume of literature has been generated on alienation and "root-lessness" or, if one prefers, *ennui*, a weariness due to a lack of interests and goals. Recently such concern has strongly revived though this time the focal term is "happiness".

The crunch issue has to be this. Why is it that the proportion of people in a given society claiming to be "happy" follows quite a steep linear progression in relation to average income up to 15,000 dollars per head per year (1995 US$; Purchasing Power Parity) by which stage a 75 to 80 per cent score is generally registered. Then a "knee of the curve" effect takes over with the happiness level claimed rising much more slowly in relation to income.[28] Latin America does well overall. None the less, it is surprising that strife-riven Colombia comes out almost on top. Maybe this result is overly expressive of a good feeling about the fall in urban crime levels there of late. Japan's relatively low rating is also debatable.

No doubt much could be explained in general on the grounds that any society will include a minor fraction who cannot readily be happy because of mental make-up, physique, life history or whatever. Hard to account for, however, is how the proportion insisting they are "very happy" seems not to rise with affluence. Indeed, American evidence indicates that the proportion rating themselves thus actually fell from 37 to 32 between 1955 and 2000.

Part of an explanation should probably be basic human ambivalence about progress towards untrammelled comfort and security, a contradiction not unrelated to our inheritance of a contingent tribal aggressiveness which looks to survive through robust endeavour. It is a trait George Orwell brilliantly explores.[29] All the same, it would beggar belief to deny that some of the likely concomitants of rising prosperity are welcomed with little equivocation. Longer life expectancy. Less clinical discomfort. Less sheer ignorance and superstition. More leisure. More leisure pursuits. More travel opportunities. More toothsome and hopefully more healthy food. Less water pollution . . .

Therefore if the net impact of economic growth on declared well-being soon proves limited or even negative, this must be because other factors contributory to the good life are receding. And surely these will mainly relate to a weakening sense of sustaining one's self within a close-knit community and in balance with ambient Nature.

These two considerations may be closely interwoven. Community bonding is often underpinned by a shared history, perhaps as reflected in common religious belief. So may it be by a distinctive and inspirational geographic background. Yet commentators vary in the store they set by such linkage. Richard Layard, in particular, seems careless of it.[30] Manifestly, however, the erosion of Nature is being felt more in our day then at the time of Smith or Marx. On the other hand, certain aesthetic positives weigh in more. The work place in countries like Britain is often more congenial than was the case even half a century ago.

What ought to occasion more concern than it yet does, none the less, is whether the availability of really challenging jobs might not be matching ever less adequately the numbers of young people and others with the aptitudes and aspirations to take them on. Do we not have to admit that, while Information Technology as a whole can be a great equalizer as regards data flow, it will become throughout a centralizer in terms of decision taking. At present, this tendency is being masked by a progressive shift of labour from primary production to the service trades which still involve a myriad of small businesses. However, in this sector, too, the trend is towards bigger units. Put this consideration alongside several others (including the decline of community and of wild Nature) and one begins to apprehend that, in the future, the great struggle for individuals and, indeed, for humankind is likely to be not for happiness seen in terms of self-indulgent consumerism. It will be for role satisfaction or self-fulfilment, for resistance to *ennui*. Frustration in this regard might all too easily (*à la* 1914) be channelled into violence against "the Other", however defined.

Simone Weil

A perspective which only too well anticipated modern rootlessness was afforded by Simone Weil (1909–43), the Leftist traditionalist whom T. S. Eliot spoke of as "three things in the highest degree: French, Jewish and Christian". Her early death in wartime London curtailed a struggle against fascism which had been heterodox but valiant — physically, morally and intellectually. The part of her posthumously published essay on rootlessness which addresses its incidence rurally makes especially arresting reading, seeing that the French countryside had customarily been seen as a solid bedrock of Catholic conservatism.[31]

She suggested that the supposed benefits modern society bestowed on the peasantry "resemble the one conferred by the financier on the shoemaker. Nothing in the world can make up for the loss of joy in one's work" (p. 781). To country dwellers, the intellectuals who championed the cause of urban workers appeared as "defenders of a privileged class" (p. 83). For in regard to everything "connected with things of the mind, the peasants have been brutally uprooted by conditions in the modern world" (p. 87). Furthermore, those young men engaged in agriculture nowadays find their "physical energy is overflowing and far in excess of what is required for the work" (p. 83).

Weil warned us, too, that if "children are brought up not to think about God, they will become Fascist or Communist for want of something to which to give themselves" (p. 91). The mission the present age must embark on has to be "the creation of a civilisation founded on the spiritual nature of work" (p. 96).

Then in a fashion strikingly reminiscent of her contemporary Eric Fromm, she senses that when "men are offered the choice between guns and butter, although they prefer butter so very much more than guns, a mysterious fatality compels them, in spite of themselves, to choose guns. There is not enough poetry about butter" (p. 96), as a rule. Study of our evolutionary history, especially since the Miocene, may be particularly helpful resolving that mystery, not too pessimistically.

Sweet Reasonableness?

In any case, many people would agree that mass society needs more poetry, more spirituality of some kind, to help it get through the multifarious world crisis now unfolding. Yet among what are sometimes dismissed as "the chattering classes" (mainly meaning those in academe, politics and the media), the concern primarily expressed is about curbing a flight from reason. The Greens and their opponents are especially keen to indict each other for it.

A recent King's College London study argues that economic and social insecurity is resulting in a constant rise in the incidence of paranoia in modern society. It tends to give rise to irrational obsession with the threat from terrorists, paedophiles or whatever rather than a balanced overview of the challenges individuals face. It is adjudged to be twice as prevalent in urban as in rural areas.[32]

So casting our purview widely, it is timely to ask whether all the ramifying tensions might not, indeed, be driving out reason in favour of brutal assertion. Very consistently, the nastiest and most dictatorial regimes in history — the most aggressive, selfish and cruel — have been characterized by wilful disdain of Reason and Reasonableness. But how might these virtues be judged? Could it ever be reasonable, for instance, to follow *à l'outrance* the logic of nuclear deterrence?

Is it not important ultimately to seek overviews which incorporate various logical exercises without being taken over by any one of them? In other words, should we not look more towards Holism: "the view that wholes have some priority over the elements, members, individuals or parts composing them".[33] A standard criticism of received wisdom by the Greens and other elements in the Counter-Culture has been that the assessments made are insufficiently holistic. Essentially the same critique was levelled by anti-industrial Romantics.

But how does one set limits to Holism? And how prone is it not to test properly its factual base and the inferences therefrom? In the final analysis must we not judge Reasonableness as a whole by the hippopotamus test? We find it hard to define but know it when we see it. Is it not the *alter ego* of Respectability, the supreme attribute of pre-1914 Liberal England as quizzically identified by Dangerfield. (See I (3)).

In May 1942, the leftist Henry Wallace (US Vice President, 1940–4) averred that "The century on which we are now entering can and must be the century of the Common Man". Since when, scores of millions of men, women and children have needlessly died in warfare or through its consequences. Torture has been resorted to extensively and ethnic cleansing massively. Several slave trades still operate. Among other barbaric survivals more locally is female circumcision. A British charity believes nearly 90 per cent of girls are subject to it still in Ethiopia, basically a Coptic Christian country.[34] Our cruelty towards other animal species is augean. Mercifully this *résumé* is far from being a balanced account of where humankind stands in 2008. But it does highlight a menacing side to it, a bitter disdain by many for holistic reasonableness.

Involvement with astrology and/or UFOs may be harmless in itself yet still bespeak disengagement. Purblind cults of political personality are, of course, downright alarming. The same goes for the proliferation of religious cults, these often with political agendae to boot. What has always to be asked is how far rank-and-file support for such departures is driven by a sense of not being rooted in a natural or social ambience one can identify with or else being oppressed, marginalized or — worst of all — simply ignored by those with influence and power.

Millenarianism is a radical articulation of this syndrome. Considered especially characteristic of Judaeo-Christianity and Islam, it has found expression in stressful times in a diversity of religious or political formats. An enveloping vision of the future is identified with an iconic leader lately arrived or soon to be expected. He tells of the imminence of History's last great struggle, a chasm of crisis which must and will be traversed to reach the celestial city, a millennium of peace. Communism has had a strongly millenarian flavour.

Democratic Response

Adam Smith and Karl Marx respectively conjured up conceptually simple solutions to the alienation problem as they perceived it. Smith placed his trust in education and civic action. Yet one has only to reflect on how literate and cultured nations like Germany and Japan went fascist in the "inter-war" period to realize that such remedies can be of little avail when a modern mass society is too riven with other discordances. For Marx, the answer lay, as so well we know, in socialist transformation via revolutionary violence. Like so many revolutionaries before and after, however, he gave far too little thought as to how things might progress after overthrow had been effected.

Simone Weil perceived the way through as operating at two levels. There must be, she strongly felt, an ambience of spirituality related to physical labour and emphatically not defined in sectarian terms. There would also need to be a whole clutch of particular mundane reforms. In principle, these two approaches could together lend themselves well to adoption throughout developing democracies.

If so, they could underpin the most basic case for democracy in the new world situation — that the information explosion is driving all polities to become progressively more open or else ever more closed and repressive. Very few people in their heart of hearts want the latter end-game unless they feel acutely menaced or otherwise set at nought. What also is for sure is that, unless democracy can address the problems of modern society on a broad front, reformist endeavours in particular communities will simply be overwhelmed. We shall then be poised to slide into the politics of negation, in one form or another, as soon as some crisis breaks.

International Regionalism

An overarching consideration has to be how far any restructuring of international political institutions can help a better world to emerge. Without begging questions about what the future holds, one can say straightaway that the UN Security Council has never looked more contributive. Regionally, too, multinational political institutions have waxed prominent. However, they lack consistency in scope and cohesion. For reasons rooted in past History and on-going Geography, the world is hard to divide into regions of broadly equivalent sizes and with similar principles of governance. Yet progress in that direction may be important for monetary management, economic development, stable democracy, cultural heritage, academic excellence, peacekeeping, arms control, and ecological preservation.

Perhaps the most remarked of such groupings is the Association of South-East Asian Nations (ASEAN) which was founded in 1967 to enhance stability in the light of the parlous situation in Vietnam. Soon after, Indonesia was admitted in recognition of its emergence from the Sukarno era. Unfortunately, however, this saddled ASEAN with a population spread extending, as of today, from Indonesia's 230 million to Singapore's four million. This must be one of the factors militating against functional integration, ethnic cleavages and pronounced income differentials being among the others.

In fact, ASEAN has largely confined its remit to collaboration in economic and cultural development. It has remained reluctant to get much involved in the internal politics of any of its members (now ten in number).[35] Though in 2007 it was highly critical of the suppression of the democracy movement by the military regime in Myanmar (a member since 1997), expulsion seems not to have been threatened. In 2008, the emphasis has been on urging the Rangoon generals to react more positively to international aid offers in the wake of the cyclone disaster. It is pertinent to recall that Myanmar's accession to ASEAN was on the basis of vague promises about liberalization.

A New World Order?

Some years ago a concerted attempt was made to widen the Permanent Membership of the UN Security Council beyond the original five: the USA, the USSR (now Russia), China, France and Britain. Germany, India and Japan were the most favoured in this regard.

Maybe one should bear in mind, however, a sombre conjunction between technology and politics. These five permanent members also fully comprise the membership not just of a nuclear club but of the thermonuclear one — i.e. the states which have mastered the peculiarly difficult task of developing not just fission bombs but the fusion ones — alias hydrogen bombs. **(See Appendix B)**. Permanent membership of the Council can therefore be read as accommodating the pronounced extension of deadliness thermonuclear warheads represent. This may do something, in fact, to sap enthusiasm for fission devices. At all events, it is a distinction we have rather lost sight of since the days of Robert Oppenheimer. **(See 13 (2))**.

In November 2003, the then UN Secretary General, Kofi Annan, recruited a High Level Panel to produce "new ideas about the kind of policies and institutions required for the United Nations to be effective in the 21st century". Reporting in December 2004, the Panel organized into six groups the potential threats to any state and therefore every state. These were war between states; violence within states; poverty, infectious disease and environmental degradation; nuclear, radiological, chemical and biological weapons; terrorism; and transnational organized crime. It acknowledged that forceful intervention might be necessary where prevention had failed; formally endorsed the burgeoning consensus that the international community had a responsibility to protect citizens when their national authorities were oppressing them; explored how peacekeeping might be improved; and stressed the cardinal importance of post-conflict peacebuilding.

What also is demonstrably needed is a solid corpus of international legislation to govern the development of Space activities. Probably, indeed, the UN should establish its own Space authority. Manifestly, too, there is a need for a more comprehensive legal regime for the oceans and sea floor.

Worthy of special consideration, also, is how to address the diversity of ecological threats faced by the north polar region, somewhere subject historically and prehistorically to climate vicissitudes that have bitten into the human psyche to an extent reflected in many works of literature. **(See 14 (2))**. Speaking as Soviet President in October 1987, Mikhail Gorbachev invited all polar countries to collaborate through treaty to manage the basin environmentally.[36] Twenty years later, however, Russian submarines dropped their national flag spot onto the North Pole. This was to underline a national claim to the sea floor beyond their shores with its huge oil and gas reserves. More specifically, it was complementary to endeavours to demonstrate geologically that the Lomonosov submarine ridge was Russian. Contrariwise,

Denmark sought to portray it as an extension of Greenland.[37] It may bear 10 billion tons of fluid hydrocarbons.

2 The Tactical Revolution

The geography of future armed conflict is harder to predict than before. A presumption one might hazard is that over several decades it will become either considerably higher or significantly lower than now. If the latter, a question afterwards could be how much this owed either to deterrence or else to arms control.

At all events, a great sea-change in military science *per se* took place not in 2001 but two decades earlier, the time when a veritable quantum jump got under way in the application of electronics to observation, guidance and automation, a jump closely associated with the progressive displacement of analogue by digital. The Twin Towers cum Pentagon outrage is better seen as a savage recourse to an old stratagem, a clandestine selective strike across a great distance by a government or a guerrilla movement or the two in collusion — a move probably born out of desperation.

Targeting the Twin Towers was not entirely unforeseen. An aerial contingency had been discussed within the site security staff. More remarkably, in 1915 somebody called J. Barnard Walker had brought out in New York and London *America Fallen. The Sequel to the European War.* One illustration depicted a skyscraper closely resembling the Empire State Building soon to be. Under air attack, its top crashes into a second such edifice. The caption luridly declaims that "in the cataclysm of the plunging towers, New York must surrender or perish".[38]

Action–Reaction

In 2001, neither New York nor the USA did either. A fortnight later, we as a couple went on a pre-arranged trip to Boston and then to Orange County California — iconic to the hard Right. In both locales the mood was resolute yet restrained.

America as a whole initially reacted with circumspection to "9/11". It was not cowed into passivity nor goaded into overreaction. Particularly welcome was the absence of any systemic backlash against Moslems. For this, some credit accrues to George W. Bush. Two days after the bombings he warmly commended the "thousands of Arab-Americans . . . in New York city who love the flag". Nevertheless, a clutch of security measures were to be introduced nationally, two of which — Guantanamo and rendition — have proved extremely vexatious. Admittedly, too, the decision to extend punitive retaliation to Iraq was ill thought through and guided at the outset by lousy if not prejudicial intelligence.

Still, the on-going debate has remained robust but calm. That compares well with how the outbreak in 1950 of the Korean war triggered a menacing resurgence of the otherwise faltering campaign, stridently led by the mendacious Senator Joseph McCarthy, against "Communists" in American public life. It was specious and profoundly illiberal. Even so, McCarthy was a force to reckon with until 1954.

It is instructive, too, speculatively to draw comparison with the aftermath of President Kennedy's assassination. A case has been made, which *prima facie* looks persuasive, for saying this was perpetrated at Fidel Castro's behest in retaliation for repeated exile attempts — made with encouragement from Jack and Bobby Kennedy

— to kill him. The Johnson administration is alleged to have withheld the evidence lest it trigger a Rightist backlash enough to keep the Democrats out of office a generation.

Likewise in Britain the London bombings of July 2005 induced no frenetic backlash. A Commons vote that November limited the time "terrorist" suspects could be detained without charge to 28 days, not the 90 the Prime Minister confusedly sought. This reaction compared favourably with, for example, that in 1947 when in Palestine two British army sergeants held by Irgun insurgents were gruesomely killed in what the Irgun claimed was retaliation for three executions for "terrorism" being carried out. Angry mobs up to several hundred strong took to the streets in a number of British cities. Not a few Jewish shops were ransacked.[39]

The Millennium Effect

Undeniably there is something awesome about the verve with which Al Qaeda staged so complex a clandestine operation as 9/11. Can one discern here a millenarian disposition, given the advent of the third Christian millennium? For a century and more, historians have debated how focal the year 1000 was for unrest within Christendom. One view has been that, as the tenth century drew to a close, the authorities often suppressed millenarian news items lest these engendered a general uprising.[40]

This time round, awareness also figured in a widespread apprehension that computer software not specifically programmed through 2000 would cause global chaos. In fact, the new century dawned with no such "millennium bug". Apparently the USA expended $43 billion in order to ensure this. Russia spent $100 million but likewise had no difficulty.[41]

Technology Differentials

Thirty years ago there was a flurry of concern about the implications for democracies under arms of emergent technical change. Anxiety was expressed about automated launch on warning. So was there lest those managing a "push button" war were either overwhelmed or desensitized; and whether military intervention would come to appear too available an option.[42] Yet a free-ranging debate about such matters never materialized.

Asymmetry

A concept that has entered the frame prominently this last decade is "asymmetric warfare". What this might mean was expressed by the US Joint Strategy Review of 1999. In the course thereof, adversaries attempt "to circumvent or undermine US strength . . . using methods that differ significantly from the United States' expected mode of operations". Technology gaps are a root consideration here.

Customarily a sigmoid or "extended S" curve has been taken to express how "specific maturing technologies" unfold. Within the given sector, performance improvement will be slow for quite a while, then become ever more rapid. In due course, it will slow right down as maturity is attained. The acceleration and then the retardation are liable to onset quite suddenly. The former will widen awhile the technical quality gaps between armed forces whereas the latter may reduce them to vanishing point.[43]

Valid though this schema may often be specifically, however, it is overridden by the "electronic battlefield" diversifying generically. While existing gaps close, new ones appear. Computer power per unit volume may not be increasing as exponentially as once it was. Even so, it still affords ever more varied and refined options.

Broadly speaking, countries like China, India, Japan and Russia might lag but a decade behind the USA and other advanced Western nations in indigenous systems development. However, this does not mean they can catch up any time soon. On the other hand, a system does not always have to be better than its equivalent on the other side. Often it simply has to be good enough to perform allotted tasks. Another caveat to enter is that biowarfare ranks as an exception in all such discussion.

Unmanned in Aerospace

Arousing special interest at the moment (in armies, navies and air forces) is the progressive application of Unmanned Aerial Vehicles (UAVs) and orbital satellites to target designation, strike, reconnaissance, eavesdropping, jamming and deception, plus wireless relay. In the Lebanon War of 1982, UAVs designed and built in Israel, with US technology input, spearheaded the air campaign in which Syria's anti-aircraft defences in the Bekaa valley were broken apart and 80 Syrian jet fighters shot down without an Israeli loss. Deception was a considerable part of the UAV contribution.

Most remarked at present, by dint of service in Afghanistan and Iraq, is the USAF's 10,000 lb Predator UAV which can remain airborne a day or two. In Iraq in 2006, Predators flew 50,000 hours and fired in anger 200 Hellfire missiles. In Afghanistan they are regularly being used to strike across the Pakistani border with the aim of killing Al Qaeda activists. One or two signal successes have been registered.

Now researchers at Harvard are developing a UAV with a three-centimetre wingspan. A case can compellingly be made for everywhere keeping all such devices strictly in government hands and subject to a tight regulatory regime. Considerable interest is being shown, too, certainly in the USA, in the Near Space deployment (for surveillance and communications) of quite or very small satellites. In this context, "quite" refers to nanosatellites weighing less than ten kilograms and "very" to pico-satellites weighing less than one. May one add that any nation attaining such virtuosity should have little difficulty developing automated Armoured Fighting Vehicles or robotic infanteers.

Command and Control

It will be important to evolve command procedures as apposite as possible for all likely circumstances, local geography included. Their importance will lie in the need to prevail while minimizing casualties all round, including one's own losses to friendly fire. The scale of these on the allied side in World War Two has been put, in one American study, at 15 per cent of the total. In the 1991 Gulf War, 17 per cent of American casualties were down to friendly fire while today British planners allow for a one in ten ratio.[44] What must be admitted is that, whatever the merits of the Internal Arms Control philosophy enunciated below, its adoption would complicate command and control that much further.

Civilian workaday experience suggests that, although real benefits accrue from more recourse to electronics in management, the avoidance of petty error is not among them. However, the bigger question *vis-à-vis* any military contingency will be whether the control structure overall is appropriate to the specific threat presented. Here the Strategic Defense Initiative debate is salutary. **(See 12)**.

Addressing this problem as it applies to the tactical revolution must largely be left to the professionals. However, one aspect invites wider discussion. How much background information should flow to troops in the field? The Western liberal disposition may be to hold nothing back. But cannot data overload be inimical to courage up front, even if the day goes well? Troops in action do need things to think about other than personal survival but not things too various and complex.

For the foreseeable future, the core objective in armed conflict will as a very general rule be not territorial conquest nor assets attrition but the paralysis of the opposing regime. Even wars about water or other material needs may be concerned with access not annexation, with compliance not elimination. In the quest for some such outcome, advances in surveillance and precision targeting will be such as to proffer large dividends from getting in the first blow. For this reason alone, we should move decisively beyond declamations about the rising supremacy of the Defence.

Air and Naval Power

In land warfare, the weaker side still uses the intricacies of geography to consolidate its situation. In the skies and on the sea, the stronger is freer to exploit its superiority, the Lanchester Square Law and the rest. In these domains, the constraints imposed by Geography are largely confined to sheer geodetics (area, distance, bearing) plus dependence on a limited number of harbours and airfields. At sea, any continental shelf is pertinent however. And still, of course, the weather can matter throughout.

Having said that, the very notion of air power is liable to be qualified considerably by the Tactical Revolution. All else apart, the status of the air superiority fighter could by 2025 be challenged by combat UAVs — more rugged and compact; cheaper to procure and fly; and demanding far less infrastructure.

A more particular advantage will be an ability to manoeuvre at perhaps 20g continually.[45] The stress on aircrew limits the manned aircraft, forwards or laterally, to 10g. An open question will always be how fully a pilot remotely located in a surface control station could exploit UAV agility. The answer will always depend on a suite of particular circumstances.

Ballistic Missile Defence could also figure, locally and regionally. Visiting Israel in August 1985, the late Edward Teller freely conceded there was no "human way" the country could thwart comprehensively a salvo of, say, a thousand missiles with warheads and decoys.[46] However, this was not to suggest that the interception rates attainable could never be worthwhile. All else apart, ballistic and cruise missiles will continue to pose a particular threat to runways.

A recent test of the manned strike aircraft was its use in 1999 against the Serbian military presence in Kosovo, a fairly well forested landscape comprised of Tertiary folding rising in places to 6000 feet and more. Freed awhile from the need to deploy against ground attack, the well-concealed Serbs survived in remarkably good shape. General Wesley Clark, NATO's Supreme Allied Commander, concluded that what in due course brought Belgrade to accept a negotiated withdrawal was the fear that the air offensive would presage ground action.

In two spheres, manned aircraft could actually come more into their own militarily.

Disaster relief is a requirement that could burgeon in the world into which we are moving. In responding to it, air and maritime assets could be complementary. Clearly aircraft may access directly continental interiors where many disasters, particularly droughts, may occur. Airheads for relief may need to be guarded by paratroops or airlanded troops.

Another possibility perhaps to revisit is the long-range reconnaissance bomber. In 1981, it was well argued that, using the B-1 airframe, it should be feasible to build a Long-Range Combat Aircraft able to carry various warloads, one option being 30 short-range cruise missiles.[47] Cruise missiles may jink and contour-hug. This can make them hard to intercept from above as well as from below. A review undertaken by the United States Navy some 15 years ago did envisage an Arsenal ship which, in one specification, would bear a crew of but 50 yet be fitted with 500 Vertical Launch System (VLS) tubes for cruise missiles, each able to home on a road bridge, a tank or whatever.[48]

During the First Gulf War, the USN fired 288 Tomahawk cruise missiles (range 1500 miles) as the lead part of an offensive against political and military command nodes; chemical facilities; and the Iraqi electrical power grid. In 1996, Edward Luttwak cited a cruise missile counter-offensive from the sea as a stratagem consistent with a yen for "post-heroic" (i.e. casualty free) war waging then understood to be prevalent in the USA.[49] Whether such engagement could always be expected to be casualty free is debatable. But it could offer the maritime West significant possibilities. For one thing, frigates firing cruise missiles as their main armament could be a good way for a small- to medium-size navy to contribute *ad hoc* to coalition task forces. And deployment offshore by ships and/or submarines remains a good posture for escalation control.

The age of missilry will make it both needful and possible for the average size of surface naval vessels to come down. Admittedly they cannot so reduce as to be effectively invisible, however supportive the Electronic Counter Measures environment. They might need to be no bigger than a dinghy for that. None the less, a more distributive panoply could make task forces more adaptive instruments of dissuasion. However, in the multilateral naval arms race currently under way across Asian waters, several littoral states are acquiring new carriers for fixed-wing aircraft and/or helicopters.

In principle, it should usually be possible to dispatch disaster relief by sea (in merchantmen and/or adaptive naval vessels) on a scale aviation cannot match.[50] Again, however, it will be necessary always to secure the place of entry which may have to be an open beach or anchorage if ports are denied by industrial unrest or warlike action. Moreover, there may also be a need to distribute supplies forcefully up country. In the Victorian era, the Royal Navy not infrequently deployed in trouble spots ashore *ad hoc* naval brigades. The Crimea, India, Abyssinia, the Sudan and South Africa proffer instances.[51] Might there not often be a need to defend relief programmes? And might not a further requirement be warships able to batten down as well as may be possible against biological threats?

Armies

The set piece "land battle" may be in line for more radical conceptual change than at any time since the herbivorous, peace-loving horse was first roped into active service,

6000 years ago, on the Ukrainian plain[52] — a region of boundless vistas and rich Black Earth grass. The separation initiated soon thereafter between cavalry and infantry today looks obsolescent. Moreover, the inverse relationship between the accuracy of fire and its range is no longer invariable, a change that militates against the premium historically placed on force concentration. Forward areas will therefore be held more lightly. So front lines will lose definition.

As the twentieth century progressed, a raft of enquiries about the future "land battle" boiled down to debate about the Main Battle Tank, the armoured cavalry as from 1916. A consensus now emergent is that MBTs as such will be deployed in smaller numbers while making, too, a quantum jump downward in size, even if manned. Some 20 to 25 tons will be representative, not 50 to 60. Like battleships, tanks have customarily been tailored, first and foremost, to fight their own kind. A progressive widening of both armour plate and gun calibre has been driven by a sectoral arms race. Now build down could similarly be reciprocal.

Arguably, armoured vehicles (tracked or otherwise) may long have been more susceptible to various antidotes (e.g. minefields) than legend acknowledges. But lately they have been decidedly vulnerable as well to new style guided weapons, infantry- or air-delivered. Field artillery, too, is poised to menace them anew with homing shells. Meanwhile, the MBTs salient attribute (namely "direct" or line-of-sight fire at short ranges) is counting for less as "indirect" or beyond-visual-range fire becomes more responsive.

There was, through the 1960s, much interest in "air cavalry", mounted forces for which the helicopter is the principal steed. Yet although helicopters are peerlessly adapted to certain mobility tasks, few would still claim they are robust enough to overwhelm *en masse* prepared defences. Take the Soviet counter-insurgency campaign in Afghanistan. The tide finally turned against it in 1986 with the West's starting to supply the Taliban with man-portable rocket launchers especially intended for use against helicopters.

All in all, the platforms prevailing on 2025 battle grounds are likely to be Infantry Fighting Vehicles in modularized form. These ground vehicles will be the mainstay of a unitary arm, the mounted infantry: a revival of the dragoon concept. No doubt IFV specifications will include some ability to close down against biological as well as chemical threats.

Residing as they do on the interface between infantry and artillery, mortars have tended to be played down by both. They could come more into their own as a means of indirect fire by mounted infantry. Their installation in light armoured vehicles is already quite regular practice.[53]

What is unclear is how big a role computers might play in the overall direction of a land battle in progress. Flows of data, very diversified in character, will be hard to process electronically. But does that mean more effort should be put into such endeavour or less? How much must an answer be shaped by the given geography?

Counter-Insurgency

Almost by definition, nearly all of any counter-insurgency commitment will be undertaken by land forces which means *inter alia* that much attention needs always be given to local factors, geographic and otherwise.

A glaring omission in the open literature has repeatedly been due consideration of the arms and ammunition insurgents acquire. Military balance compilations have paid little or no attention to small arms transfers, licit or otherwise. Moreover, academics were averse to considering small arms inventories. Lack of hard evidence would be the given reason. But it has not been difficult to discern a preference for explaining the origins and outcome of insurgency war in terms of political, economic or social realities or else of charismatic leadership.

Times have lately changed. A 2006 Geneva study indicates exports worldwide (1999–2003) as reported by the UN Comtrade, compilations which derive primarily from customs data. The annual value was around two billion dollars for small arms and light weapons; and 0.6 to 0.7 billion for small arms ammunition, often a critical factor.[54] But what of deliveries that circumvent customs returns?

Colombia still stands out as a country besmitten by internal violence, an endemic state of affairs originally derivative from the break up of the Greater Colombia (including, too, Venezuela, Panama and — in due course — Ecuador) that Simon Bolivar had been instrumental in founding in 1819 to be a building block for the unitary Latin American state he naïvely hoped for.

The population of Colombia is 46 million. Her military say 800,000 unregistered weapons are in circulation in the country. The police reckoning is at least three times higher. The accepted jargon locally is that there are Leftist guerrilla groups and Rightist paramilitary ones. There are also many criminal gangs without defined *leitmotifs* save the celebration of lucre, secrecy and violence. Arms flow in illicitly from all directions, across open sea and jungly land. Those destined for the Right come largely from Central America and the USA; those for the Left considerably from the Middle East and Eastern Europe. Nowadays Venezuela is closely watched. Meanwhile, the "illegal drugs business is to a large extent a mirror image of the gun trafficking business" especially as regards the paramilitaries.[55] Violence in the countryside is largely political; that in the towns (appreciably reduced the last few years) mainly criminal.

The world over, political and criminal violence can readily overlap. Such conflation may be getting stronger but it has a long pedigree. Al Capone (1899–1947) — the terror of Chicago between 1920 and 1931 — was wont to provide gambling dens and alcohol "speak-easies" along with medical supervision of prostitutes and abortionists. To that extent, he offered a social alternative, notwithstanding his masterminding hundreds of murders and his exploitative aims throughout. Obversely, Hezbollah and the Tamil Tigers have been among the insurgent movements deep into illicit drugs dealing. Diamonds have been another insurgent interest, notably in Angola, the Democratic Republic of Congo and Sierra Leone. Underworld togetherness has been uncovered within the Russian federation as between Russian crooks and Islamic radicals.[56]

The penultimate convergence between the two genres is surely to be seen in how the drugs baron, Khun Sa (d. 2007) unilaterally created a rogue state in the "Golden Triangle" northern border zone across Myanmar, Thailand and Laos. At one time he had a well accoutred army 20,000 strong. Moreover he was supplying up to 80 per cent of, for instance, New York's consumption of street heroin. By 1996, however, the production thereof was shifting to Afghanistan; and the several sovereignties affected were tracking his revenues better. So he ceremoniously surrendered to the Rangoon government in exchange for legal immunity. The UN Office on Drugs and Crime (UNODC) in 2007 estimated the global illicit trade in drugs at $320 billion annually; and even that in human trafficking but one order of magnitude less.

War Without End?

It is, in the longer run, vital that counter-insurgency strategies encompass not just the military means to prevail against insurgent "terrorism" but the political and economic reforms antidotal to it. The rub is, however, that reform does not easily override folk memories of events too long ago to be redeemable. Take two Islamic territories respectively located to east and north of the volatile Fertile Crescent.

In British India, the Raj waged war continually against Pathan tribes in what became, in 1901, the North-West Frontier province. In 1897, Winston Churchill, then a subaltern in the 4th Hussars, attached himself to the Malakand Field Force deploying for reprisal out of the Noswhera railhead.

In due course, they passed by the Mamund valley, the warlike inhabitants of which had lately been in a live and let live relationship with Calcutta. Sporadic fighting developed, culminating in a night action in which the army incurred several dozen casualties. Churchill described its reaction as follows: "We proceeded systematically village by village. We destroyed the houses, filled up the wells, blew down the towers, cut down the great shady trees, burned the crops and broke the reservoirs in punitive devastation." He further noted that, in the broader context, our "tendency to road making was regarded by the Pathans with profound distaste". This applied particularly to "the great road to Chitral" south from the Hindu Kush.[57] How far might such folk memories bear on the apparent ability of Osama bin Laden to find sanctuary somewhere in this general area ever since 2002?

Lately, this question has come rather to a head. For Yousaf Raza Gilani, Pakistan's new Prime Minister, has declared himself anxious to abolish the Frontier Crimes Regulation (FCR), the colonial hangover that still allows of collective punishment in the frontier tribal areas. A 2008 poll has shown 39 per cent of the tribesmen want it amended while another 31 per cent want it abolished.

Chechnya can be the other thumbnail case study. After generations of conflict between Moscow and the Chechens, things turned really ugly when, in 1942, Stalin deported the entire people. Supposedly this was to end their alleged collaboration with the Wehrmacht, 600 km away though the latter still was. Scores of thousands died in burning villages or in cattle trucks headed for frozen steppes or when in limbo out there. The survivors returned in the fifties, utterly devastated.

Come the break-up of the USSR, the Chechens did not gain independence. Hostilities ensued but, over several years, Moscow regained effective control. But Chechen mistrust remained implacable. Yeltsin was suspected of using agent provocateurs to commit atrocities which could be blamed on the Chechens. Then, post-9/11, Moscow was seen as concocting a putative Chechen sliding away from their moderate Sufist Islam towards the involuted Wahhabi fanaticism historically associated with the Arabian peninsula.

Late 2006 saw the covert murders in Moscow of the journalist and Chechen sympathizer, Anna Politkovskaya and in London of her friend, Alexander Litvinenko — formerly in the Russian secret service. British intelligence found Litvinenko had died at Moscow's behest. The evidence was compelling.

Currently the residual armed resistance is closely contained within the mountain fastnesses of south Chechnya. Even so, the Chechens *per se* may pose more of a challenge today than in 1991. A clinching argument against separation then was that half the population was of Russian stock. Now those people have largely left.[58]

Meanwhile, the West and its allies can claim a fair measure of success militarily in the post Cold War situation, albeit through the expending of 15 precious years and at a cost in life, treasure and sometimes liberality. Some big insurgent plots have been foiled post-9/11. So have many small ones. Some sizeable peacekeeping operations — e.g. in Northern Ireland, former Yugoslavia, Sierra Leone, East Timor and the Lebanon — have broadly achieved at least their immediate objectives.

The Twin Campaigns

Even so, a number of situations of internal conflict give serious cause for concern — Dharfur, Palestine, Somalia . . . Moreover, the outlook for the larger scale military commitments in Afghanistan and Iraq still hangs in the balance. In Afghanistan, the initial dispatch of relatively light ground forces was backed intensively by air power. Within two months, a Taliban field force initially 40,000 strong slumped from controlling 90 per cent of the country to retaining no urban strongholds, not even Kandahar.

This *volte face* owed much to the Taliban overestimating ludicrously the resilience, in the face of precision air strikes, of the vast spread of fortifications they had built into the Tora Bora, a mountain node — steppic and tundra — rising to 12,000 feet close by the Khyber Pass. Much was also down to the active support of a Northern Alliance of warlords. Instrumental, too, had been how, since before the turn of the millennium, the National Security Council in Washington had well identified Afghanistan as the regional centre of a "lethal cocktail of drugs, weapons, Islamic militancy and terrorism".[59]

In certain fundamental respects, however, the achievement was ill consolidated. With only a quarter of the agricultural land in use in 2002, a fourth consecutive year of drought, it proved impossible to prevent the country's re-emergence as the world leader in opium production, an estimated 90 per cent of the global total in 2006.[60]

On 26 August 2002, *Newsweek* published online an account of how that January a thousand or so Taliban were accorded positive surrender terms by a Northern Alliance warlord but then crammed into cargo containers and left to suffocate to death. Washington accepted that this account was materially correct (plus or minus one or two hundred) but took no overt action. This omission will have encouraged certain warlords to carry on much the same old way, not least as regards sleaze.

As of late 2008, the Taliban is very much a force to reckon with, militarily and otherwise. Its appeal will have been further strengthened by how piecemeal arrangements for the distribution of foreign aid have helped corruption to thrive. On the other hand, free education for girls (two million at school already) was being accepted more readily than might have been feared. It should eventually prove a major bulwark against a Taliban return unreconstituted. The Afghan outlook will also much depend on how Pakistan progresses.

The Coalition liberation of Iraq was accomplished with a cavalier flourish, viewed as a mobility operation. But it derived from a strategic appreciation that was unsound in cardinal respects. Regarding Weapons of Mass Destruction (WMD), Saddam had effected a truly classic ruse. Apparently go to great lengths to conceal an asset you never did have or have no longer. This is what his impeding of the Hans Blix inspectorate was about. A close parallel lies in the pains taken by the Allies in the run-up to D-Day in 1944 to give Berlin the impression that they had in East Anglia an additional "9th Army", the intimation being that the main invasion thrust would come through the Pas de Calais. But in 1944, neither the Germans nor anybody else had the benefit of continual reconnaissance by orbital satellites to check things out.

Had the Security Council proved able to resolve that Hans Blix should be given more time with an expanded inspectorate, Saddam might have been put truly on the spot and world opinion against him consolidated. The upshot might even have been his departure without all the bloodshed and that brief but very damaging vacuum of power hard upon the Coalition victory.

At all events, the absence of synoptic planning for post-conflict restitution was a desperately negative circumstance. An authoritative British assessment has been that Donald Rumsfeld simply "discarded the detailed plans for the post-conflict period prepared by the State Department".[61] Granted, the 2007 surge in American and Iraqi military operations in Baghdad, against both Shi'ite and Sunni insurgents, was deeply impacting. Yet with Iraq still so soft a state and so broken a society awash with firearms, one cannot yet feel all that confident about its future as a democracy in so unstable a region.

The Current Parameters of Conflict

As implied above, the West and its allies look rather stronger today than in 2001, *vis-à-vis* revolutionary violence. But there have been further big terror strikes — Bali, Madrid, London, Islamabad , Mumbai . . . And even if one can see such stratagems as owing something to desperation in the face of such measures as the international pooling of intelligence, they can still be hard to retaliate directly against. In the Bali explosion of October 2002, quite a few of the 190 lives claimed were Australian. Two months later, the then Prime Minister in Canberra, John Howard, threatened to strike at militants outside Australia if they seemed to be preparing a repeat. Indonesia, Malaysia, Thailand and the Philippines all expressed their aggravation.

The need these days to address all aspects of a warlike situation accents the truism that territorial gain can now be a rather mixed blessing. Correspondingly, vehicular mobility will often assume more importance as a way of evading precision attack by a requisite few yards than for advancing 70 miles a day. The continued advocacy of penetration swift and deep owes something to the blitzkrieg's heyday. Yet so does it to a lingering belief that the Roman Empire was torn apart by galloping hordes. Alas, this legend is one which historians have radically revised the last half century.[62] Even the luscious Hungarian plain has been adjudged able to support but 15,000 cavalry in Late Antiquity, allowing for other pasturing requirements.[63]

The case for cautious consolidation can most readily be made where cities are concerned. For offensive or defensive warfare within them, a detailed appreciation of the landscape (natural, artefactual and social) is usually even more imperative than in open country. The same applies to police work and everyday civil administration.

Among the seminal contributions of late from the US Army War College has been one drawing lessons adjudged pertinent to the modern urban problem from an earlier Classical setting — namely, the salutary experience of the Roman army essaying peace enforcement in Germany in the first century AD. Three legions (altogether 20,000 men) were lost in their entirety in the Teutoberger Wald in AD 9. Five years later, another such foray was triumphant. This is put down to the new Roman commander, Germanicus, recognizing that the crux of the problem had been not some mental block against close order combat in dank, shadowy forest. It had been a failure to read aright the political setting.[64]

A current instance of a dangerously imbalanced approach to urban trauma is presented by Mogadishu, the capital of war-torn Somalia. Two-thirds of the population have become refugees in the face of an Ethiopian military take-over, many now living rough. Yet reconstruction aid, international or otherwise, has been largely absent across the country and totally so in Mogadishu itself.[65] Maritime piracy has made things worse, 30 ships being seized in the first nine months of 2008. It would be hard to find a grimmer example of the need for a joined up survival strategy. The June 2008 peace accord just might make room there for this.

Preserving Heritage

A plea will be made below for a comprehensive philosophy of limitations in regard to war waging. A corollary must be to avoid the attrition of cultural heritage. In World War Two, this happened frightfully. Witness the bombing of Japan even granting some invaluable sites — most notably Kyoto and Nara — were scrupulously spared.[66]

Afghanistan and Iraq have become disturbing case studies in this regard, too, the latter being a major fount of early Eurasian culture and the former a cultural crossroads of very similar longevity. In Afghanistan in February 2001, the Taliban ordered the destruction of all Buddhist statues; within weeks televiewers worldwide saw the giant Buddhas of Bamiyan being shelled to pieces. And so it went on. This *kulturkampf* had begun the previous autumn, ostensibly as a response to assertions that an infidel had prayed before a Buddha in Kabul Museum.[67] Soon some Northern Alliance members and moderate Taliban (plus, of course, looters) began to transfer movable artefacts out of the country.[68] So could this precaution have further exacerbated tension? Was the whole business staged by Al Qaeda to cover its 9/11 preparations?

For many years, the Saddam regime in Iraq had remained careless of an obligation to conserve the heritage. However, the 2003 campaign brought a sad situation abruptly to a head. Sixteen thousand objects were looted from the Iraq Museum by outsiders but also insiders; and only 4000 have yet been recovered.[69] Other repositories have been ransacked too. Even worse, unrecorded field sites were ravaged, usually by uninformed diggers who might retrieve in good state only one per cent of what a trained archaeologist would. The fraction returned or regained has been very low all round.

What we are talking about here is the savaging of archaeological treasures of incalculable significance. It is not just that Mesopotamia was, more than anywhere else, the cradle of the ancient civilizations of Europe and South Asia. It is also that the writings on tough imperishable cuneiform tablets have proved amazingly informative. They may enable us "to know far more about the social and intellectual life of a Babylonian mathematician than ever we will about Euclid or Archimedes".[70]

Two precepts must surely be inscribed in stone in the light of these bad experiences and, may one add, to come operationally into line with UNESCO policy on cultural heritage. The one is that material from our distant past anywhere must be treated as the inheritance of all humankind. The other is that forces assigned to active operations have an obligation to protect historic sites against degradation from whatever quarter. The comment reputedly made by Donald Rumsfeld that "in war things happen" was desperately inadequate. Future historians may see archaeological impoverishment as quite the worst upshot of this long Iraq conflict, worse even than the terrible death toll all round.

Internal Arms Control

Nuclear proliferation apart, arms control is currently on the agenda a sight less than it was half a century back. Not a few of us see the non-weaponization of Space as difficult but feasible and eminently desirable. Yet it figures little in intergovernmental discourse save as an agreed objective in the trilateral exchanges Moscow, Beijing and New Delhi have evolved the last several years. Meanwhile, biowarfare still tends simply to be seen as the giddy limit in ghastliness, too awful for inclusion in broader strategic perusals. Were matters that resoluble.

The multipolarity of the contemporary world is quite an obstacle to international arms control. So is the absence (save in Korea) of well-defined fronts since this makes it hard to feed in geography in a measured fashion. Therefore we should more actively explore unilateral initiatives by countries or alliances. Such an approach could be made compatible with the kinds of armed conflict likely to arise whatever way the world scene goes. Multilateral arms control cannot be so adaptive.

Historically, various weapons or actions have been identified as only to be used with specific authorization. A school of thought in medieval times said Greek Fire should never be employed against fellow Christians. Likewise in the Catholic Church there was a disposition early on to outlaw the crossbow because of the jagged wounds it inflicted. Since 1918, poison gas has widely been proscribed. NATO has ruled out since the fifties the use of theatre nuclear weapons without the consent of the White House plus the government of the country to which they belong and the one in which they will be delivered. Similarly, civil police or military engaged in internal security have on countless occasions required high level authorization regarding the coercive measures allowable. International law, too, has been very much concerned with constraints on how force may be applied.

Not that a philosophy of limitations in war is without its difficulties. Non-lethal devices may seem *a priori* more appropriate to internal security but what if they blatantly effect humiliation or suffering. Torture comes into this category. Undeniably free recourse to it will do one's own side terrible moral damage. But to duck the dilemma by denying it ever yields hard information that may urgently be sought can be downright disingenuous. In World War Two, the French forces of the interior — *les Maquis* — reckoned they had two days to reorganize after one of their key people had fallen into Gestapo hands. So a measure for highly exceptional circumstances, in a world generally more fraught than what we yet know? (See 13(4).

The unrestrained use for surveillance of ultra-miniature UAVs could likewise be deeply controversial. With both torture and surveillance, a primary concern should be not to create an open precedent for regimes (allied ones included) less scrupulous than we believe ours in the West still to be. The same goes for such juridical issues as detention without charge.

Lasers and Landmines

Regarding the battle scene, various kinds of bomb or shell might best be subject to internal arms control as here defined. Cluster bombs are surely a case in point. Two other genres merit particular attention. These are lasers and land mines. The former have exercised the International Red Cross a good 20 years. Yet this subject political parties across the West seem never to address at all.

From 1400 to 300 nanometres on the electromagnetic wave spectrum (see Appendix A) is a "window" of penetration through the atmosphere. High power beams transmitted within it can therefore threaten vision acutely, a sombre reality especially applicable to that part of it (800 to 400 nm) our eyes see the world through. Wilful blinding, one should note, has almost never been a stratagem in human conflict. Another concern is that certain lasers mounted high (probably airborne or Space-borne) might sweep wide areas of the Earth's surface, very possibly occasioning multiple fires and other damage.

One obstacle to arms controlling lasers by formal multilateral agreement is their already being used for measuring distances and guiding bombs and shells. Another is the impossibility of neatly distinguishing between beams that are lethal and those which are not. Too much depends on range, time on target, absorption rates and atmospheric clarity. In addition, pulse compression is now routinely facilitating very high power peaks.It seems that about the only hope we have even for the internal arms controlling of terrestrial lasers is an understanding that they will be directed against surface targets only for ranging and precision guidance unless and until a wider remit is authorized from above. Such a proscription would not be easy to enforce. But it would gain some extra credence in the context of an international ban on the weaponization of Space, lethal lasers naturally included.

Land mines likewise lend themselves at best to unhappy compromise. Undeniably land mines or "roadside bombs" are prone to inflict hideous injuries on the lower body, a plain truth seared into the consciousness of all who have been anywhere near such a device exploding. Moreover, an estimated 50 million mines strewn around from past conflicts still kill or maim 10,000 people a year, many of them children. On the other hand, minefields are stabilizing in that they can blunt offensive action. In World War Two, the very flat Western Desert roughly between Alexandria and Benghazi was regarded as a terrain "made for war", meaning open mechanized war. Yet in the year of decision, 1942, first the Axis armies but then their British-led adversaries found their offensive action significantly constrained by deep minefields.

The Ottawa Convention of 1997 represented an attempt to avoid dialogues of the deaf between those who found mines too disgusting to be imperative and those who found them too imperative to be disgusting. The way through lighted on was a general proscription on anti-personnel mines. Once again definition was difficult. In any case, there was never going to be universal compliance. So here, too, the best answer may involve internal arms control, no dissemination of mines except as sanctioned by the highest appropriate authority. Moreover this may be a remedy appropriate to the kind of military crises likely most frequently to arise. Thus for NATO forces to lay mine-fields as part of the drive for peace in former Yugoslavia would clearly have been inappropriate.

Declarations of Conflict?

Since 1945, the Declaration of War, a formal diplomatic procedure that still mattered to some people in 1939, has gone right out of fashion. But could this be unfortunate? In 1964, Herman Kahn was asked what might one do confronting an adversary one felt unable to wage an all-out war on. His reply was "Declare war", thereby reserving all intermediate options. It is a proposal worth reviving in relation to Internal Arms Control. Undertake no intrusive mini-surveillance against a country, for instance, unless you have declared yourself to be in a State of War or, perhaps more appropriately, a State of Conflict with her? All else apart, such a proscription might help prevent technological advance resulting in unbearable levels of international suspicion and anxiety, paranoia if you will.

Thermonuclear No First Use

Contingent exercise of the prerogative of turning a military conflict nuclear need not perforce lead to boundless escalation. It might finally result in fewer casualties than many other episodes the last two centuries or so.

Even so, such decision taking would be a singularly awesome responsibility. After all, the High Explosive expended by all the combatants in World War Two could well be surpassed, in terms of sheer energy release, by a single thermonuclear warhead. To which should contingently be added a whole raft of qualitative nuclear effects, clinical but also ecological. Even well short of a "nuclear winter", nuclear recourse would badly impoverish and distort biogeography.

None the less, a 1985 study by the late Sir Anthony Farrar Hockley and myself opposed any adoption by the Alliance of a strategy of no nuclear first use regardless.[71] Since when, the geopolitical map has changed profoundly. But the general thrust of our argument remains for me every bit as germane. After all, warlike crises threaten to become more frequent and less foreseeable. Even so, one can hardly feel comfortable about how this stance jars with one's general pitch in favour of a free, peaceable and beautiful world. Nor with the attendant divisiveness within one's national body politic.

One move which could make these contradictions less acute could be a return to the distinction between thermonuclear and fission warheads as made doctrinal by Robert Oppenheimer (1904–67), a doyen of American nuclear physics and science management. One does not have to see as well-founded his inference that the USA should not seek to develop thermonuclear warheads since — so he thought — the USSR was unlikely to. Nor, to go on the other tack, does one have to admire the rancid politics surrounding the withdrawal of his security clearance. Without accepting either, one can still feel that the West would lose nothing and might gain much by adopting a policy of "no thermonuclear first use". There is enough of a natural break between the respective spectra of firepower to warrant a distinction being preserved between "thermonuclear" and "fission".

3 Biowarfare

Yet much of what has just been said seems destined quite soon to be subsumed by another military revolution. Awareness will dawn that another big war could well be dominated by microbes. An apposite point of departure for a response strategy might be the unconditional condemnation as "terror" any recourse to them as instruments of war. For such a prospect does seem singularly evil, irrespective of whether those responsible are governments, insurgents or common criminals or any mix thereof. So was it if our ancestors did use disease to eliminate the Neanderthals.

A demonstration of how biowarfare may be facilitated by trends towards globalization is afforded by the Black Death. It erupted in or soon after 1331 from Yunan-Burma or else the Manchurian-Mongolian steppes.[72] In 1347, the contagion reached a Tartar army besieging the Genoese port of Kaffa in the Crimea. The Tartars catapulted some infected corpses over the battlements, vaguely expecting some local advantage to accrue. Alas, the bubonic bacillus (or was it E coli?) duly progressed by sea to Genoa whence it spread around Europe. The lucid French chronicler, Jean Froissart (c. 1337–1410) wrote that "a third of the world died". That reckoning may have been on the low side, judging from the English evidence.[73]

In modern times, offensive germ warfare has been the preserve of governments rather than insurgents. In 1932, the Japanese established Unit 731 near Harbin, Manchuria. By August 1945, several thousand Chinese and Soviet prisoners had been put to death there in pursuance of a biowarfare capability. In 1942, a cocktail of diseases was directed against Chinese villagers in Chekiang causing hundreds of deaths and many long-term afflictions.

In February 1952, first the North Koreans then the Chinese alleged the USAF was resorting to germ warfare in Korea. After Stalin's death in March 1953, Moscow then Beijing dissociated themselves from this charge. What was disconcerting, certainly in Britain, was how ranking Leftist intellectuals had given it credence awhile. Lately, South Africa's well respected Truth and Reconciliation Commission has elicited evidence that, from 1978, some within the Apartheid regime did wage biowarfare against its opponents within the republic and Rhodesia. Over time dozens of people may have been killed or incapacitated.

Quite the most flagrant example yet of a covert programme was the one the USSR proceeded with despite having signed the Biological Weapons Convention of 1972. The defection to Britain in 1989 of Vladimir Pasechnik (1937–2001), a laboratory chief, exposed a Biopreparat network an order of magnitude stronger than Western intelligence suspected. An "archipelago" of 18 major facilities employing 35,000 people was spread across the USSR.[74] Soon this revelation was being elaborated on by other defecting specialists. Also out in the open now is that in 1979 a facility exploded at Sverdlovsk, this being followed by a local outbreak of anthrax.

Perhaps a dozen countries sustain biowarfare programmes at the present time. However, direct surveillance is beset with difficulty. Much or often all of the focussed development work can take place in small nondescript premises usually indistinguishable from high overhead from the rest of an urban sprawl. Quite a lot of the information comes from "blue skies" research. The distinction between offensive and defensive is harder to draw than normally applies in war. Nor can one tell beforehand, unless by espionage or through defections, what kind of microbes will be used.

4 Offensive Supremacy

What has just been said points towards the Offence having commanding advantages over the Defence. So do other considerations. Grim dividends attach to a successful first strike; and quite possibly there will be uncertainty on the part of the Defence as to where a particular threat emanated from or even whether it was Man made.

Nor is it reassuring to ponder the biological realities. Among them is how advanced life forms find it veritably impossible to keep abreast of microbes in terms of evolutionary adaptation to unfolding situations. Typically the latter reproduce by binary fission every quarter of an hour or so. They therefore go through a million generations in the time humankind takes to go through one. Gaining naturally a particular immunity is, for a species as for an individual, very much a matter of challenge and response, defence on a learning curve. A vaccine is an attenuated form of a disease which serves to "teach" an involuntary host how to cope.

Lately progress has been made with producing vaccines which expose more the inner cores of microbic cells responsible for such diseases as influenza and cancer. These will be harder to negate by the germs in question mutating spontaneously or through human manipulation. But it will be totally impracticable, physiologically and logistically, to vaccinate against the scores of pathogens that could be weaponized. A decade ago Ken Abilek (the Biopreparat first deputy chief who defected from Russia in 1992) put the number at not less than seventy.[75]

Also to reckon with are the large number of available toxins. These are lifeless but lethal substances derived from living organisms — microbes, animals and plants. Until 1985 or so, they had to be extracted laboriously from often large amounts of the relevant living matter. Nowadays "the gene that directs the production of the toxin can be isolated and inserted into bacteria that can be grown in fermenters" in quantity. An alternative can be to engineer the toxin gene directly into antibiotic-resistant bacteria or viruses. Thus scorpion toxin has been engineered into a virus which then attacks caterpillars to deadly effect.[76]

Genetic Manipulation

In principle, the artificial "tweaking" of an individual's immune system should, once the *modus operandi* is sufficiently advanced, effect some specific protection in advance. Certainly the early indications are reasonably encouraging for run-of-the-mill medical conditions. Even so, application of the technique is still some way off, and when achieved will remain expensive.[77] Besides, it is evident that radical tweaking could be the more readily effected with microbes. Indeed, it is conceivable that in, say, 15 years' time entirely new microbial forms will be laboratory fabricated fairly widely. Polio viruses are being synthesized already.

Partial Cover

As good a protection as any against aerosol particles might often be respiratory masks.[78] Moreover, their acceptance as a precaution by society at large might be made easier by their relevance to some of the spontaneous global pandemics we are warned could become more frequent. However, the masks would have to be much more efficient than heretofore to be effective against a threat not from inert gases but from virulent microbes able to multiply menacingly once inside the human body. Nor are masks any defence against diseases that are contagious or (like cholera and typhoid) waterborne.

Nor, of course, is protection afforded to crops or animals. A former Director General of Britain's Chemical and Biological Defence Establishment at Porton Down has been among those warning that biowarfare strikes against livestock and crops may result in "major socio-economic damage".[79] One is talking here of targeting populations which may be confined by geography in a way humankind itself is not. Among the plant diseases identified as especially eligible for bio-attack are the viral tobacco mosaic and sugar beet curly top; bacterial rice and corn blight; and the fungal potato blight. Some 8000 fungal species are known to cause plant disease. So are more than 200 bacterial species and 500 viral ones.[80]

Proliferation and Application

Already the indications are that biotechnology as a whole will spread round the world every bit as insistently as other technical revolutions. Thus the Zen zone of East Asia has entered the business in quite a big way. It may be the locale for over a thousand companies, much like Europe and North America respectively. Though Japan has been seen as the biotechnology leader regionally, China is coming up fast. So, indeed, is Singapore. There is no indication that a wish to attain a state of art biowarfare option lies behind this East Asian awakening. But something fairly close to that will follow *ipso facto*.

As regards operational employment, the delivery of biological warheads via ballistic missiles would never be easy. Tests during World War Two showed that the temperature extremes and shock an aerial bomb was subject to during descent and impact might kill 95 per cent of the microbes in solution within its casing; and that the survivors would tend to be in droplets too large for human inhalation. If instead biological warheads exploded well above the surface, there could be a conflict between the aerosol released being fine enough to be inhaled yet coarse enough to descend before undue dispersion.

A more functional threat could come directly from the surface via a compact micro-spray or a suicide bomber who sets out already incubated. Micro-sprayers can be designed to diffuse droplets as small as those generated, 1915–18, with gases such as chlorine. Some 45 years ago, experiments were carried out to evaluate the threat of spraying from the sea. In one, a ship steamed for 156 miles just ten miles off the American coast, disseminating the while 450 lbs of fluorescent material. Traces were detected across 35,000 square miles of land.

A most worrisome aspect of this whole problem is the difficulty of confirming responsibility. Take Palestine in May 1948. There was a serious outbreak of typhoid in Acre and Arab villages around. It does seem to have been started deliberately. But by whom and with what authorization, if any?[81]

Another characteristic of biowarfare is that it will probably make guerrilla movements less dependent logistically on external sponsors — state governments or whoever. This may not encourage extreme behaviour in all cases. But it could well do in some.

Yet having said that, and thinking in terms of territorial compass, there are two classes of polity which might just have an outside chance of withstanding a major attack. The one would have a large area with maybe a generally low population density — e.g. Canada. The other would be compact — e.g. Singapore, Taiwan and Israel. But even the preparation for such a scenario could involve any free society in a terrible corrosion of its values. Yet actually to go operational could be to make things much worse in this respect. Could it, for example, ultimately involve quarantining of any affected communities? And how viable would any state be, in any case, if much of the outside world was ruined?

Unusable?

Scepticism is often expressed about biowarfare being resorted to except perhaps in some carefully circumscribed way. For one thing, we are presumed to have emerged from those mindsets of 45 to 50 years ago in which the deaths of scores of millions in a strategic exchange could be cogitated dispassionately.

However, in the febrile world into which we are moving, such genocidal previsions could return. After all, we can hardly be satisfied that they have ever gone so very far away. The present leadership of China could not possibly be described as ghoulish or fanatic, viewed in the round. Yet in 2005 a dean at her National Defence University considered in a lecture how a Sino-American showdown over Taiwan might lead to a major nuclear exchange. Should the USA make a high technology non-nuclear strike on Chinese territory, he believed "we shall have to respond with nuclear weapons . . . We Chinese will prepare ourselves for the destruction of all the cities east of Xian. Of course, the Americans will have to be prepared for hundreds of cities being destroyed by the Chinese".[82] The geography of such a conflict as thus described takes a bit of fathoming. At all events, it would be a war of pronounced asymmetries culminating in a dreadful symmetry.

What has to be borne in mind is how readily people can slide into a frame of mind in which punishing to the maximum extent a stronger protagonist matters more than one's own welfare or anything else. This, after all, is what fights to the finish are all about.

Fortunately, most political and military leaders including most guerrilla leaders do set their opponents within defined limits, as often as not geographically defined limits. But suppose one was up against someone or some grouping fanatically convinced that the whole world was against them; and suppose that party had biowarfare capability. Might this not be used *àl'outrance* if things got desperate? What if Hitler had had this option in his closing weeks? What if Osama bin Laden was in a similar situation? And what if biowarfare did begin with microbes not too virulent being directed against localized objectives — e.g. local crops? Would not such a gambit pave the way to rapid escalation? Is not the dissemination of addictive drugs as a stratagem in insurgency war a precursor?

A biowarfare exchange could constitute the gravest threat since Toba 72,000 years ago. It is one that will have to be met not reactively but in anticipation. It is said that the USA alone has spent 50 billion dollars on biodefence since 2001, several hundred installations being involved. So how clear a sense is there of where all this is leading or could lead?[83]

Notes

1 Sam Knight, "The Bomb in the Box", *World Today* 59, 2 (February 2003): 17–18.
2 *The Economist* 385, 8555 (17 November 2007): 11.
3 Bob Rowthorn, "Numbers Matter", *Prospect* 125 (August 2006): 10.
4 Fridtjof Nansen, *Through Siberia. The Land of the Future* (London: William Heinemann, 1914), p. 202.
5 New Local Government Network survey cited in *The Economist*, 388, 3597 (13 September 2008): 42.
6 "It's grim up North say life quality statistics", *Daily Telegraph*, 11 July 2003.
7 Matthew Arnold, *Culture and Anarchy* (London: Smith, Elders, 1869), pp. xviii–xxi.
8 N. J. G. Pounds, *An Historical and Political Geography of Europe* (London: George D. Harrap, 1947), Figs. 72 and 73.
9 D. Trevor Williams, "Some problems of the strategic geography of London and the Thames Estuary", *Scottish Geographical Magazine*, XLVIII, 5 (September 1932): 274–9.
10 Basil Collier, *The Battle of the V-Weapons, 1944–5* (London: Hodder and Stoughton, 1964), pp. 151, 165.
11 Special Report "The World Goes To Town", *The Economist* 383, 8537 (5 May 2007): 13.
12 "Rapid city growth means China faces rubbish crisis by 2020", *The Guardian*, 10 January 2007.
13 Richard Norton, "Feral Cities", *Naval War College Review* LVI, 4 (Autumn 2003): 97–106.
14 Robert W. Snyder in Kenneth T. Jackson (ed.), *The Encyclopedia of New York City* (New Haven: Yale University Press, 1995), pp. 297–9.
15 Manuel Castells, *The Internet Galaxy* (Oxford: Oxford University Press, 2001).
16 Audrey Kurth Cronin, "Cyber-Mobilisation: the new Levée en Masse", *Parameters* xxxvi, 2 (Summer 2006): 77–85.
17 John Humphrys, "Virtual Nightmare", *Daily Mail*, 5 January 2008.
18 "NATO considers response to cyber warfare", *International Defence Review* 40 (July 2007): 5.
19 Alfred E. Zimmern, *The Greek Commonwealth* (Oxford: The Clarendon Press, 1922), pp. 84–5.
20 *Chiltern News* 182 (December 2006): 21.
21 *Pocket World in Figures* (London: Economist, 2007), p. 99.
22 Donald Greenless, "Stagnation Marks Anti-Corruption Fight", *International Herald Tribune*, 6 April 2006.
23 The pursuit of these aims is carried forward helpfully in John Minkes and Leonard Minkes (eds.), *Corporate and White Collar Crime* (Los Angeles: Sage, 2008).
24 Robert J. Samuelson, "The Upside of Recession", *Newsweek* CXLIX, 20 (14 May 2007): 60.
25 Noam Chomsky, "Profits of Doom", *New Statesman*, 3 June 1994.
26 D. S. Lewis (ed.), *The Annual Register: World Events, 2006*, Vol. 248 (Bethesda: ProQuest, 2007), pp. 242–3.
27 Teodor Shanin, "How the Other Half Live", *New Scientist* 175, 2354 (2 August 2002): 45–7.
28 Richard Layard, *Happiness* (London: Penguin, 2006), p. 30.
29 George Orwell, *The Road to Wigan Pier* (London: Victor Gollancz, 1937), Chapter XII.
30 Layard, *Happiness*, pp. 68–9, 179–80.
31 Simone Weil (Arthur Willis, trans.), *The Need for Roots* (New York: Harper and Row, 1952), pp. 78–99.
32 Daniel and Jason Freeman, *Paranoia: the 21ˢᵗ Century Fear* (Oxford: Oxford University Press, 2008), pp. 53–4.
33 Nicholas Bunnin and E. P. Tsui-James (eds.), *The Blackwell Companion to Philosophy* (Oxford: Blackwell, 1996), p. 751.

34 London, *Womankind*, March 2008.

35 David Martin Jones and Michael L. R. Smith, "ASEAN's Imitation Community", *Orbis* 46, 1 (Winter 2002): 95–109.

36 Jeff Ballot, "Northern Vision", *Toronto Globe and Mail*, 10 October 1987.

37 "Russia pushing its claim at North Pole", *International Herald Tribune*, 2 August 2007.

38 Reproduced in I. F. Clarke, *Voices Prophesying War, 1763–1984* (London: Oxford University Press, 1966), p.160.

39 David Leitch in Michael Sissons and Philip French (ed.), *The Age of Austerity, 1945–51* (Harmondsworth: Penguin, 1964), pp. 74–5.

40 Henri Focillon, *The Year 1000* (New York: Harper and Row, 1971), p. 60.

41 R. S. Lewis (ed.), *The Annual Register 2000* (Bethesda: Keesing's Worldwide, 2001), p. 443.

42 Paul Dickson, *The Electronic Battlefield* (London: Marion Boyars, 1977), Chapter 9.

43 E. Jantsch, *Technological Forecasting in Perspective* (Paris, OECD, 1967), pp. 156–63.

44 Editorial, *Daily Telegraph*, 7 February 2007.

45 g, the acceleration due to gravity or 32 feet per second per second.

46 Stanley A. Blumberg and Louis G. Panos, *Edward Teller, Genius of the Golden Age of Physics* (New York: Charles Scribner, 1990), Chapter 24.

47 Jacquelyn A. Davis and Robert L. Pfaltzgraff, *Power Projection and the Long-Range Combat Aircraft* (Cambridge: Institute for Foreign Policy Analysis, 1981), p. 30.

48 Scott C. Turner, "Floating Arsenal to be 21st century battleship", *International Defense Review* 29, 7 (1 July 1996): 44–7.

49 Edward N. Luttwak, "A Post-Heroic Military Policy", *Foreign Affairs* 75, 4 (July/August 1996): 33–44.

50 Jeremy Blackham, "The Potential of UK Maritime Forces in Palestine", *RUSI Journal* 153, 2 (April 2008): 60–3.

51 Arthur Bleep, *The Victorian Naval Brigades* (Dunbeath: Whittles, 2006), pp. vi and vii.

52 Jared M. Diamond "The Earliest Horsemen", *Nature* 350, 6416 (1 March 1991): 275–6.

53 M. P. Manson, *Guns, Mortars and Rockets* (London: Brassey's, 1997), pp. 52–3.

54 *Small Arms Survey, 2006* (Geneva: Graduate Institute of International Studies, 2006), Fig. 31.

55 *Ibid.*, p. 227.

56 Dr Mark Galeotti, "Crime Pays", *The World Today* 58, 879 (August–September 2002): 37–8.

57 Winston Churchill, *My Early Life* (London: Thornton Butterworth, 1930), Chapter XI.

58 Vanora Bennett, "Chechnya's Theatre of War", *The Liberal International* X (April–May 2007): 20–3.

59 Ahmed Rashid, "Epicentre of Terror", *Far Eastern Economic Review* 163, 9 (11 May 2000): 16–18.

60 "A World awash in Heroin", *The Economist* 30 June 2007, pp. 71–2.

61 Mike Jackson, *Soldier* (London: Bantam Press, 2007), p. 320.

62 E. A. Thompson, "Early Germanic Warfare", *Past and Present* 14 (November 1958): 2–29.

63 Rudi Paul Lindner, "Nomadism, Horses and Huns", *Past and Present* 92 (August 1981): 3–19.

64 Vincent J. Goulding, "Back to the Future with Asymmetric Warfare", *Parameters* XXX, 4 (Winter 2000–01): 21–30.

65 Sally Healy, "Broken City", *World Today* 64, 1 (January 2008): 23–5.

66 Yu-Ying Brown, "Private Libraries in Japan", *Alexandria* 7, 1 (1995).

67 "Afghanistan: Iconoclastic Fury Unleashed Again", *IIAS Newsletter* 27 (March 2002): 45.

68 Carlo Power, "Saving the Antiquities", *Newsweek*, 14 May 2001, pp. 64–6.

69 *International Journal of Contemporary Iraqi Studies* 1, 1 (2007): 110.

70 Eleanor Robson, "Cradle to Grave", *Oxford Today* 16, 1 (Michaelmas 2003): 12–14.

71 Professor Neville Brown and Sir Anthony Farrar-Hockley, *Nuclear First Use* (London: Buchan and Enright for RUSI, 1985).

72 William H. McNeill, *Plagues and People* (Garden City: Doubleday, 1976), pp. 161–5.

73 J. Z. Titow, *English Rural Society, 1200–1350* (London: George Allen and Unwin), Chapter 5.

74 "Vladimir Pasechnik", *Daily Telegraph*, 29 November 2002.

75 Ken Abilek with Stephen Handleman, *Biohazard* (London: Hutchinson, 1999), p. 281.

76 Wendy Barnaby, *The Plague Makers* (New York: Continuum, 1999), p. 134.

77 Dan Jones, "Superimmune", *New Scientist* 193, 2592 (24 February 2007): 42–5.

78 Barnaby, *The Plague Makers*, pp. 148–9.

79 Graham S. Pearson, "Why Biological Warfare Matters", *The Arena* 3 (October 1995): 3.

80 Malcolm Dando, *Biological Warfare in the 21st Century* (London: Brassey's, 1994), pp. 36–44.

81 Barnaby, *The Plague Makers*, pp. 121–2.

82 Major-General Zhu Chenghu quoted in *The Guardian*, 16 July 2005.

83 Editorial, "The Best Defence", *New Scientist* 199, 2668 (9 August 2008): 5.

14 | The Age of Ecology

1 Global Warming

Climate change and its impact have been themes since Classical Greece and Rome. There has likewise been a long tradition of considering how climate moulds a society's general character. It wells up in Late Renaissance to Enlightenment times, encouraged by the voyages of discovery. Its herald was Jean Bodin. His lead disciple *vis-à-vis* climate was Abbé du Bos; and his was Charles de Montesquieu 1689–1755). David Hume, too, (1711–76) stands out in this context as in others. So still more does Edward Gibbon (1737–94).

Through the first half of the twentieth century, this interest waxed strong again, certainly among geographers. Names like Ellsworth Huntington and Frank Markham come to mind. The usual presumption was that the most energizing climates were those in mid-temperate or somewhat higher latitudes, the tropics and their surrounds still being seen as too torrid, much in line with the Classical understanding. This owed a lot to ignorance about historical sub-tropical empires. Besides which, its relevance to the future had been sharply curtailed by the advent (in Buffalo in 1902) of the air conditioner. Today Singapore is a spectacular example of vigorous adaptation to high humidities and/or temperatures. It is prosperous, imaginative, orderly and virtually corruption free.

All the same, if our shrinking world is getting less comfortable and secure in manifold other ways, we shall *ipso facto* be all the more susceptible to any additional stresses, climate change among them. Take an historical comparison of European winter severity. A standard measure is the cumulative shortfall of monthly temperatures, December through February. On that basis, the winter of 1946–7 was demonstrably in much of Europe the worst since 1840 with acute political consequences especially in the Germanies, Britain and Greece. But the winter of 1962–3 was the worst since 1740. Yet in the annals of world politics, this receives no mention. For then, both sides of the Iron Curtain, Europe was at a post-war high of stability and confidence with sustained growth, low unemployment and little social unrest.

The comparison just drawn depends on assessing what the direct consequences of freeze-ups may have been. More often, in the future as in the past, temperature change will chiefly be important as a measure of climate variation which may impact more in other ways. Altered rainfall patterns may be critical. Trends in the distribution of ice on land and sea can eventually be too. As Archimedes would have said, changes in the volume of sea ice do not affect sea levels. But fluctuations in the volume of land ice can do so appreciably after a while. Besides, ice on land or sea reflects the sun's rays strongly (see below).

The Mechanics of Change

Fundamental to climate trends in general are rhythmic variations in the Earth's orbit and axial inclination, their respective periodicities being measured in tens of thousands of years. The review carried out by the Serbian physicist, Milutin Milankovitch (1879–1965) showed that, for some millennia, the underlying trend worldwide has been towards cooling. This has essentially been because the "astronomical" cycles (slow alternations in the Earth's orbital path and axial inclination) have been limiting the insolation contrast between summer and winter, a circumstance conducive to icefield expansion. Cooler summers mean less snow melt at high latitudes while warmer winters (with the air therefore able to retain more water vapour) yield more fresh snow.

At present, warming due to the atmospheric accumulation of "greenhouse" gases is overriding decisively the Milankovitch effect. However, our awareness thereof can be lessened by apparently random seasons and also by an assortment of weather/climate cycles (short term or long term; transient or enduring) which may or may not have been elucidated. Some will be intrinsic to the ocean and atmosphere circulation. Of these, some will have evident "teleconnections" with climatic tendencies further afield.

At the heart of a virtually worldwide teleconnection web is El Niño. Across the tropical Pacific, the displacement of surface sea water is heavily from east to west by dint of a great gyre each side of the equator. Yet between these is an equatorial counter-current moving the other way with its axis parallel to, though several degrees north of, the equator. Once or twice a decade, a regular Christmastide surge to the Peruvian coast is especially strong. Hence the term "El Niño" or boy child. The El Niño mechanism modulates *inter alia* the Indian monsoon and the flow of the Nile.

Otherwise a big volcanic explosion may eject into the rainless stratosphere a volume of fine particles liable to persist sufficiently to sustain cooling up to seven years. Moreover, it is not too unusual to get two or three strong eruptions within several years. But most evident among the extrinsic influences are the fluctuations in solar activity, the basic sunspot cycle averagely of 11 years being a visible expression thereof. More and/or bigger sunspots collectively indicate more solar activity even though individually a spot is a coolish feature. Recent cycles have shown a solar irradiance difference between peak and trough of one part in a thousand. In itself, that differential might transiently alter terrestrial surface temperatures 0.25 degrees Celsius.

The Specifics of Greenhouse

The peak wavelength of electromagnetic radiation from a given surface will be close to being in linear inverse proportion to its absolute or Kelvin temperature. The mean temperature of the Earth's surface is currently just under 290 Kelvin whereas that of the Sun's is 6000. Therefore simple multiplication predicts the former will have a peak wavelength for the emission of electromagnetic energy some 21 times longer than does the latter: close to 10,000 nanometres (nm) as opposed to 485. **(See Appendix A)**.

That comparison may be tested against the energy absorption profiles of the two most significant greenhouse gases. Water vapour has several absorption peaks between 940 and 6300 nm. Carbon dioxide has a moderate one just below 5000 nm and a strong one at 15,000 nm. So both gases do indeed block longer-wave radiation from the Earth more than they do the shorter solar wavelengths. They thereby make significant contributions to rises in terrestrial surface temperature to levels at which a thermal balance is restored. In the case of water vapour, in fact, its sheer volume ensures it always contributes a rise of about 10 degrees Celsius or Kelvin. Nevertheless, water vapour has very generally been excluded from greenhouse gas projections. This is because the connection between any variations in its incidence and human economic activity is a second order derivative.

So ignoring water vapour for the purposes of argument, as analyses regularly do, one can say that 60 per cent of this anthropogenic greenhouse effect is being induced (mainly through fuel combustion and deforestation) by added carbon dioxide (CO_2). Methane (NH_4) contributes 30 per cent, largely via cattle and rice paddy. The remainder comes from nitrous oxide (N_2O), chlorofluorocarbons (CFCs) and low altitude ozone (O_3). The average level of carbon dioxide in the atmosphere has risen from 280 parts per million (ppm) in 1850 to 315 in 1958, the year direct global monitoring began; 350 in 1988; and 379 in 2005. The incidence of methane rose from 0.85 ppm in 1850 to 1.32 in 1958; 1.70 in 1988; and 1.77 in 2005. So methane's prevalence rose proportionately faster than carbon dioxide to 1988 but has been the slower since.

From the plausible permutations (including the introduction of mitigation strategies), a raft of global predictions to 2100 has been made, notably by the UN Intergovernmental Panel on Climate Change. The variant it currently designates A1 T seems particularly pertinent. It accepts current economic trends as broadly a given, while leaving room for mitigation. It anticipates "very rapid economic growth"; the world's population peaking in mid-century; "the rapid introduction of new and more efficient technologies"; a levelling up of inter-regional differences in per capita income; and emphasis on non-fossil energy sources. The median prediction for the global rise in mean temperature the rest of this century is 2.4°C; the limits of the likely range are 1.4 and 3.8.

A report commissioned in 2005 by Britain's then Chancellor of the Exchequer accepts between 2 and 3°C as the general median prediction. It noted that this meant the Earth would reach overall "a temperature not seen since the middle Pliocene around three million years ago", a level of warming on a global scale "far outside the experience of human civilization".[1] It is worth recalling that the late Pliocene was characterized by truly searing droughts on the African plains our pre-human forebears were evolving on. **(See 2 (1))**. It is an arresting demonstration of how temperature change may signify more general climatic variation, this quite possibly having dire ecological consequences..

The Oceans

A crucial dimension in the climate debate is afforded by the oceans. Their thermal capacity is a good thousand times that of the atmosphere. Their currents are presently responsible for 40 per cent of the net heat transfer from the equator to 70°N. They react very conservatively to changing circumstances.

However, ice fields are another matter, largely on account of their high albedo — i.e. how reflective they are of the Sun's rays. At high latitudes, snow and ice may reflect as much as 70 to 90 per cent of the sunlight incident upon them. In other words, an ice field acts as its own refrigerator. This causes it either to expand more or contract more than would otherwise be the case in given circumstances. That applies very particularly out at sea where the whole circulation favours the ice being extensive though thin in contrast with the huge nodes of ice which may form ashore. The ready extension but now retraction of Arctic sea ice explains why predictive modelling now consistently shows high northern latitudes warming more than anywhere else.

One inference is that there will be, in the northern hemisphere at least, a reduction in the thermal gradient between low latitudes and high, the gradient that basically drives the circulation of the sea as well as the air. The Gulf Stream, for instance, is part of a North Atlantic gyre which is wind driven though also "thermohaline" — i.e. driven by contrasts in sea density as determined by temperature and salinity. Both drives derive from the tropics to pole thermal gradient.

In the near future, oceanic inertia (thermal and mechanical) will likely preclude any part of the North Atlantic gyre being basically reoriented by climate change. However, the corollary is that, as and when the pattern does much alter, the said alteration will be rather abrupt and very radical. In 2002, Sarah Hughes of the Aberdeen Marine Research Laboratory reported that the Gulf Stream had already slowed 20 per cent since 1950. Nicholas Stern writes of sophisticated modelling predicting a weakening of the North Atlantic gyre by up to a half this century.[2] That is tantamount to predicting radical reorientation of flow perhaps by 2100. Air temperatures could then be several degrees lower than would otherwise be the case in Western Europe and eastern North America. And, of course, other disjunctions would arise elsewhere in the Northern Hemisphere including — one would have thought — the Arctic. However, recent work at the University of Southampton, a European leader in oceanography, has reportedly suggested a less drastic slowing of the Gulf Stream. More research is needed, including about any effects on precipitation patterns. Meantime we do well to eschew media hype about Western Europe flipping right back to an Ice Age.

Hurricanes and Erraticism

Large tropical revolving storms (known in the Caribbean as "hurricanes", in the Pacific as "typhoons", and the Bay of Bengal as "cyclones") have been adjudged likely to increase in frequency and, indeed, average strength as the equatorial seas warm. The reasons were indicated in 3 (2). It is what does seem to have happened the last 40 years or so.[3] Enquiry continues into how much this trend may be overridden by El Niño fluctuations.[4]

Yet this is part of a more general quandary as to whether there is an inclination towards more short-term variability or, as *The History of Climate Change* study prefers it, "erraticism". Is the weather fluctuating more day-by-day, week-to-week, season-to-season . . . ? Since the onset of the present "global warming" debate, in 1988, popular belief pretty much the world over has been that the weather is somehow becoming more freakish. That variability can vary has long been accepted. The tough question has been how this relates to secular weather trends. As far back as 1912, the Swedish oceanographer, Otto Pettersson, well observed that "Part of the 13th century and the whole of the 14th century show a record of extreme climate variability". That is to say, it was as the climate turned colder that the weather turned more erratic.

Important at this juncture is to recognize that what comes across to dwellers in a particular place as more erratic weather may lend itself to either of two contrary explanations. The one is that the pre-existing climate regime is indeed producing more erraticism. The other is that this regime is being displaced by a quite different one.

Abnormally recurrent droughtiness in such regions as Mediterranean Europe, Southern Africa and Australia will often be attributable to a broad polewards movement of the subtropical anticyclonic belts. These zones of blue skies, descending air and steppic/desertic landscapes are geologically traceable across hundreds of millions of years. But their strength and disposition are subject to global climate trends. A collateral tendency, if the atmosphere's global circulation is weakened by reduced tropics to pole thermal gradients, is for more rain to fall near coasts exposed to moist oceanic winds while less does in continental interiors.

Then again, the historical experience of Britain, in particular, has been spells of one or more decades during which high winds have been more frequent (*c.* 1300; before 1600; *c.* 1700 . . .). The first of these three spells were near a turning point towards secular cooling; and the last near one towards secular warming (see below).

Greenhouse Sceptics

Some 45 years ago, Thomas Schelling, now a Nobel Laureate at the University of Maryland, made a singular contribution to the delineation of strategic studies. But by 1985, he had moved on to urban and environmental economics. Though not sure the White House should have discarded Kyoto as brusquely as it did in 2001, he would himself have preferred to prepare methodically for deep emission cuts later this century.[5] However, the trouble with this is that a crunch could come much sooner. Analysis has suggested that the Arctic could actually be ice-free in summer by 2080.[6] Some even say 2040.

Richard Lindzen, a ranking atmospheric physicist at MIT, has been prominent as a greenhouse sceptic these dozen years and more. Many of his reservations have been in the realm of the geophysical. But like others, Schelling included, his disposition has been to minimize the societal impact of the most plausible levels of greenhouse warming. Thus in 1994 he was claiming to discern a "general consensus" that a global mean temperature rise of 0.5 to 1.2°C consequent upon a century of global warming "would present few if any problems".[7] Likewise in 2005, giving testimony to Britain's House of Lords, he averred that a doubling of CO_2 would lead to about 0.5°C warming and a quadrupling (should it ever occur) to about 1.0°C. Neither would, he thought, pose a particular societal challenge.

However, when temperature is being used as a global marker for broader climate change, a one degree Celsius alteration cannot be lightly dismissed. This especially applies when it is taking place fast and when a spatially averaged reading masks big differentials as between major regions. Mean air temperatures worldwide are esti-mated to have fallen but one degree between AD 1250 and 1700, from the zenith of the Little Climatic Optimum to the nadir of the Little Ice Age, a nadir determined by a quieter Sun and vulcanism. Then it took to 1940 to recover this loss more or less. Throughout these seven centuries, there were repeatedly manifestations of the social, political and geopolitical impact of climatic tendencies.[8]

Lindzen has written to better effect when warning against the crude politicization of this whole debate: the bowdlerizing, stereotyping, motive questioning and general venom involved. An especially disturbing instance was a 2003 ruling by Denmark's Committee on Scientific Dishonesty against Bjørn Lomborg, then a statistician at the University of Aarhus and the author in 2001 of a remarkably comprehensive critique of environmentalist conventional wisdom. This ruling was that Lomborg had been scientifically dishonest. However, the professional indictment was not proceeded with in the face of a gale of disapproval, not least from the Copenhagen government. But this McCarthyism of the Left had sullied awhile the peerless contribution Scandinavia has made to natural science this last century and more, particularly to oceanography and climatology.

Certain of Lomborg's arguments could have been honed better. More specifically, his book should have taken due account of a keynote graph he himself had prepared, inferring global temperature means for the millennium just past. All the five series it brought together show a rise of temperature through the twentieth century which is quite unprecedented within the said time frame: a uniquely rapid rise to a uniquely high level.[9] Moreover, the human impact of climate change is much accentuated by its pace being so contrasting inter-regionally. But this the 2001 study ignores almost completely. Thus it contains no indexed reference to the Arctic. There is but the briefest allusion to changing patterns of rainfall deficiency in the subtropics. And so on.

Still, Lomborg did raise some germane concerns. How well do strategies designed to mitigate the greenhouse effect stand up to cost-benefit analyses (pp. 300–316)? Are there any theoretical grounds for believing global warming induces extreme weather events (pp. 292–97)? Has not the threat posed near term by rising sea levels been hyped in the media (pp. 289–91)? None other than Al Gore has, under the mantle of Reason, seriously aired as a putative possibility tectonic activity in Greenland causing its huge ice cap to break up and slip into the sea![10]

Cosmic Stimulation

Work led by the Danish National Space Centre has lately carried forward an argu-ment that fewer "seeds" for the formation of water droplets may be created when cosmic radiation through the atmosphere is weak because of the resistance proffered by a strong solar wind. Low clouds may therefore form slightly less extensively and, being averagely composed of larger water droplets, tend to be expended more readily as rainfall. The upper clouds composed of ice crystals (i.e. the cirrus family) are said to be affected much less. This is significant. For whereas water droplets reflect incoming solar radiation, ice crystals have more of a greenhouse effect.

As it so happens, the current decrease in cosmic ray flux has proceeded progres-sively since about 1900. So the proposal has been that this factor has been considerably responsible for the global warming registered across the twentieth century — a rise of about 0.8°C.[11] It is a hypothesis worth testing thoroughly. It should be, first and foremost, in relation to putative cloud generation by aerosol "seed" particles. How readily can cosmic radiation allow of particle aggregation to a sufficient size to act thus?[12] And how critical may seed supply be to cloud formation?

The proposal is supported by a graph which depicts a neat correlation between cosmic radiation and cloud cover. It does so on the questionable assumption that the latter parameter, too, can be extrapolated back to the year 1700 with sufficient accuracy. Still it does show some correlation with the relatively cool interlude from 1940 to 1965 or so. However, its opening years, 1940–2 were very acute, certainly across Europe; and witnessed, too, a very strong El Niño.[13] Scanning more widely, one rather gets the impression that the current lessening of cosmic ray flux may soon be troughing out as solar radiation weakens.

What all this could mean for global warming by 2100 is by no means certain. But it is unlikely to involve a decrease of more than half a degree from present temperature projections. Yet this could tip the balance between a global warming crisis that remained just about manageable within acceptable political norms and one that looked hopeless. In other words, this Danish hypothesis might be read as a call to action as opposed to lapsing into either defeatism or euphoria.

Easement and Impact

The sundry strategies for greenhouse easement these days receive plenty of attention. What still needs stressing, however, is that three of the most radical will probably have to be set within a modified international legal or quasi-legal framework. Structures in orbit intended to shield the Earth or else to harness solar energy for use on it may require radical development of Space Law. Likewise, utilizing the oceans to recapture more carbon dioxide by stimulating the blooming of phytoplankton may necessitate modification of the Law of the Sea.

Above all, a revival of nuclear power ought to involve the development of regional bodies for uranium enrichment and plutonium separation. To have the developing world draw closer to nuclear autonomy in a way thus constrained may be a needful ingredient of a strategy for the avoidance of that most horrendous of prospects — the proliferation of biological bombs. Studying regional climate change along with its impact could be among the other functions of such multilateralism.

What one cannot generalize about is the nature of the impact climate change might have on international security overall. The world depression of 1929–32 compromised in manifold ways the prospects for peace. The credit crisis of 2008 could do likewise, particularly if it proves less resoluble than is hoped as of December. The former episode was regularly spoken of at the time as an "economic blizzard". This latest is being termed a "financial tsunami" with potential for "meltdown". Real tsunamis or meltdowns or, if you like, blizzards could likewise have a mix of adverse consequences for the world order.

2 Malthus Recast

As is these days well appreciated, every variant of Marxism-Leninism has performed badly in the ecological domain. Several reasons might be adduced. One could be that the early Marxists, not least Karl himself, endorsed the technology-driven goal of "socialist abundance" via the "conquest of nature". By so doing, they were overreacting to a Romantic movement that had often disported itself as not just anti-industrialization but fundamentally anti-science, in abstract principle if not for material comforts. Yet they were also reacting against Malthusian pessimism, this rationalized with the proposition that population tended to increase by "geometrical progression" (i.e. a constant percentage or exponentially) whereas production tended to register only "arithmetic progression" (i.e. a constant increase in absolute terms). Thomas Malthus cited war as among the inevitable "natural checks" — mindless of how (not least in the industrial age) warfare has tended to destroy production capacity more than people.

By responding the way they did, however, the Marxists missed the chance of developing one of the most cogent critiques that can be levelled against capitalism operating along classically *laissez-faire* lines. Its costings exclude what are now termed "external diseconomies of scale". Every new car-owner in Madrid, say, is an added burden for every existing one and, at one stage removed, for the city at large. A new oil-fired power station adds to the carbon dioxide level in the atmosphere around the world. Having been ignored by the Marxist founding fathers and not properly addressed by Thomas Malthus, such external diseconomies received systematic attention only post-1918. Arthur Pigou, a Cambridge political economist, led the way.

Not Climate Alone

We may yet be some way off a clear understanding of how our global climate is responding to incidental human interference. We are even further from comprehending how this is panning out regionally. A fundamental complication is that the species within a given ecosystem are likely to be affected to varying extents.

Yet what must also be conceded is that, even if the Earth's climate was set fair for stability, we would still face a manifold crisis in ecology and resources. Uncertainty about non-ferrous metals shortage should again be on the agenda because many factors affect it and about most of them we are chronically knowledge deficient. Undue anxiety in the early fifties and again in the early sixties does not excuse complacency now.[14][15]

Output of oil and then natural gas (i.e. methane) are likely to peak globally within two or three decades. Supply will get more costly as less accessible deposits are tapped. Those lately discovered off Brazil's north-east coast will likely make it a major producer but this by sinking wells on seabeds as much as 2300 feet below sea level. The obverse of this globally is that demand will be further constrained by the greenhouse factor. However, this applies to methane less than to oil. The chemical formula of methane, CH_4, shows its combustion will largely produce water vapour. This is less greenhouse than carbon dioxide, molecule for molecule.

For obvious reasons, coal is the worst carbon dioxide emitter. But as world energy prices rise considerably higher, it could become viable to capture power station emissions and store them underground. Though no such facilities are operational at present, some pilot plants are under construction. All the ramifications of the hydrocarbon problem make a broad international dialogue at once more difficult and more needful.[16]

A natural resource ever more prone to chronic as well as short-term short-falls is water. The UN estimates this already applies where a billion people live and that by 2025 some 1.8 billion could be thus exposed. One remedy well in the frame at present is desalination. Thirteen thousand plants are operational; and this number is scheduled to nearly double by 2025. However, the current output represents only 0.5 per cent of global use, whereas the billion already exposed to scarcity constitute 15 per cent of the world's population. Huge extra investment is needed.[17]

However, environmental lobbyists are unhappy about how desalination consumes energy. Peter Blomberg of Friends of the Earth, Middle East is quoted as saying it should be a technology of last resort. As usual, a basic difficulty is that big investment can easily take two decades to decide on and implement. What is clear is that not a few countries will have to conduct a saddening and difficult movement out of agriculture. A Hebrew University estimate has been that water productivity in Israeli agriculture has increased a creditable threefold the last half century. Yet farming still accounts for 60 per cent of the country's water consumption while contributing but 2 per cent of its GDP.

The water crisis reminds us that causation by climate change and that by other factors can never be neatly separated. Witness, too, the concern over how polar bears are threatened by the incipient meltdown of their habitat though also, according to Canadian and other studies, by the accumulation in their bodies of fire-retardant chemicals. These could well compromise their immune and reproductive systems.

High Northern Latitudes

In fact, polar bears are a totemic indication of the multiple crisis of pollution the North's fragile polar and sub-polar domain is facing. Melting of the permafrost (soil at a depth at which, at a given latitude, it naturally stays frozen) is one insidious aspect — e.g. in Alaska and Siberia. Overfishing is another. So have been acid rain and "Arctic haze", the latter comprising blankets of smog largely from Eastern Europe. The Arctic basin has been warming twice as fast as the globe as a whole, thanks to the "ice albedo feedback". (See 14 (1)).

The Great White North is still for Canada (as for Russia) focal for national identity. Unfortunately, however, the dilemma of how to define and apply a benign policy towards indigenes was to prove as hard to resolve around the Arctic as in other areas subject to Western incursion. Back in the 1950s, Canada felt constrained to consolidate its sovereign bounds by relocating Inuit families from northern Quebec to unfamiliar territory in the high Arctic. Many of these "human flagpoles" sickened and died.[18]

Ancestral Evocation[19]

What then of a particular concern informing this whole study? One alludes to an apprehension that our survival will be acutely at risk if, against a background of social and psychic rootlessness, there is a general flight from reason. Many people with North Eurasian backgrounds will feel a profound resonance with a far North whence came the successive glaciations that were for our Palaeolithic ancestors so testing an experience. Literary responses abound. Witness the writings of Fridtjof Nansen and of Vilhjalmur Stefansson (1879–62), a Canadian-Icelandic explorer-cum-scientist suffused with sensibility.[20] Witness, too, the experiential poetry and prose of Robert

W. Service and the empathy William Morris felt towards Iceland's pre-industrial society and artistic culture.

In World War Two the Germans put weather stations in Greenland. Duly, the Allies sent small raiding parties against them. But the brave and resourceful men on each side felt constrained from coming to mortal grips because, in that primordial setting, internecine human conflict seemed trivial.[21]

Granted, not all the evocations have rung true. It would be hard to find a contribution more jejune than the rendering by the poets W. H. Auden and Louis MacNeice of their 1936 pilgrimage to what the former had foreseen as the "holy ground" of Iceland. In fact, Auden comes over as particularly discomfited. Those vulcan landscapes, sternly enchanting to a sympathetic eye, he likened to "the useless debris of an orgy".[22]

No matter. Through the nineteenth century those the Amerindians scorned as "Eskimos" (e.g. eaters of raw flesh) but whom we now know as Inuit came to occupy a "conspicuously privileged" place in the imagination of a Western world fancifully "fascinated by their position amidst the ice".[23] Then again, three leading astronomy scholars of the Early Modern period (Nicolaus Copernicus, Tycho Brahe and the philosopher, Immanuel Kant . . .) spent their lives around the Baltic shores. To be denied any part of the Northern ambience could be to forego a significant part of our sensibility. A neo-Malthusian assault on our psyche?

The End of Nature

Still, any struggle to protect the polar regions can but be part and parcel of a wider endeavour to preserve Nature. Not that it is ever easy to define how Nature is bounded. One must acknowledge forthwith that we "children of the Wisconsin" and of the other glacial spans have always sought, to an extent singular among animal species, to free ourselves from Nature. We live in artificial caves. We wear fabricated skins. We contrive access to the whole biosphere and beyond. Even so, a large part — one could say an older and deeper part — of our psyche yearns to keep in touch with Mother Nature. Failure to do so sufficiently may aggravate that sense of emptiness already too prevalent in post-industrial society.

Referring particularly to man-induced climate change, Bill McKibben, an Anglo-American columnist, put the climate aspect with stinging lucidity in 1990: "By changing the weather we make every spot on Earth man-made and artificial. We have deprived Nature of its independence and that is fatal to its meaning . . . A child born now will never know a natural summer" (pp. 54–5). A premise regularly stated by environmentalists at the time of Henry Thoreau (1817–1862) was that the sky at any rate was forever safe from human spoliation. This no longer rings true. Moreover, those who, like McKibben himself, seek to resolve their crisis of religious belief "to a greater or lesser extent by locating God in Nature" (p. 66) could therefore be "in for a siege of apocalyptic and fanatic creeds. A certain way of thinking about God — a certain language to describe the indescribable — will thus disappear" (p. 74).[24]

Biodiversity is clearly a salient issue. Although its diminution is often discussed in global terms, the local scene is what impacts most on individuals. The same applies to loss of quietude. The great Leiden historian, Johan Huizinga, wrote to effect in 1924 about the passing of the medieval world.[25] He had no illusions about how the fourteenth century alternated between "tearful piety and frigid cruelty" (p. 40). None the less, he averred that a small country town could then experience a nocturnal peace more profound than ever obtained currently: "The modern town hardly knows silence or darkness in their purity, nor the effect of a solitary light or a distant cry" (p. 2). Yet in such respects the Western European landscape of 1924 was much closer to that of 1324 than to that of 2008. Ultimately, this situation could also compromise Peace of the kind extolled on armistice days.

Closely akin is how artificial light impacting on atmospheric dust obscures the night sky. One finds that, for a fifth of the world's people, the Milky Way is no longer visible to the naked eye. In the USA and the EU, the proportion respectively exceeds two thirds and a half.[26] Aircraft condensation trails significantly worsen obscuration by night and day.[27]

In the final analysis, such psychological and aesthetic aspects of environmental disturbance must stand alongside the more material, more classically Malthusian, considerations. Bill McKibben stresses the need for any American contribution to a containment strategy to derive from the renovation within of a national society that is becoming "hyperindividualized". As yet, "very few understand with any real depth that a wave large enough to break civilization is forming; and that the only real question is whether we can do anything at all to weaken its force."[28]

Malthusian Basics

Taken on its own terms, the classical Malthusianism of several decades ago is turning out to be a self-defeating prophecy. As recently as 1988, the executive director of the United Nations Population Fund (UNPF) said she feared the 2100 population total would top 14 billion.[29] On the basis of current received wisdom, however, the prospect is that the said total is set to rise by two-and-a-half billion to reach nine billion, on median projections, by 2050. Then growth would proceed more slowly another decade or two prior to a sustained turn down.

A caveat to enter concerns radically greater longevity, a prospect the scientific press now ponders. By 2020, mimetic calorie-restricting drugs may be pushing our general understanding of a good age from 85 towards 95. Soon thereafter, genetic manipulation of stem cells within the foetus or later may proffer the prospect of several more decades averagely. This could revive birth rates in that the emergence of an extra generation could induce young couples to broaden the family base. Much might depend on how the reproductive life span was being affected. A worrisome complication is that, for a generation or so, the benefits of this perceived boom would accrue unevenly within many national societies and as between them all.

Another caveat has to be that commentators may have taken too little account of the part played in the pressure on population by rising prosperity and evolving life styles. Such oversight could extend to a consideration as banal as the amount of extra living space the members of a rising generation may averagely require to have around themselves most of the time so as to accommodate all their artefacts and aspirations, the latter including privacy and quietude. In Britain the last half-century or so the population has risen by a fifth but the number of households has by over two-thirds. Air passenger mileage worldwide seems destined to grow between 2 and 5 per cent per annum the next half century unless "green" taxation impacts very acutely. Besides, the actual population is still rising fast across much of the developing world — notably in Black Africa, Andean America and parts of the Middle East.

Still, much of the more developed world has already fallen well below a Total Fertility Rate (TFR) of 2.1 births per prospective couple, the level thought necessary to keep a population in equilibrium, assuming no net migration or quantum change in longevity. Indeed, this threshold of decline was passed several decades ago over much of Europe. At present, the TFR stands at 1.28 in Italy and Spain; and 1.25 in Poland. Japan's TFR is 1.27 and South Korea's 1.25. Social disjunctures are foreseen as a result of the downward trend. Tax breaks intended to enhance their TFR are operative in various countries: France, Poland, Singapore, Australia . . .

Also to be remarked is the divergent demographic experience as between Russia and the USA. In the former, faltering health care coupled with alcoholism, poor nutrition, toxic exposure and low morale mean a male life expectancy of 59, twenty years below prevailing Western norms. Moreover, housing shortages, low wages and poor job security discourage pro-natal attitudes. With a TFR of 1.28, the population is falling by 700,000 a year. It is currently 142 million. With a TFR of 2.5, high immigration and a population almost at 300 million, the USA is expected to reach 420 million by 2050.[30] In 1941 the USSR had 175 million and the USA 135.

The USA is obviously well attuned to this evolving prospect. Indeed, the danger is that it will prove too much of a good thing in that it will be associated with an inward brain-drain in science, technology and other aspects of modern culture on a scale big enough to engender anxiety and resentment internationally. Meanwhile, those countries still under population pressure so immediate that they are failing to generate adequate welfare face another contradiction. They will be especially vulnerable to lethal pandemics, to savagely disproportional Malthusian checks. In August 2007, the World Health Organization warned that, thanks to globalization, pandemics could spread faster than ever before.

A contemporary manifestation is how a number of diseases which modern medicine appeared to have routed for good have been fighting back. Tuberculosis and salmonella are familiar examples. So, too, is malaria though in this case the criticality can often be the enhanced resilience of the mosquitoes. What is more, some maladies (e.g. SARS, ebola and HIV) appear to be afflicting our species for the first time, transferring in part from their customary hosts. Two consequences follow from that particular effect. The one is that our immune system may not have had time enough to develop by trial and error an adequate response. The other is that the disease itself may be genetically less inhibited about inflicting a high mortality rate if it does not ultimately depend on ourselves for accommodation. The same can apply with the threat that looms behind all else, that of biowarfare.

The Urban Syndrome

The inadequacy and fragility of urbanized mass society has been too prominent a theme in fiction and social science now to cast aside. Though regularly sidetracked by Leftist establishments, literary and political, it has often figured in socialist thought: Robert Owen, the Anarchists (especially Russian and Spanish), G. D. H. Cole, William Morris, Julius Nyerere, and, if you will, Mahatma Gandhi . . .

Nor should we dismiss as facile convention a recurrent disposition in literature in general to conjure up a supposedly idyllic rural–urban balance a generation ago: George Orwell, Thomas Hardy, William Cobbett and Oliver Goldsmith being familiar exemplars. Arguably, the most vexatious aspect of urbanization is not its extent at a given time but the rate at which, and the manner in which, it is further extending. As late as 1800, only 3 per cent of the world's population lived in urban agglomerations. The UN finds that today 50 per cent do; and that by 2030 the percentage will be 60. Needless to say, much of the current increase is in the poorest and, thus far, least urbanized parts of Afro-Asia. Between 2000 and 2030, the number of urban dwellers in the "less developed regions" is expected to double to four billion.[31]

In 1950, the world had 83 cities with a population in excess of a million. A decade from now it could have 500. The UN expects that by 2020 the following conurbations will exceed 20 million: Delhi, Dhaka, Jakarta, Lagos, Mexico City, Mumbai, São Paulo and Tokyo-Yokohama.

Many of us instinctually recoil at what Lewis Mumford famously condemned as such "purposeless gigantism".[32] Still, mega-conurbations even in the developing world may not be afflicted with social malaise as inexorably as the more shrill critics of urban existence would have had us believe. The tensions associated with city life have by no means always been uncreative. Nor must urban frustration invariably find release in the same set patterns of deliberate violence. At 35 million, Tokyo-Yokohama is easily the largest of the mega-cities. Yet it is well laced with positivism and still remarkably free of such individualized crime as mugging and vandalism; and this is despite manifold other manifestations in recent years of alienation among the young.[33] There have lately been throughout Japan 100 to 400 thousand young people who wilfully locked themselves in their rooms for weeks or even years on end.[34] More generally, Japanese youth is in a state of passive rebellion against "the system".[35] Some thirty years ago, an upbeat interpretation of the longer-term urban outlook globally was published from within the Mediterranean pro-"cives" tradition. It still commands attention. The gist was as follows. Urban expansion will continue insistently though stressfully well into this century. Then it will slow down markedly. As it does, the quality of urban life will much improve. Then sometime next century an "ecume-nopolis will come into being, binding together all the habitable areas of the globe as one interconnected network of settlements". Penultimately, this state of dynamic equilibrium will proffer lasting peace through international melding. Even so, our survival will further depend on our establishing a "genuine partnership with Nature", a balance to be struck "at every level from the single home with its garden to the entire globe".[36]

To the British pioneer of town planning, Sir Patrick Geddes, bringing the countryside into the cities as authentically as possible (gardens, natural parks, treelined avenues . . .) was an essential part of the re-engaging with Nature that he, like many of his generation (e.g. in the Boy Scout movement) saw as essential to the bodily and mental health of people growing up in the city.[37] [38] Effecting such a transformation across the developing world will, alas, be enormously expensive. There will also be acute shortages of usable space. A strong revival of interest has been evident of late in building very high though this owes something to intra-state competitiveness on non-lethal terms and a return to a long tradition of buttressing civic authority by making city centres very conspicuously the elevated hubs of grandly concentric ground plans.[39] The anthropological lineage is a quest for the Mandate of Heaven.[40]

Contrasting Similarity?

The UN finds that 70 per cent of sub-Saharan Africa's urban population was living in slums in 2005; the number thus encompassed was rising at 4.5 per cent per year.

For South Asia, the percentages were 57 and 2.2. Elsewhere across the developing world, the proportion of urban dwellers in slums varied, region by region, between 24 and 34 per cent; the regional growth rates thereof were very generally spread between 1.25 and 1.75 per cent.

What this can mean for the human condition is demonstrated in all its sordidness by Kibera, the slum in the middle of Nairobi, Kenya. Depending on the estimator and on seasonal migration, the Kibera population is put at between 600,000 and 1.2 million. Shacks are packed so tightly within its 630 acres as often to be accessible only on foot across streams of putrid ooze made dusty or else all the muddier by seasonal weather. Sewage runs freely. People resort at night to the "flying toilet", polythene bags thrown from the doorway. The ubiquitous stench is overpowering. Rape and other crimes against the person are commonplace.[41] Corruption engenders a feeling of hopelessness. This in turn sustains corruption. Yet many teenagers and young adults have preferred Kibera, in prospect at least, to rural stagnation.

Till recently, it seemed as if Kibera was too worn down by suffering to consider political violence. But it saw its share of post-election strife in 2008. There as elsewhere in the country, however, the killing and maiming was tribe on tribe, using spears and clubs. There has not been as yet any more systematic insurgency. As and when there is, it will be mindlessly insensitive to write it off as "terrorism", however much it may owe to outside stimulation. So much for the late Spiro Agnew's acid comment, "When you've seen one slum, you've seen them all". New York's immigrant tenements never got as bad as that. Luanda, Kinshasa and Lagos are among other African cities with slums that more or less do.

A crucial long-term requirement (which all concerned will find hard to meet this century) is that the emergent cities of the developing world effect urban progress more along the lines Geddes and other visionaries the first half of last century looked to.

Even such contemporary success stories as New York or London still fall short. When the then Mayor of London, Ken Livingstone, lately said "Its worse being poor here than anywhere else" in Britain, he had in mind a pronounced widening of income differentials within the capital, this then accentuated by yet another property boom. Meanwhile subjective feelings of congestion are being stirred by how crowded the South-East of England as a whole is becoming. Then again, though one may discern brittleness in the multi-ethnicity of the old East End, the familial and societal bonding often proffered could engender security and motivation for many individuals, some of the more aspiring excepted. The knife-wielding and gun-toting gangland warfare that has been so evident in English cities, but especially London, in 2007–8 must owe something to an especially pronounced weakening of an inter-generational social structure. Correspondingly, the collapse of manners and friendliness wherever people interact within "lonely crowds" is peculiarly marked in Britain's capital. It bespeaks a prevalence of anomie that could sometime be mobilized by political extremism. The British National Party and the almost as distasteful leftist Respect Party each have significant footholds in east London.

A cardinal strategic principle has always been the security of the home base. No medieval prince would embark on a Crusade without feeling sure everything was aright at home. In the modern world, the concept considerably relates to ensuring the internal stability of nodal urban areas. Generically they can flip all too readily into social and political violence, often because of the interaction between local and wider issues.

Notes

1 Nicholas Stern, *The Economics of Climate Change* (Cambridge: Cambridge University Press, 2007), p. 15.

2 *Ibid.*, p. 19.

3 P. J. Webster *et al*, "Changes in Tropical Cyclone Number, Duration and Intensity in a Warming Environment", *Science* 309, 5751 (16 September 2005): 1844–47.

4 "Have Hurricanes met their Match in El Niño?", *New Scientist* 194, 2605 (26 May 2007): 21.

5 Thomas C. Schelling, "What makes Greenhouse sense?", *Foreign Affairs* 81, 3 (May–June 2002): 2–9.

6 Matthew Sturm *et al.*, "Meltdown in the North", *Scientific American* 289, 4 (October 2003): 43–9.

7 R. S. Lindzen, "On the Scientific Basis for Global Warming Scenarios", *Environmental Pollution* 83, 1 and 2 (1994): 123–34.

8 *HCC*, Chapters 9 to 11.

9 Bjørn Lomborg, *The Skeptical Environmentalist* (Cambridge: Cambridge University Press, 2001), Fig. 154.

10 Al Gore, *The Assault on Reason* (London: Bloomsbury, 2007), p. 205.

11 Henrik Svensmark, "Cosmoclimatology: a New Theory emerges", *Astrophysics and Geophysics* 48 (February 2001): 18–24.

12 U. Dusek *et al.*, "Size Matters More than Chemistry for Cloud-Nucleating Ability of Aerosol Particles", *Science* 312, 5778 (2 June 2006): 1375–8.

13 S. Brönnimann, "The global climate anomaly, 1940–42", *Weather* 60, 12 (December 2005): 336–42.

14 See the author's *Future Global Challenge* (New York: Crane Russak, 1977). Chapter 7 (b).

15 David Cohen, "Earth Audit", *New Scientist* 194, 2605 (26 May 2007): 34–41.

16 Matthew Burrows and Gregory F. Treverton, "A Strategic View of Energy Futures", *Survival* 49, 3 (Autumn 2007): 79–90.

17 "Tapping the Oceans", *Economist Technology Quarterly*, 7 June 2008, pp. 20–2.

18 "Anxiously watching a different Arctic", *The Economist* 383, 3831 (26 May 2007): 56–8.

19 *HCC*, pp. 4–5.

20 Vilhjalmur Stefansson, *The Friendly Arctic* (London: Macmillan, 1921), Chapter 2.

21 Bernt Balchen, *War Below Zero* (Boston: Houghton Mifflin, 1944).

22 W. H. Auden and Louis MacNeice, *Letters from Iceland* (London: Faber and Faber, 1937), p. 139.

23 Francis Spufford, *I May be Some Time. Ice and the English Imagination* (London: Faber and Faber, 1996), p. 187.

24 Bill McKibben, *The End of Nature* (London: Viking Penguin, 1990).

25 Johan Huizinga, *The Waning of the Middle Ages* (London: Edward Arnold, 1955).

26 P. Cinzano, F. Falchi and C. D. Eldridge, "The First World Atlas of the artificial night sky brightness, *Monthly Notices of the Royal Astronomical Society* 328 (2001): 689–707.

27 David J. Travis *et al.*, "Contrails reduce daily temperature range", *Nature* 418, 6898 (8 August 2002): 601.

28 Bill McKibben, "How Close to Catastrophe?", *New York Review of Books* LIII, 18 (16 November 2006): 23–6.

29 Dr Nafis Sadak, *The State of World Population, 1988* (New York: UNDF, 1988), Table 1.

30 Paul and Anne Ehrlich, "Enough Already", *New Scientist* 197, 2571 (30 September 2006): 47–50.

31 *Bulletin of the Atomic Scientists* 62, 2 (March/April 2006): 34.

32 Elizabeth Baigent, "Patrick Geddes, Lewis Mumford and Jean Gottman: divisions over megalopolis", *Progress in Human Geography* 28, 6 (2004): 687–700.

33 Makoto Watabe, "Youth Problems and Japanese Society", *The Japan Foundation Newsletter* XXVII, 3–4: 1–20.

34 Alan Macfarlane, *Japan Through the Looking Glass* (London: Profile 2007), p. 228.

35 Akira Kogima, "Robbed of Dreams", *World Today* 160, 3 (March 2004): 26–7.

36 C. A. Doxiades and J. G. Papaioannu, *Ecumenopolis. The Inevitable City of the Future* (New York: William Norton, 1974), Chapter 38.

37 Elizabeth Baigent, *Progress in Human Geography* 28, 6 (2004): 687–700.

38 Patrick Geddes, *Cities in Evolution* (London: Williams and Norgate, 1949), pp. 52–7. A somewhat abbreviated version of a text first published in 1915.

39 Anthony Aveni, "Bringing the Sky down to Earth", *History Today* 58, 6 (June 2008): 14–21.

40 For an arresting introduction to this theme see Frances MacDonald, *The Vertical Frontier: William D. Howells and Frederick Jackson Turner*, a 2008 Master's Dissertation, Oxford.

41 "The Strange Allure of the Slums" in *The World Goes to Town*, Supplement to the *Economist* 383, 8527 (5 May 2007), p. 6.

15 | Critical Regions

1 A Fertile Crescent?

Curving through the Levant and Mesopotamia from Gaza to the head of the Gulf, the Fertile Crescent is the primary source of those Neolithic "economic revolutions to which we owe our cultural progress".[1] Much of Iran lies outside it as formally defined. Strategically, it is all part and parcel, especially at present.

Known as Persia before 1935, Iran retains a nostalgic pride we do well to allow for. Cyrus the Great (d. 526 BC) established an Achaemenidian empire extending from Cyrenaica to the Indus and to the Aral. Not least could it be commended for religious inclusiveness. Three religions originating in ancient Persia were Zoroastrianism, Mithraism and Manichaeanism. The empire, its capital Persepolis included, was trashed by Alexander the Great in 334 to 331 BC for no reason but vainglory.

A general conversion to Shi'ite Islam from AD 648 could connote doctrinal belligerency. But there was a gentler side. By the early Middle Ages, Persia was the main fount throughout Islam of *sūfism*, an ascetic-cum-mystical approach to knowing through love an Allah inaccessible to mere reason. Best remembered of her *sūfi* poets is Omar Khayyam who was also a peerless mathematician.

Across Southern Asia, *sūfism* has long been disposed towards syncretism with other faiths. It also encourages *wali*, local saints, more than does the Islamic mainstream. This may relate to the survival within Iran of diverse faith and tribal identities. Tribal worship can assume a fervency awesome to observe.

Modern Imperialism

As in Afghanistan so in Persia, Russia and Britain competed for influence in the nineteenth century. Some Islamic clergy favoured Tzarist Russia as less of a modernizing influence. In 1907, the year of decisive London–Moscow *détente*, diplomacy divided the country into three spheres: Russian, neutral and British. Soon the oil factor came into play.

In August 1941, the USSR and Britain jointly invaded Iran putatively to curb growing Axis influence therein. But that objective had already been secured by British invasions of Syria and Iraq. The aim now was to use the Trans-Iranian railway (from Bandar Shahpur to Bandar Shah) as a supply route to the USSR. Iranian radicals were never going to be relaxed about that. Nor were they about the coup staged with CIA backing in 1953 to oust the militant prime minister, Muhammad Mussadegh.

With Mussadegh's departure, Reza Shah Pavlevi sought to impose his own authority, his focus being to promote reforms designed to head off any revolutionary alliance of "the Reds and the Blacks" — i.e. the Marxists and Islamicists. He may have come to feel, however, that the West was saddling him with geopolitical tasks which were altogether too diverting. The Central Treaty Organization (CENTO) which, after 1958, comprised Britain, Turkey, Iran and Pakistan, linked NATO and the South-East Asian Treaty Organization to complete the Dullesian ring of containment. Then the British encouraged, from 1965, an Islamic Alliance between Iran and Saudi Arabia. Soon, too, the Iranian navy was being invited to assume a policing role in the Indian Ocean.

In October 1974, three years after huge celebrations at Persepolis in honour of Cyrus the Great, the Shah gave a press conference about nuclear energy. He received sympathetically a suggestion that one way to avoid "nuclear blackmail" was to build reactors fuelled by natural uranium. He revealed plans to install 23,000 megawatts of generating capacity. Reactors would be bought "both from the Western countries and from the East if they are so willing. A diversity of suppliers is what we envisage".[2]

Eventually, in February 1979, the Red-cum-Black revolution did take place, spear-headed by Ayatollah Khomeini. Like Reza Shah, Khomeini was to justify his autocracy as promoting the public good. But this was now interpreted as the Islamization of the common weal, not development Western style. He distrusted bureaucracies. He declaimed "economics is for fools".[3] His regime would be a populist theocracy. Early on, it shot some leaders of the Baha'i and Jewish communities.

Iraqi Aggression

Eighteen months into this revolution, Saddam Hussein gratuitously launched a full-scale invasion, using a panoply of advanced weapons supplied by the USSR and France. Inspired by Shi'ite millenarianism concerning the imminent return of the "hidden Imam" to herald a new world order, the Iranians fought back hard, launching many human wave attacks. By 1984, they were poised to take Basra but were then to be thwarted by repeated Iraqi recourse to poison gas.

A peace accord restoring the *status quo ante* was signed in 1988. A million Iranians had been killed or maimed. Much damage had been inflicted on urban areas by Scuds. And so on. The rest of the world had done nothing, save that Moscow and Paris had sent Baghdad more military wherewithal.

The June 2005 Presidential

The turn-out in Iran's 2005 presidential election was 62 per cent for the first ballot and 60 for the run-off. In the first stage, the vote was spread very evenly between the six candidates cleared to contest by a clerical Council of Guardians, an evenness expressive of diverse regional loyalties. The victory of Mahmoud Ahmadinejad, Mayor of Tehran, was a surprise. Cast in the Khomeini mould ideologically, his appeal was to the traditionalist and underprivileged masses. Since Iran remains the most solidly Shi'ite country in the *umma*, the Community of the Faithful, his political persona could conflate religion, nationalism and social justice. Anti-corruption was a keynote theme.

His assumption of office was never going to ameliorate the harsh rejection of liberal values. In May 2008, Ahmed Batabi escaped to the USA. He had been sentenced to 15 years imprisonment in 1999 because a picture of him protesting had appeared on *The Economist's* front cover. A summation of what he had been through in the interim makes chilling reading.[4] In 2008 a number of convicted thieves have had their right hands and left feet removed.

Israel and Holocaust

On various occasions, Ahmadinejad has declaimed that Israel must and will be erased from the history books. If one believes he is seeking WMD, these declamations must be read as a genocide threat. They could owe something to his "rather reckless contempt for American military power", an attitude which may notch down as the situations in Iraq and Afghanistan belatedly improve.

His reiterations that the Holocaust never happened or has yet to be confirmed are "seen by many Iranians as rather peculiar at best; and at worst not only morally wrong but dangerously ill-advised". Generally the Holocaust "is dismissed in Iran only in so far as it relates to the idea that the Palestinians have been punished for the sins of another continent".[5] Holocaust denial has been Mahmoud Ahmadinejad's way of defying with pure unreason the *sūfist*-led intellectual tradition valued by educated Iranians and of which he is no part.

Nuclear Ambitions

More assured of patriotic support with or without Ahmadinejad is the development of a civil nuclear programme, support well consolidated when President Bush included the country in an "axis of evil" in 2002. As regards bomb development, the current spiritual Supreme Ruler, Ayatollah Ali Khamenei, has said building or using nuclear weapons is against Islamic law. However, a study at the Princeton University Center for Human Values in 2002 illustrated how uncertain such judgements were within Islam given the "inevitable gap between precedents set in the 7th century and the questions of subsequent generations".[6] As much could be said, of course, for other ancient faiths.

The politics of this business initially revolved around whether IAEA inspectors should be allowed back into Iran to monitor the uranium enrichment by gas centrifuge basic to its whole nuclear programme. Negotiations with Iran have been led by an EU troika — Germany, France and Britain. In November 2004, Iran negotiated with them a deal to suspend enrichment altogether partly in exchange for the EU's resuming trade talks.

However, within days of taking power in August 2005, President Ahmadinejad announced Iran would reactivate its Isfahan plant for the gasification of natural uranium, the first stage in the enrichment process. **(See Appendix B)**. Even should this facility be completely or partially underground, it might not be difficult to immobilize by air or missile strikes. Otherwise, and if Iranian protestations that they are interested only in the peaceful application of nuclear energy do prove disingenuous, Tehran could have enough weapons-grade U 235 to make its first bomb sometime between 2009 and 2015.

Actually fabricating such a device should not be beyond its ability. Indeed, this was one conclusion drawn in the US National Intelligence Estimate report of December 2007. However, the main message taken from that study, which represented a broad consensus among 16 agencies, was that Iran had stopped work towards a bomb in 2003. On the other hand, IAEA personnel reported in 2008 indirect findings appearing to indicate high explosive triggers for nuclear bombs being developed still, along with work to adapt the nose cone of a Shaheen rocket to carry nuclear warheads. The Iranians themselves acknowledge receiving from A. Q. Khan a document relevant to "gunshell" triggering. (See also Appendix B). Much extra monitoring might be required in the event of an IAEA re-entry.

Alternative Pathways

Quite possibly, Ahmadinejad will not be re-elected for a second term in 2009. Unemployment had climbed steeply by mid-2008. Sanctions were biting, notably in regard to banking services. High oil prices may not correct matters. And foreign policy issues could also be raised by critics within the conservative establishment, the Supreme Ruler likely among them.

If the present incumbent does exit, his successor is almost bound to be conservative but perhaps less likely to be as fired up as he so far has been. Either way, it will be necessary to insist on IAEA inspectors returning in strength. Whether those concerned should stick out for a dismantling of Iran's uranium enrichment facilities is another question. Such a stipulation any Tehran government might have to stand too proud to accept; and could be superfluous to requirement.

A military strike, even one confined just to the Isfahan site, could have unpredictable repercussions. These might be limited, as appears to have been the case with those from Israel's strike against Iraq's Osirak reactor in 1979. More likely, instabilities would be set in train, immediately or later, which could lead to indigenous WMD in some shape or form. Iran has nearly four times the area of Iraq. Population-wise it is two and a half times the greater. Its social geography is more complex. So there could be a major "what follows" problem. May one add that the same applies to the covert operations the Bush administration is now undertaking in Iran, with or without inter-allied consultation.[7] Even at this stage these might impede a worthwhile nuclear agreement.

Holy Land Foci

The conflict in the land holy to three great monotheisms has been addressed in its military and strategic aspects in 9 (5). One should now seek to trace certain causal threads in the background to this situation.

The Archbishop of Canterbury warns of the Islamic world's perception of continual aggression towards it on the part of the Christian West. The Crusades are a core theme. The first was launched when Pope Urban II, himself a French aristocrat, addressed a council convened at Clermont in 1095. Such contemporary accounts as we have are partial in both senses. But his punch line apparently was that the land of France was "too narrow for its own population; it furnishes scarcely food enough for its cultivators. Hence you murder one another". He duly urged the knightly class to liberate the Holy Land instead.

Historians question whether France was experiencing Malthusian pressure as chronically as this rendering implied. But undoubtedly endemic knightly mayhem was being aggravated by a several-year drought. Another aim Urban had was the reunification under Rome of the Latin and Greek churches.

A well-found army 30,000 strong left for Jerusalem the next year. Displaying immense fortitude, it marched across Europe and Anatolia to storm the Holy City in July 1099. Its departure had been preceded by that from Cologne in April 1096 of tens of thousands of *sans culottes* from the manifestly overpopulated and socially disaggregated Lower Rhine-cum-Meuse region. They had been galvanized into a mission next to impossible by the millenarianism of Peter the Hermit, a charismatic phoney. Soon they were massacring Jews in other Rhenish towns. Most of these *sans culottes* themselves disappeared in one way or another before ever reaching the Levant. The survivors came together there as an ill-accoutred coterie styling themselves the *tafurs* — i.e. vagabonds. They were well to the fore in the killing of all Jerusalem's Jews and nearly all its Moslems hard upon the city's fall. Neither of the peoples in the current dispute have much to appreciate the Crusades for.

Outside observers see other commonalities in the mental make-up of Palestinian Arabs and Zionist Jews. One is resilience in the face of adversity. Another is a disinclination to examine their own contribution to a miserable and menacing impasse. A third is apprehension that past injustice will repeat itself, not least as regards denial of land or water.

A Churchillian Lapse

In 1921, Prime Minister Lloyd George moved Churchill to the Colonial Office. One task awaiting him was implementation of the 1917 Balfour Declaration on Palestine. He had been, at least since 1906, "very well disposed towards Zionism".[8] It was a pitch not unrelated to his empathy with the Boers, alias the Afrikaners. In his Commons maiden speech, he had praised the Boer as a "curious combination of the squire and the peasant" and told how ashamed he had been to see such men ordered about "as if they were private soldiers" by young subalterns. The Boers were a "People of the Book", the Old Testament. He therefore saw them as possessed of industry, integrity, enterprise and dependability. The same applied with the Jews. Lloyd George thought likewise.

All concerned should have recognized that the crux was preparing the Palestinians (through education reform and in other ways) for an accommodation any people, including the British then or the Israelis now, would have found hard. It was inadequate for the Colonial Secretary just to aver that, under the rule of law, every inhabitant of Palestine stood to gain from the proven ability of the Zionists to spearhead development, rural as well as urban.[9] Mutuality could never be effected entirely spontaneously. For one thing, the Palestinian political culture lacked any refinement after centuries of soporific Ottoman rule. Witness a memorandum presented to Churchill in March 1921 by the Executive of the Haifa Congress of Palestine Arabs. One passage reads, "The Jew is a Jew all the world over . . . He encourages wars when self-interest dictates and thus uses the armies of the nations to do his bidding".[10] So a fascistic backlash had already started.

False Economy

The omission of a Palestinian Arab development programme may have been Winston Churchill's but it was consonant with the British colonial tradition with its heavy reliance on "indirect rule" (i.e. by pre-existing indigenous authority) coupled with economic *laissez-faire*, this perhaps within the parameters of imperial tariff preference. France's philosophy of cultural assimilation might have been adapted more easily to this singular political experiment. Still, all governments felt too short of money to be very imaginative during the post-war economic slump.

That money remained a major constraint is evident in the grossly inadequate size of the security forces in Palestine. Configurations and nominal rolls fluctuated. But in mid-1929, there were just 1600 in the Police, not 200 of them British; and 85 in an RAF armoured car detachment. That year 130 Jews and 120 Arabs died in political disturbances.[11] As World War Two broke, Palestine was nowhere near conflict resolution.

Post Holocaust

In the aftermath of the Holocaust, things ineluctably turned more fraught. Violent action–reaction became endemic. A car bombing by Jews was quickly matched by an Arab one. The British authorities also turned angrily reactive. They came to see all Jewish fighters as "terrorists", especially after the Irgun's bombing of the King David hotel in Jerusalem in July 1946. British troops in-country numbered 100,000 by that year's end.

In a calmer atmosphere, London might not have brusquely rejected the UN partition plan of 1947. One objection raised was that more Arabs would be living under Jewish control than Jews under Arab. Yet this reflected how the apportioning of land allowed for substantial Jewish immigration. The really critical provision was the internationalization of Jerusalem. Successfully implementing that would have put Israel/Palestine well on the way to becoming a binational confederation albeit one in which each component had its own security forces.

No doubt the Irgun Revisionists would have sought violently to wreck the whole scheme. Whether their zealotry could have prevailed against Ben-Gurion, the British, the Americans and world opinion in general is another matter. The greater likelihood is that Revisionist rejectionism would have given such a partition respectability in Arab eyes.

A Hashemite Lapse

Of the wars waged between Israel and her Arab neighbours after 1948, quite the most impacting on strategic geography was the 1967 one. It was so because King Hussein elected to join in. He did, it seems, on the basis of two cardinal errors in situation appreciation.

The one concerned the tactical air balance. In Amman that September one was told on good authority that the Jordan Arab Army's mind set had been that it would fight under conditions of air parity. It sounds like yet another instance of warriors in light blue simply being discounted by those in khaki. Jerusalem apart, it was never remotely likely that tactical aviation would thus be written out of the assize of arms.

Air campaigns tend not to proceed *pari passu* for long, particularly where the geography is tightly delimited. Usually they soon yield a decided advantage, one way or the other.

On this occasion, it was going to be Israel's way, given that her qualitative ascendancy would rapidly be brought to bear through pre-emption or however. Besides, the proficient Jordanian air arm could never have been remotely a match for the Israeli numerically. Nor, even on super-sanguine assumptions, could Egypt's air arm expect to extend protection over the Hashemite realm. Syria, too, would have found total cover infeasible. Moreover, there was for Damascus a motivational glitch. Right up to the line, it had been waging a splenetic propaganda offensive against Amman. The king himself had been depicted as a "jackal" whose mortal remains should be "nailed to the door of the British Embassy" as soon as possible.

Nor would it be easy to argue convincingly that the Hashemite regime would have been overthrown by insurgents had Jordan not gone to war. As it was, the guerrilla movements waxed stronger in the wake of Jordan's military defeat. Furthermore, there was some shift leftwards within the insurgency domain away from Fatah and towards the Popular Front for the Liberation of Palestine, a new movement which emphasized that a revolutionary transformation of the Arab world must precede the final showdown with Zionism.

However, the high profile the guerrillas therefore assumed in downtown Amman as of early 1970 concealed operational weakness. They were incurring heavy casualties in their operations against the Israelis in the Jordan valley; and this would soon induce them to resort to aerial hijacking. But that made King Hussein decide in September the time had come to have his army reimpose royal authority. This it did very professionally. (See 6 (1)). Since when, insurgent movements have been absent from the Jordanian political landscape.

Oxford's leading specialist on the Israel–Palestine imbroglio is Avi Shlaim, very much an Israeli peacenik. His verdict is that "Hussein made the mistake of his life by jumping on Nasser's bandwagon. The price he paid was the loss of half his kingdom, including the jewel in the crown, the old city of Jerusalem".[12] In the light of the above, it would be hard to disagree.

A further consequence of Amman's abstaining might have been the collapsing once and for all of the Nasserist vision of an Arab weal truly united from the Atlantic to the Gulf. The disparate geography never allowed of this; imagining it could only inhibit genuine progress — Islamic, Arab, Palestinian . . . Originally there was a notion that the *umma* created after the death of the Prophet could be run as one caliphate from Damascus or Baghdad. This did not survive two centuries. Later the Ottomans proved unable to run well so over-extended an imperium. Nor has the Arab League been effective since 1945 as a regional planning body.

Israel's Demographic Gridlock

In July 1967, Yigal Allon (an Israeli government minister soon to head the foreign office) enunciated a plan for the future of the newly occupied West Bank. The text was never published and never became official. None the less, it afforded the framework for policy thenceforward.

A key stipulation was that the Jordan river must be regarded as Israel's strategic border to eastwards. Indeed, Allon's view was that, in order to secure it, work must immediately start on a string of settlements along the valley. Their locations were to relate topographically to (a) providing early warning, (b) blocking mechanized incursions and (c) counterattack.

That September, a ranking Jewish author of leftist inclination and I talked at some length with Allon. My colleague argued forcefully that regaining the Wailing Wall was Israel's historic opportunity to break out of her Zionist *leitmotif*. Allon appeared absolutely unmoved. Nor did he evince the slightest hubris about recent military triumphs. His dourly implacable pitch throughout was that "we have had enough" of Arab rejectionism as expressed in guerrilla incursions from Gaza and bombardments from the Golan.

A dependable 1988 summation of the plan told how Allon had acknowledged explicitly the national aspirations of the occupied Arabs. True in this to his social democratic background, he had assured the government early in 1969 that his scheme did not "require the annexation of areas densely populated by Arabs". Nevertheless it was "anchored in the Jewish peoples' historic right to the Land of Israel from the moral standpoint". Moreover, the "Jewish character of the State of Israel had to be maintained". He envisaged the complete incorporation into Israel of an expanded Jerusalem. How much expansion the Holy City could take without its personality being compromised was a subject of debate in itself.[13]

A plan so characterized by conflictual loose ends lent itself to slanted interpretations. Soon there was Jewish settlement well outside the guidelines Allon had laid down. Apprehending this, David Ben-Gurion earnestly opposed all settlement in the occupied territories. That he could no longer prevail was the more regrettable in that, at middle management level, the Israelis were pursuing some enlightened policies in the occupied territories that first decade — a freer press, health care, agricultural development . . .

Things were made no easier by the binary geography of the occupied Palestinians. By 1975, it was evident that Jordan was making ready to disclaim responsibility for the West Bank and Gaza. Not only in Israel were doubts expressed about how these two entities could together constitute a statehood able to survive more than a couple of years.[14] After all, the population of crowded Gaza had effectively derived from the Jaffa evictions in 1948 and was very radicalized politically, a mood regularly recharged by the "long hot summers" effect. The situation *in toto* called not for a quick fix but for a major exercise in social engineering. It still does.

A Fresh Start?

Is it not high time to ask what the current "Middle East peace process" amounts to a decade and more on? A serious indictment has to be that it resonates too little with the respective peoples. Disillusionment and distrust run too high. The extremists on each side (ethnic-cleansing Revisionists and hard-core Hamas) seem currently at a discount. But it would take little for them reciprocally to recover.

If this reading be correct, it was incongruent for President Bush to anticipate a peace accord being reached in 2008 or President Sarkozy to claim a break in the deadlock as a first fruit of his Mediterranean multilateralism. We remain a long way from enough Israelis feeling able to make a comprehensive offer enough Palestinians would feel able to accept. And time is on nobody's side, whatever extremists intimate.

From *c.* 1955, much was said, in relation to central European security, about

Confidence Building Measures (CBMs). To some of us these smacked of affectation since that situation locked into a wider West–East balance which was essentially nuclear. The Palestinian–Israeli conflict is more detached but also very confined. Several miles either way can often be critical which makes mental blocks harder to dissolve. It further means that confidence building and substance generation must continually interweave.

A radically fresh approach may therefore be required by the two nationalities and by the outside world. Should there not be a new "road map", one which sets out s blueprint for a viable Palestinian state (meaning, most basically, a functioning democracy with a stake in a unified Jerusalem) which is to be finally realized, not this year or next but by a target date perhaps a full decade hence. *En route* there would need to be not contrived CBMs but an interactive peace building process. Palestinian bitterness might be assuaged by sustained gradualism better than ever it could by trite immediacy; and this could invoke a positive feedback in Israel and elsewhere. So may one speculate about the particulars of a viable accord.

Apropos Jerusalem, can one not envisage dual nationality for the Palestinian residents of the old city, this within a special municipality, and presuming — of course — open access to religious sites? A Palestinian parliament within the city bounds (as already envisaged by all) could also build intercommunity confidence. Israeli retention of the new Jewish suburbs on the east side of the city ought to be negotiable, subject to compensation for land sequestered since 1967.

There is wide agreement that deeper into the West Bank the great majority of Israeli settlements will need to be withdrawn as peacebuilding proceeds. They are too much a driving force for trends to ruinate Palestinian landscapes, force up Palestinian property prices, and compromise the social democratic identity of the Zionist mainstream — e.g. transit roads for settler use only. Moreover, the settlements themselves are edifices made ugly by tight confinement to the hill tops.

It would, of course, be fair to point out that such a situation is not unprecedented. The British army's hill stations in India had similar connotations.[15] But they were set within wider horizons and could adopt more congruent layouts. Having said that, if a proportion of the settlers were willing to remain, though under Palestinian rule, that could assist confidence building.

A raft of other issues would need also to be addressed. Easing Malthusian pressures would not be easy against the background of a looming water crisis aggravated by climate change, a crisis Israel's exceptional prowess in desalination could be instrumental in managing. Both the Jewish diaspora and the Palestinian could become more vulnerable if, as is by no means impossible, the world ambience becomes chronically more tense. In which case, they will look for a right of return to their respective territories to be preserved in this regard at least. Both communities will need to contract their agricultural sectors. A Palestinian programme of family planning will be important. So will addressing the situation of the Lebanese Palestinians, together with the democratization of Jordan. Also a resolution of the Golan question might detach Damascus from what currently is the less rational authoritarianism of Tehran. Then by no means least is the need to involve the Arab oil sheikhs heavily in funding development. It seems a *sine qua non* in principle as well as in practice.

An acid test of growing togetherness would be collaboration on security, including thwarting threats presented or instigated by authoritarian regimes unhappy about democratic harmony progressing thus. Additionally, internal arms control and external monitoring would likely be needed to secure the mutual border in each direction.

Israel's 12,000 Palestinian prisoners successively being freed or transferred would be one measure of confidence gaining. Another would be a progressive opening up of the barrier wall. For the latter, we have a present instance which is quite relevant. It is the beautiful Renaissance-style stone bridge built then by the Turks across the Neretva river at Mostar, the chief city of Hercegovina. During the strife hard upon the break-up of Yugoslavia, it was sundered by shelling: an action which sharply separated the Moslem part of the city from the Christian. Now the bridge's faithful reconstruction betokens reconciliation. Negation has turned positive. Unfortunately though, the same cannot yet be said of Bosnia-Hercegovina as a whole.

An outcome along the lines here aspired to would have commended itself to the greatest Zionist of them all. In 1940, Albert Einstein envisioned forging "an advantageous partnership which will satisfy the needs of both nations".[16] One must further aver that the Holy Land, the region and the wider world will be in still greater peril if we enter with this dispute unresolved what is liable globally to be the menacing second quarter of this century. A regional susceptibility to climate change underlines this truth.

Levantine Climate Change

The period *c.* AD 700 can instructively be considered in relation to the Levant and thereabouts. That it witnessed a deterioration in the human condition has long been recognized. But there used to be a disposition to attribute this basically to depredations by conquering Islamic horsemen, a contrast sometimes being studiedly drawn with how the early Zionists were making the desert bloom. A sensibly restrained example will suffice. Walter Clay Lowdermilk was an American soil conservationist who became Chairman of UNESCO's Commission on the Arid Zone. A devout Christian, he saw Zionism as a way to revitalize the Holy Land via Jewish–Arab partnership, culminating perhaps in a Jordan Valley Authority. He adjudged the Prophet's nomads to have been responsible for destroying "many cities" and "cultivated areas" but concluded that Palestine was not finally "plunged into the age of darkness" until the Crusades.[17]

However, a negative link with martial Islam was called into question by Rhys Carpenter, a Classical archaeologist at Bryn Mawr. He pointed out how in nearby Anatolia, which those nomads never reached, many churches were built in the fifth and sixth centuries. Come the seventh, no more were; and existing ones fell into decay. Nearby Syria they did occupy but as something akin to liberators. No towns had to be stormed. Yet here, too, church building died away.[18] Nor should we overlook the matchless progress of Islamic agriculture[19] and, indeed, architecture, in Spain.

Most conclusive, however, is the scientific evidence of a climate change from moist cool to arid warm as medieval warming began to set in generally. In 1998 Arie Issar, a ranking Israeli hydrologist at the University of Negev, went firmer than ever before about such a switch in the Levant between 500 and 700.[20] Likewise, Amos Frumkin and colleagues at the Hebrew University of Jerusalem, having measured drainage channels in bedrock in the northern Negev, concluded that *c.* AD 600 the water cycle got weaker remarkably quickly.[21]

With climatic zones tending to move polewards under the influence of global warming, the Holy Land may soon become appreciably more arid, this relatively suddenly. So far this time there has not been a secular decrease in regional rainfall but showers have got shorter and more intense, a trend conducive to more water running off straight to the ocean. More seriously, by 2100 total precipitation could well reduce by a third. What should probably be discounted this century is salt pollution of the main Israel–Palestine acquifer by rising sea levels. Yet overall the geography of this region is turning more fragile, harder to make a centrepiece of planetary togetherness rather than of international discord.

2 The Orient Ascendant

Pandit Nehru, India's first Prime Minister, was among the many who misread Maoist Communism. He divined China to have the inner strength to reinterpret Marxism-Leninism but presumed she would broaden it forthwith. From whence came his faith in building Afro-Asian neutralism on a New Delhi–Beijing axis, a faith savagely mocked by his troops coming off worst in the 1962 Himalayan War.

Through the twentieth century, the resilient toughness of East Asia in general has been expressed in surges of martial ardour or economic dynamism. Crucial to this has been a collectivist spirit arguably strengthened by the absence of Church-versus-State dichotomies. Not that these East Asian attributes are wholly beneficial. With them can go an inability to relate to other cultures except via absolute rejection or wholesale ingestion. Evidently the development of pluralist modes within the body politic does not come easily. The concept of constitutional opposition has yet to sink deep.

Confucianism

Confucianism can still be seen as a philosophy which unites the region. Though some-times treated as straight religion, it owes its unifying power more to its lack of reliance on metaphysics, saving the notion that "the Mandate of Heaven" is somehow conferred on good rulers.

As for Confucius himself, he lived (*c.* 551–479 BC) at the same time as Buddha (*c.* 563–483 BC), Zoroaster (*c.* 628–*c.* 551 BC) and the founders of Jainism. He was also a contemporary of Sun Tzu, rated (by virtue of his text *The Art of War*) the first strategic theorist known to History. He, too, stressed psychological and ethical concerns. Witness his averration that it is better "to take intact a regiment or company or squad than to destroy it".[22] Embracing Geography as well, he highlighted the importance of "focal" ground. He was a field commander in the state of Wu, formed around the pivotal Lower Yangtse.

Confucius came from the Shantung peninsula. He lived through the worsening instability after the calm of the Shang and early Zhou dynasties, a time leading to what would later be termed the "Warring States period", 481–221 BC. His overriding concern was to promote peace, albeit through philosophy not generalship.

The cardinal attribute identified in his writings is *rén*, a very special empathy. Rightfully it should apply to five binding relationships: friend and friend; wife and husband; younger and elder brother; child and parent; and, above all, subject and emperor. At each stage, the viability of the nexus depends on both parties being educated to behave selflessly. The whole matrix functions properly only when all five linkages do.

Confucius sought to promote as best he knew dynastic stability. Since when, Confucianism has usually acted thus regardless of whether a ruler be good, bad or dreadful. Long supportive of this proclivity in China was the development during the Tang and Sung dynasties of the three-tier state examination system designed to test stringently someone's basic suitability for official positions from local magistrate to imperial adviser.

Buttressing government thus has often been seen as involving the entrenchment of rigid conservatism. Max Weber (1864–1970), the pioneer political sociologist, identified it as a prime reason for the backwardness of the China of his day.[23] Sometimes, however, tight centralization allows of radical policy switches. The history of China and of Japan has afforded successive examples, by no means all of them for the good. Some have been overtly anti-Confucian. But this does not gainsay their being the products of a political culture Confucian thought has done much to generate.

The heavy educational emphasis on the Confucian and other classics never precluded Chinese science from being highly dynamic until that traumatic era, the early fifteenth century (see below). The imperial weather records go back continuously 2500 years. There was the "early conception of an infinite universe, with the stars floating in empty Space".[24] What we call the "solar wind" was recorded in AD 365 through the slewing of cometary tails. There is data adequate to discern the perturbation of Halley's comet during its appearance in AD 837.[25] Printing developed early in the Middle Ages. Into the fifteenth century China was also ahead in such technologies as shipbuilding, hydraulics, iron smelting and fibre spinning.

The First Emperor

The creation of China by Emperor Qin (see 2 (2)) ranks high among History's radical achievements. Without it, Geography could well have allowed this land of the three great latitudinal rivers to develop as valley empires, all three of them so elongated as to stand desperately exposed to neighbourly depredations and liable to disintegrate periodically into strings of mainly landlocked fiefdoms. A much remarked example of how thoroughly Qin consolidated the unification is his regulating the axle width of all vehicles using the main roads out of Xian.

Through the High Middle Ages

Two beneficent adjustments made by medieval China (described in 3) were facilitated by an imperial administration that was unitary, strong and adaptive down to village level. The economic lift-off led by the introduction of Khampa rice via imperial edict was the one. The other was Kublai Khan consolidating nationwide his position as Yuan Emperor. His was a very Confucian approach — one singular in Mongol terms, Persia perhaps apart. Since Yuan China was so large and thriving, it imbalanced the Mongol world and thereby contributed to its break-up.

Global Excursions

In 1405, Admiral Zheng led out of the Yangtse 300 ships with 28,000 men aboard.

This and subsequent expeditions bespoke Chinese dominance within the Indian Ocean and possibly beyond. An arresting hypothesis has been advanced by an ex-Royal Navy submariner with an Oriental background. His inference is that, between 1421 and 1423, there were four Chinese task forces which traversed between them the High Seas, reaching the Arctic and the Antarctic.[26] At all events, these navigators found on return the imperial authorities turned off completely. It marked the end of such adventures bar one voyage to Arabia in 1435. This rejection even entailed the destruction of many fleet records and ship designs.[27]

Some of the forays had incurred heavy losses. Famine was causing unrest along China's northern border. Jealousy of Zheng as a Moslem eunuch came into play. So, too, will have mandarin anxiety about cultural pollution. So may have a shortage of quality timber.

A further factor has lately been revealed more clearly by chronicle studies. The Emperor since 1402, Yong Le was a usurper within the Ming dynasty. Insecurity and aggressive pride generated within him boundless ambition and manic cruelty. Feeling safer in Beijing than Nanking, he made the former his seat of government. To which end he built the Forbidden City (using a million forced labourers, there and elsewhere) between 1406 and 1421. After weeks of opening celebrations, he saw fit to massacre his large harem. That very night his Forbidden City was ravaged by fire caused by lightning. Petrified by this heavenly punishment, Yong Le collapsed into depression. He died broken hearted in 1424. So a single lightning stroke may have set the world up to be turned Eurocentric. China progressively lost room for manoeuvre and zest for initiatives.

Confucian Demise?

The imperial examinations were abolished in 1905. Then in 1911 the Manchu dynasty was overthrown by Republicans, many of whom had spent years abroad. There was duly a backlash seeking to entrench Confucianism as received religion. But those opposed included most of the leading Confucians.[28]

Chinese Communism has not infrequently been seen as a Confucian derivative although its cadres might have cited Sun Tzu more. A bad distortion was engendered during the Cultural Revolution (1966–74) with Confucius being ritualistically blamed for every ill post-1949. Since when he has been accepted into the approved pantheon of sages gracing China's illustrious past, sometimes as *primus inter pares*.

Elsewhere in East Asia, sundry attempts have been made to portray Confucius as encouraging devolved democracy as a prospective counterpoise to Marxian centralism. But this can be oxymoronic. His concern was to extol familial and especially social ties, not political activism.[29]

Continued Switching

At all events, the tradition of sharply radical policy switches was not going to be cast aside post 1949. The Great Leap Forward in rural affairs from 1958 was to be a demonstration of the contrary, calamitously so.

Aridity acutely affected much of inner Eurasia through 1960. Well understood even at the time was how it helped bring Nikita Khrushchev down in 1964, this by compromising the plan he had launched in 1955 to cultivate the Virgin Lands of the

USSR's steppic south. Collateral rainfall deficits during the Great Leap Forward accentuated the tragedies ecological iconoclasm was generating. Purblind campaigns had been launched against grain-eating birds, the massacre of sparrows reputedly passing the billion mark. Crop planting densities were recklessly doubled. "Much of the accessible timber around tens of thousands of villages and towns was cut to provide charcoal for the primitive 'backyard' furnaces producing useless pig iron".[30] This quote is from a University of Manitoba study that tells of ecological savaging throughout the Maoist era, relying almost entirely on Maoist sanctioned sources. As regards the Great Leap Forward, we now know 20 million died of famine over three years.

Another catastrophe was the Cultural Revolution (1966–74), Mao's manipulation of the Chinese flank of the worldwide youth revolt. It was characterized by rabid xenophobia, industrial chaos, shattered universities, closed research centres, desecrated churches, teachers terrorized by crazed children, thousands of casualties in street violence, labour camps, and the gleeful destruction of cultural relics. From a Hong Kong observation post, one could see squads of farmworkers marching to and from the paddy, each and every one clutching at high port a copy of Mao's "little red book" of analects. It was therefore surprising afterwards to find a ranking Cambridge economist suggest that "The Cultural Revolution makes the accusation of aggressiveness less plausible than ever".[31] In 1960, Karl Wittfogel had seen Beijing as having already "developed a total managerialism and a degree of economic and personal control never exerted by any hydraulic society".[32]

Those who endured the Cultural Revolution were left very solidly resolved nothing like it should ever recur. Within two years of Mao's death in 1976, China was moving towards its present market economy set within Capitalist Communism — five year plans and tightly guided democracy.

At 1300 million, the mainland population is twice the 1958 figure. With over half still country dwellers, the crux is the creation of 20 million new urban jobs annually. Also a big drive is now under way to modernize the western provinces, meaning just over half China's area and a quarter of its people.

Nor does the economic and technological agenda stop there. Take transport development by China abroad. Direct links to Tashkent and to Rotterdam are being planned. Pakistan's internal network is being enhanced; and a deepwater port at Gwador in Pakistani Baluchistan has been agreed in principle. Collaboration on modernization in Myanmar has proceeded since "Beijing played a crucial role in shutting down" a Communist insurgency there is 1989.[33] China has invested heavily in and traded ever more extensively with Latin America the last 10 to 15 years, filling a void left by diminished American and Russian interest.[34] Since 1965 it has ramified progressively its economic presence in Africa, adhering still to the declaratory Maoist precept of "non-interference in internal affairs". A third of China's oil supplies are from that continent.[35] Neither should Beijing's Space ambitions be forgotten.

On the military side, its build-up of ICBMs proceeds but gradually: of 46 deployed, six are Dong Feng-31s — land-mobile and MIRV-bearing. One suspects that, were these fully satisfactory, more would be in service now. Three strategic missile-firing submarines have also been observed though two may not be fully commissioned as yet. The first attempts to develop this genre began in 1965.

Additionally, 750 Intermediate and Medium Range Ballistic Missiles (IRBM/MRBM) are in service, all of them able to be trained on Taiwan, opposite to where a high proportion of this missilry is positioned. Meanwhile the country's local war forces are large but of uneven capability. Chinese aircraft designers are striding ahead aerodynamically. Electronically, their progress seems more halting, Russian and Israeli inputs notwithstanding.[36] Aerodynamic virtuosity let down by electronic mediocrity was the fate of Soviet warplanes throughout the Cold War.

The Human Factor

Concern about the human stress associated with material development surfaced vocally at the 10th National People's Congress held in Beijing in March 2007. Prime Minister Wen Jiabao acknowledged "long-standing and deep-seated" problems including slow rural economic growth; excessive growth elsewhere; educational and medical demands; safety at work; and deficiencies in local government. Other such matters could also be adduced, conditions in the prisons and labour camps for instance. In 2003, the officially admitted number of executions for all offences topped 7000.

As regards the leadership cadres, my own impressions particularly derive from editing a conference held in Beijing in 2000 between officials, military and academics from China and from Britain, France and Germany.[37] The Chinese conferees struck us Europeans as not only reasoning but reflective. But these people were second or first echelon within the public service. One suspects too many party functionaries at village and precinct level were and are obtuse or smoothly manipulative, fairly corruptive either way.

Nor should one suppose sweet reason and reasonableness always reign supreme higher up. The previous July a crackdown on Falun Gong dissidents had been launched. Eventually 100,000 would be summarily sent to prison and perhaps another thousand to mental homes. Our Chinese colleagues advised us the whole movement was mad. Agreed, its crackpot cosmology confirmed that. But one should ask next what specific factors were driving 30 million mad. And if one is looking for a better alternative in the form of functional democracy at grass-roots level, a good approach could be through issues ecological.

As things now stand, China has set herself a daunting agenda. One could have said as much were its population but the size of Norway's, instead of 300 times it. A continued risk is economic overheating, a tendency analysts in the West suspect the central government plays down by rather understating growth, inflation, exports and unemployment.[38] Lately the world at large has been judging China by the Beijing Olympics and, to an extent still, Tibet. On the latter score, it is as well to recognize that Tibet's population is so divided between the religious traditionalists and the pro-Beijing modernizing elite (Tibetan and Han) that unqualified autonomy near term would likely precipitate vicious internal strife.[39] As regards the former, the world agrees the Games' opening ceremony started with an absolutely epic co-ordination of aesthetic inspiration and managerial aplomb. It arguably betokened a more general virtuosity in the face of formidable challenges.

Whether or not the Chinese miracle is destined to turn sour will not be revealed in 2009. Should it eventually do so, the outward manifestation will be a negative nationalism, fuelled by recall of the long decades of quasi-colonial humiliation.[40]

One China, Two Voices?

An obvious target for an overtly forceful diversion will then be Taiwan, itself a veritable economic miracle. The Taiwanese could not uphold their sovereignty unilaterally, given growing economic interdependence but also the IRBM/MRBM threat. During the island's first presidential election in March 1996, Beijing ordered that four missiles be fired to near her coast and that confrontational naval exercises take place. To this bystander, the public mood on Taiwan seemed utterly calm but there was a run on the Taiwan dollar, while the index of economic activities showed its biggest monthly drop for seven years. The USA deployed a naval task force regionally but declined Taipei's request to buy some submarines. The 25,000–strong US presence in Okinawa also strengthens regional deterrence but does so somewhat controversially since it involves the lease for military facilities of what *in toto* amounts to a fifth of that island's land. And even with American back-up, Taiwan's precarious situation would preclude its withstanding prolonged pressure. The choice would soon be between backing down and escalating significantly.

The March 2008 Presidential was won by Ma Ying-jeou of the Kuomintang (KMT) by a margin unexpectedly large. Many voters finally accepted Ma's argument that growing economic interdependence was the basis for what he had called "one China with different verbal expressions". He was thus departing radically from the KMT pitch pre-1982 which had been to maintain martial law largely in order to suppress the Taiwanese identity, dialectic and otherwise, ostensibly in order to create an archetypal national base for the liberation of all China. However, this new approach drew criticism on the grounds that putatively the mainland's economic miracle was levelling out.[41] If that is really borne out near term, it will be because of world recession.

The Korean Peninsula

Nothing yet seen in real life bears more resemblance to *Nineteen Eighty-Four* than the way the North–South division of Korea in 1945 produced two oppressive regimes leaning on one another in reciprocal justification. Unfortunately there have been repercussions elsewhere ever since the Korean War armistice. Witness the state terrorism Pyongyang has waged. There have been the many tunnellings under the demilitarized armistice line; submarine intrusions into South Korean territorial waters; the seizure of the *Pueblo*, a US surveillance vessel in 1968; the murder of the Republic of Korea's (ROK) first lady that same year; hacking two American officers to death in the Demilitarized Zone (DMZ) in 1976; ROK Cabinet Ministers blown up in Rangoon in 1985; and repeated kidnappings, these admittedly including (1977–83) 15 Japanese children bathing on the Sea of Japan coast. All have manifested a chilling awareness of how to disconcert without invoking effective response. The 1968 incidents cannot have been unconnected with its being a year of decision in Vietnam.

However, in response to overtures from the North in 1998, Seoul has evolved a "sunshine policy" of engagement in a variety of mundane ways. Episodically this has gone forward. In 2005, for instance, 400,000 visitors crossed the DMZ. Even so, reunification will be very hard to achieve, save in the long-term on a confederal basis. More immediately, the nub is whether North Korea can be induced by multilateral regional diplomacy with economic incentives to give up its military nuclear programme. North Korea conducted a nuclear explosion half-successfully in 2006, and is thought to have amassed enough plutonium to make several more bombs. (See Appendix B).

The contrast between the two Koreas is today starker than ever. The South's elective government has achieved economically an "Asian tiger" lift-off. Lately it has been vying with Iceland to be a world leader in broadband internet. The North remains subject to a fascisto-Marxist personal dictatorship; and in 2005 faced widespread famine as floods ravaged its fragile agriculture. The South agreed to send 150,000 tons of rice. Then Pyongyang seriously harassed two of the delivery ships. The situation is all the more tragic and menacing in that nowhere has been better placed by Geography to reaffirm its national unity than has Korea. The unification of Korea was first effected by its Silla kings in the ninth century AD. Between 1592 and 1597 the Koreans defended themselves successfully, especially at sea, in a devastating war against Japan.

Japan's Resurgence

In 1945, Japan's ruination was such that it was generally expected it would have to settle for living standards little above the Afro-Asian average. Instead the country got onto an economic growth curve averaging 10 per cent a year. Forecasts made *c.* 1970 medianally projected Japan's GNP for the year 2000 being close to the American.[42]

Come that millennial year, the latter was still (at Purchasing Power Parity) two-thirds as much again. Nevertheless until 1990 or so, Japan rather than China had been seen as the next economic Superpower. Among the explanations for the former's impressive emergence were the immanent sociological ones identified in advance by Ruth Benedict in her classic 1946 essay on cultural anthropology, *The Chrysanthemum and the Sword*. One says "essay" advisedly because one criticism levelled against her was that she had put aside numerical analysis.[43] Anticipating this objection, she had insisted that resolving the seeming contradictions in the Japanese *leitmotif* could not be done via opinion polls geared to occidental parameters.[44] More crucially, the said essay foresaw that, when circumstances permitted, the Japanese would again seek a "place of honour among the nations of the world" but this time through peaceful economic growth.

Japan had been through a number of such switches historically. Faced with a marked thermal range between summer and winter, the early medieval Japanese felt obliged to choose. Mitigate either monsoonal summers or chilly winters. The former was given priority. (See 3 (2)). Architecturally this involved recourse to wooden structures well screened yet well ventilated. Among the characteristic features were floors raised off the ground. Walls comprised of sliding panels (*amado*), wide verandas (*engawa*) and steep roofs with broad overhangs. Stoicism was called for come wintertime.

The main thrust in this cultural direction came late in the Heian period (794–1185). This was also when the *samurai* were first waxing prominent in central and provincial government. It was, too, the first great era of the Japanese sword as a weapon — functional, aesthetic, spiritual and precise. Archery, too, had these qualities bestowed on it.[45] The Japanese martial tradition was well and truly born.

Another sea-change was *sakoku* (1639–1854). This seclusion posture involved limiting the foreign presence in Japan to small Dutch and Chinese enclaves in Nagasaki and a Korean in Tsushima. Also no Japanese were allowed to travel abroad, those already overseas in South-East Asia or elsewhere could not return. The purpose was generally to strengthen the authority of the shogun and particularly to stamp out Christianity, a creed seen as subversive, particularly after a major Christian rebellion in 1637.

The era was a peaceful one. Literacy spread remarkably; and as the country, especially the merchant class, prospered, the population of the capital Edo (Tokyo) grew to near a million. The secret lay in a geographically uniform culture of contentment; and in tight internal security, the most potent element of which was an efficient secret police. Still, a century and more before Commodore Perry and his US naval squadron arrived in Edo bay in 1853 to present an ultimatum to open up, "Dutch" (i.e. western) learning was spreading from Deshima, the Dutch factory in Nagasaki bay, in fields such as medicine and Copernican astronomy.

Nevertheless, the "unequal" Treaty of Kanagawa in 1854, giving the Americans some port access, was a big shock to the political system. It led on to across-the-board modernization, Western style. Progress was far from smooth but was insistent. The upshot was that, come 1904–5, Japan could astound the world by defeating Tzarist Russia on land and sea.

The next diametric adjustment was the lurch, expedited by army-backed terror, into fascism (1931–4). The world war involvement that followed engendered within Japan a strong reaction against the military ethos, *bushido*. One of several reasons was that the treatment of prisoners by the imperial army consistently contrasted with the correctitude and chivalry it had regularly evinced during the Russo-Japanese war.[46] In both conflicts, its troops displayed extraordinary courage in battle.

As an offshore archipelago facing problematic continental neighbours, Japan today retains under arms close to 500,000 personnel. But inhibitions buttressed by the constitution still constrain overseas deployment. During the 1994 confrontation over Pyongyang's nuclear programme, Tokyo declined a request from Washington that Japan earmark minesweepers and patrol planes for possible hostilities.[47] But it does now commit troops to non-combatant aspects of peacekeeping and has extended her oceanic surveillance.

The Next Departure?

Japan is at a crossroads again. Very many of her people retain a downright inspirational penchant for friendliness, politeness, competence, reliability and — not least — making that which is functional, beautiful. But in 1989 the country had something of its own precursor of the near global credit crunch of 2008. Against a background of slowing growth nationally, the banks diversified their investment portfolios, not least into property. When the bubble burst, there was a welter of bad company and personal debt and underresourced pension funds.

This meant the national economy would be hit hard by the "dot.com" boom and bust which followed. Between 1990 and 2002, the Nikkei index of leading share values fell to what had been its 1984 level. The economy has been taking a long while to recover. Nor has its situation been helped by national politics being too consensual, too polite. Hence the simmering youth revolt which finds expression in diverse ways. Sooner or later it may acquire a clear identity for good or ill.

One way through may lie in the promotion of an East Asian regional consciousness via the pursuit of shared values. Respect for Nature is a core theme within the region's religious traditions — Buddhist, Shinto, Taoist . . . In regard to this, however, today's East Asian nations have not unblemished records. Nevertheless there are positive signs. Salient is China's new-found willingness to reckon with global warming. In 1986, one could only discover one small institute within the Beijing scientific community prepared even to recognize the problem. Now the Greater China director at the Beijing office of the international NGO Climate Change confirms that China is committed to (a) closing down all coal-fired power stations of 300 megawatts or less, (b) developing new coal-fired plants "supercritically" efficient as regards greenhouse emissions, and (c) increasing from 8 to 15 in the 2006–2020 timeframe the energy percentage coming from renewables.[48] If all this works out, it will tip the balance in the country's political culture towards environmentalism — a switch likely to have positive resonance at local level.

Confucianism Today

What then of Confucius? Around 250 BC *Tao te ching*, the foundation text of Taoism, warned of his lending encouragement to excessive governance. A latter-day significa-tion of that was the celebration of his birthday the Beijing authorities staged close by Tiananmen Square weeks after the 1989 massacre. Still, his influence is more posi-tively discernible in the readiness of Chinese philosophers to debate with the West their approach to guided democracy[49] along with more esoteric questions like the theory of knowledge.[50]

Perhaps the time is ripe to explore as well the other side of Confucius, the upholding of family and personal bonding: something he saw the basic importance of but took rather as a given. As Prime Minister of Singapore, Lee Kuan Yew turned his unitary city state from a British naval base to an "Asian Tiger" democracy. Part of his strategy was having Confucianism taught in schools.

The philosophy's original formulation is obsolete as regards the family stratifica-tion it endorses. But its most basic precept has never looked more germane. It is that close and enduring person-to-person networking within the family and without can encourage a more compelling sense of obligation to society at large. Witness the response of local people to the 2008 Szechuan earthquake; and, indeed, to the 1995 Great Hanshin earthquake at Kobe, Japan, an event that caused $100 billion of damage.

A movement in this philosophic direction could also be healthy in that it would represent an authentically non-Western contribution to modern social philosophy. Contrariwise, some would argue that the reassertion of Confucianism regionally could revive customary East Asian notions about geopolitical hierarchy. Samuel Huntington has foreseen its reinforcing Chinese hegemony in East Asia. Allowing that this could augur well for regional peace, he feels the implications for relations with the wider world are less certain.[51] In practice, however, the difference between East Asian models of international hierarchy and Western ones of the balance of power may be more apparent than real. Some in the West see the USA is *the* hegemon.[52]

3 The European Pillar

The Atlantic Alliance was effectively launched in 1947–8 with the Marshall Plan. To administer this programme from the American side, an Economic Co-operation Administration (ECA) was established. Legislation authorizing it passed the Senate by 69 votes to 17 and the House by a like proportion. Written into the Act was the goal of European integration.

Congress passed resolutions in favour of European federation in 1947. Two years later, Allen Dulles, J. W. Fulbright and others sponsored a Committee for a United Europe. Congressional exasperation at Britain's refusal to join the nascent European Coal and Steel Community (ECSC) in 1952 further showed how European unity was now an American mainstream interest, with such Republican leaders as John Foster Dulles well on board. Dean Acheson and Averell Harriman had been advocating on behalf of the Truman administration what in the Kennedy years became the "twin pillars" concept for an Atlantic community.

Thereafter, bipolarity remained robust through the Cold War and beyond despite sharp discord: mass protests in Western Europe in favour of "nuclear disarmament" *c.* 1960; against "Vietnam" through the late sixties; against "cruise missiles" in the early eighties; and, to an extent, "Star Wars" (i.e. BMD) soon afterwards. Though the intergovernmental differences over "Suez" in 1956 (London and Paris versus Washington) had been bitter, a strong London–Washington accord was rebuilt within a year. Paris moved towards Gaullism instead.

An overt Congressional challenge to Atlanticism had come from the "Pacific Firsters" in Congress in the course of 1951, with Joseph McCarthy riding high and "Korea" turning into a deadlocked war of attrition. But no more Senators voted that year against sending another four army divisions to Germany than had against ECA.

Along with their geographic reorientation, the Pacific Firsters placed more emphasis than did Eastern Republicans or Democrats on the deterrent or punitive potential of air and naval power. They placed less store by multilateral alliances or conflict limitation. Some were close to McCarthy. Behind it all lay resentment at the grip perceived to be exercised on their party and country by Ivy League and Wall Street. They themselves tended to come from the Middle and Far West. However, their impact on the national scene ensured the accession to the White House in 1952 of Dwight Eisenhower, a thoroughgoing Kansan Atlanticist with arms control predilections.

Early 2001 saw the arrival in the Oval Office of a George W. Bush not then much inclined to enter into military commitments abroad. As and when warlike crises broke, Colin Powell as Secretary of State was disposed to work within the Atlantic Alliance and the UN. But other members of the administration (Dick Cheney, Donald Rumsfeld, John Bolton . . .) were impatient of this. In a tightly fought presidential, Bush had secured a solid swathe of middle America. His Democrat opponent, Al Gore, had won three more peripheral but very sizeable segments (New England, the northern prairies/Great Lakes; and the far West) plus, overwhelmingly, Washington DC. Soon the dialectic about interventionism would be galvanized by 9/11.

Europe Goes European

The pristine concern of much of the American political elite with European unity interacted with the emergence of a dynamic European Movement within non-Communist Europe. Its lead personalities included Paul Henri Spaak of Belgium plus Robert Schuman and Jean Monnet of France. Monnet claimed "we are not making a coalition between states but a union of peoples", a decidedly utopian pitch in relation to Europe's tortuous geography and the historical legacies thereof. In opposition, Winston Churchill was zealous for European unity but declined to involve Britain when Premier again, 1951–5.

In 1901, H. G. Wells had foretold how, once Germany had been curbed in successive wars, the destiny of Western Europe would hinge on federation between the states extending about the Rhine, states sure to become a single economic entity within the next 50 years.[53] Add Italy to his grouping and you have the European Economic Community (EEC) as it came into being under the Rome Treaty in 1958. The other two bodies similarly formed to make the European concept operational were the ECSC and the European Atomic Energy Community (Euratom). The three were to merge in 1967.

For the next three decades all sides saw closer integration and further enlargement as alternative priorities. Nowhere was this understanding more explicit than in Britain; and nowhere was a wider Europe more favoured. There the term "wider" was casually presumed also to connote global vision, never loss of cohesion.

Apprehension of the latter effect was one reason why Charles de Gaulle was hostile to Britain, Denmark, Ireland and Norway acceding to the EEC, another being his belief that "perfidious Albion" would just be a Trojan Horse for Washington. In 1973, following his departure, Britain, Denmark and Ireland were admitted. Norway declined, her electorates having rejected in a referendum the advice of their political elites. A disinclination to pool Norway's fisheries had weighed heavily.

Greece, Portugal and Spain entered in due course. Then in 2004, ten more states joined concurrently. Meanwhile, in the wake of the 1993 change of name from EEC to European Union (EU), a tendency had developed within Brussels to treat integration and expansion as less in conflict than once felt. Across the EU at large, a popular reaction against this perspective has been developing since.

The Enlargement Debate

Antipathy to boundless enlargement has been adjudged to lie behind the decisive "no" votes against a European constitution in 2005.[54] A report to the European Parliament in March 2006 by MEP Elena Brok cautioned ministers that enlargement required support resources; and maybe an arrangement whereby active applicants for admission might join on an interim basis. A desire to develop a concept of "absorption capacity" was aired.

However, the commissioner responsible for enlargement negotiations, Olli Rehn, stressed that the preconditions for entry routinely included human rights safeguards, constitutional and judicial, plus economic competence. He deemed it inadvisable to interrupt what he saw as an endeavour to build solidity in the least stable parts of the European continent. "For the sake of Europe, let us not shake this", he pleaded. Conversely, during the fraught debates hard upon 9/11, a distinguished British columnist had contended that, impelled by its Right wing, the USA has always sought "a liberal free market Europe, securing capitalism and democracy but with no capability to become a partner in the exercise of Western political and military power". Hence, he argued, the consistent pressure on the EU to enlarge to the point where it "collapses into dysfunctional paralysis".[55]

Turkey, a Special Case?

Certainly the current Bush administration, though before it the Clinton one, too, has backed the Turkish EU application very overtly. Originally one argument was that Turkey could be invaluable for continental missile defence. General grounds for dismissing this concern are further considered below. Suffice for now more particularly to remark that a great circle track from Tehran to Berlin, say, bypasses Turkey widely. In any case, these days Washington likely sees support for Ankara over the EU as a needful offset for differences with her over other questions. Palestine can cause tensions contingently. Cyprus does continually. Meantime the USA has not as yet been prepared to act decisively against Turkey's Kurdish rebels based in northern Iraq nor been ready to countenance Turkey's doing overly much herself.

Yet Washington never loses sight of Turkey's nodality *vis-à-vis* force projection to Iraq and Afghanistan. In the Spring of 2003, as the invasion of Iraq was being launched, this question precipitated a deep crisis within Turkey, the crux being the unpopularity of the breaking war among the public. The outcome was a parliamentary decision on 20 March to open Turkish air space to transitting coalition planes subject to strict limitations. It was a move which received the army's lukewarm endorsement.

Turkey Emergent

Notwithstanding advice that if it does accede, this cannot be before 2014, Turkey remains as yet in the frame. Nevertheless, the action–reaction between her and the EU has turned more negative these past several years. Correspondingly, while the military still see themselves as pro-active guardians of constitutional secularization, they are doing so more in the mien of a revived Ataturk nationalism and less in that of Euro-consciousness. Respect for the army remains high, and scepticism prevails about EU-induced reforms.[56] In April 2007, the army sought to influence the choice of President by publishing on its web site a threat to stage another coup, its fifth since 1960. Moreover, in June 2008 the army supplicated in the constitutional court to have Islamic political parties (including the present ruling party, the AK) banned.[57] They failed but only just.

The current Prime Minister, Recep Tayyhip Erdoğan, is a fervent though liberal AK Moslem, one long keen for EU membership. On that score, the rub starts with the country's Gross National Product as yet being (despite incipient signs of the long-awaited lift-off) a twelfth that of Germany's although demographically Turkey is already almost as big and area-wise is nearly twice so. At 73 million, the population is up ten million on a decade ago. What with the migratory pressure that this situation would engender and its eccentric location geographically, Turkey could imbalance the union drastically. The upshot might be a worsening all round of relations between Islam and Judaeo-Christianity.

The Turks might best fulfil their destiny through simply being a cornerstone of the Atlantic Alliance and a keystone of the Middle East. Recent years have seen a veritable explosion of contacts between the Arab world and Turkey. But there is a deal of Turkish interest in Israel's developmental achievements; and customary pride in how the Ottomans received Spanish Jews after the *reconquista*. Recently Turkish diplomacy has sought to further the Palestine/Israel "peace process". Positive, too, is the government's recent decision to sponsor exigesis of the *hadith*, non-Koranic statements attributable to the Prophet. It bespeaks an open-mindedness worthy of broader emulation.

Further Enlargement?

So when and where will EU enlargement stop? Norway and Switzerland evince little desire to join. Before its acute banking crisis in 2008, Iceland was divided. Morocco has been told Geography does not commend her, none of her land mass being in Europe. However, Belarus, Moldavia and the Ukraine are often mentioned as eligible. Before her 2008 war with Russia, Georgia avowed its desire to join.

Few would feel able to object to the eventual incorporation of all the legatees of former Yugoslavia plus Albania. But should it not now be made explicit that things will be left there? This would confirm the EU boundary up to the western border of the USSR as it was early in 1940. To extend further east wherever could be to create a posture provocative of Moscow yet ill-placed to withstand a riposte from that quarter.

Internal Management

It does seem that the current Europe of the 27 takes longer to make decisions than did its slimmer predecessors. But might the problem go deeper? Take Romania, one of the most recent entrants: credited with an elite suffused with new-found confidence, one perhaps poised to revive the longstanding sense (somewhat mythic but quite impacting) of Latinist/Romance affinities with Western Europe. Yet still its income per head is not a sixth the EU average. The social exclusion of Romas is as frightful as elsewhere in the Balkans EU.[58] Two million of her citizens live abroad. Sleaze remains rampant, not least in those upper echelons. The international borders are porous.

This instance shows how difficult it may prove to uphold EU common standards as things now are. A derivative problem is that EU defaultings on social policies might encourage more unilateralism within its Euro-zone on monetary objectives. Lately the euro has risen sharply relative to the dollar, to which the membership has responded well enough. One can also acknowledge that the regular summits the USA and the EU are together promoting on international monetary collaboration could prove to be the axis on which progress is made in this field. But for a whole clutch of reasons we must still anticipate the next few decades will be ones in which the world economy will be unsteady and sometimes stormy. Should the euro and/or the EU crumble at any point, the continent might be more at risk from violent unrest than if European unity had never been envisioned. The increased number of land-locked polities could prove an added aggravation.

A Russian Menace?

Sundry indications of malaise concern one when applied to contemporary Russia. Alcoholism and corruption are among them. So are the suspicious deaths of 20 ranking journalists during the Putin years, with no convictions yet secured.[59] Hope that such blemishes can be erased by prosperity has lately drawn succour from economic growth led by a hydrocarbon boom. But the production of oil at any rate may now be near a long term peak. A lot may depend in the short term on what happens to hydrocarbon demand internationally; and on how far President Dmitry Medvedev is able and willing to move beyond Putin. On the latter score, events in Georgia in 2008 do not reassure.

Some observers are uneasy, too, about a revived sense of the worth of the military though thus far this looks not unhealthy in itself. Most worrying of late have been the overreaction to the South Ossetia problem and the threat of oil and gas shut offs being used to pressurize neighbours. Disturbing, too, was the 2007 cyberattack on Estonia. (See 13 (1)). So again was Putin's surmise that a corridor through Lithuania could link Mother Russia to her Kaliningrad outlier on the Baltic. However, this last may well have been a ruse to divert attention from the Russo-German gas pipeline under the Baltic, a facility which could free Moscow to focus precisely any gas embargo aimed at Poland and/or Belarus. All of which should contingently be seen within the context of Limited World War. A Russia much withdrawn compared with the USSR might be the more tempted to take action along her European borders in reaction to unwelcome Western moves elsewhere she cannot directly inhibit.

As the Scandinavians regularly comment, it would be better if the EU spoke with something like one voice on such matters. No less desirable, however, is that it and the Atlantic Alliance firm up on a stance towards the Russian Federation which reassures as well as confronts. Extensions of NATO beyond the bounds proposed above for the EU would certainly be read by the Kremlin as designed to disturb its hemmed-in country's peace of mind, albeit with a gambit which could readily be upturned. Yet at NATO's Bucharest summit in March–April 2008, the Eastern Europeans were reportedly in the main eager to extend the alliance, ultimately perhaps to Georgia.[60] Still, Poland's Foreign Minister did ask whether the concept of "mutual defence" was at risk. Surely nothing would tax it more than Russian harassment of NATO's entering the Caucasus. Such a projection would stand even more exposed than West Berlin was.

Another question to revisit is the installation of strategic BMD in Poland and the Czech republic. In this case, too, the location proposed could be tested geodetically in relation to Middle Eastern threats. More fundamental, however, is the inadequacy of surface-based emplacements for area defence at strategic level. Experience to date suggests that if a regime can build missiles able to project warloads 1500 miles or so, it will fairly soon be able to incorporate in them multiple warheads and maybe, too, credible decoys able to saturate surface defences.

Why then bother? The answer appears to be, alas, that a BMD military industrial complex is looking for persuasive precedents. If comprehensive missile defence is accepted as viable by Japan, Israel, NATO Europe or whoever, this could clinch politically the case for its introduction across the USA. Given its territorial extent, the said coverage would need to be provided by weaponry (presumptively lasers) mounted in orbital satellites.[61] Thus would be negated by indirect approach a taboo against Space weaponization not a few of us earnestly seek to uphold.

European High Technology

How best to forge a European technological community has been the subject of debate these 40 years past. A prime aim has been to counter *le défi américain* in high technology, civil and military, thereby securing freer choice for Europeans and the world at large. An additional objective now might be to preserve something of the rich heritage not a few European countries can lay claim to in Science, pure and applied.

Western Europe as a whole thrives in certain up front sectors. For instance, the Galileo advanced navigational satellite (under development by the European Space Agency) should enter service in 2020; and hopefully generate 100,000 new jobs. By then, too, a very high speed rail network (at least 200 km per hour) will extend over the region.

The CERN particle accelerator facility at Geneva has long contributed powerfully to fundamental physics; and the Large Hadron Collider will carry this further. France is hosting the International Thermonuclear Experimental Reactor (ITER), a key stage in the endeavour to have controlled nuclear fusion contributing to energy supplies by 2060 or so. This accolade was won against the Japanese, supplicating with well-founded American support. Their main facility at Naka is exhilarating to observe. Japan is duly guaranteed 20 per cent of the ITER posts though contributing only 10 per cent of the total budget.

Europe has a weak record as yet for innovation in IT and some branches of biology. More generally, one must say that science and technology policy has been too characterized by a lack of strategic foresight. In 2006 the EU introduced the world's first scheme for trading permits for carbon dioxide emissions only to have it precipitately collapse. Similarly, the emphasis lately placed on biofuels already looks mistaken. EU energy policy in general is being confusedly reconsidered in the light of Moscow's penchant for economic war in this domain, a trait that might have been foreseen. Expectations *c.* 1966–7 of the sale of 300 Concordes, the Anglo-French supersonic airliner, showed no awareness of how the advent of the 747 "jumbo jet" in 1969 would cheapen subsonic flight. Actual sales were not a tenth that number.

The nub of the general problem is that reaching agreement about any multinational project is so difficult those concerned do not venture to think too radically lest difficulty is thereby compounded. The few think tanks Brussels lays claim to reputedly comment but tamely on the ongoing status quo. The many in Washington DC are more inclined to look towards the next big problem, the next big idea.

The A380 and 787 Dreamliner

At present, the big test of European technologies is the Airbus A380, the largest airliner ever built and the most direct successor to the Boeing 747 "jumbo", 1400 of which have been sold. Unfortunately, the A380's prestige has been dented by delays not inordinate in themselves yet indicative of management problems. The development and production of different major sections in different countries has been questioned on efficiency grounds. The two biggest shareholding communities, the French and the German, have clashed repeatedly. The British are not participating after all. Charges of insider trading have been levelled. Meeting production costs in euros then selling the machines in dollars has thus far been disadvantageous.

Nevertheless, deliveries have been due to start the winter of 2008–9 and 185 orders have been received. A marketing estimate has been that 1500 aircraft in this "very large" genre will be needed by 2025, half of them for Asia-Pacific service. The A380 still looks well placed to capture most of whatever the market is. Airbus needs to sell about 500 to break even.

Boeing, Airbus's transatlantic rival, has taken the view that the main growth sector in air traffic will not be as between the great hubs — Chicago, Copenhagen and so on. It will be in flights directly between cities rated secondary in this context. This is the requirement its twin-engined 787 Dreamliner is designed to meet. Deliveries should start in early 2009. Orders or quite firm commitments top 700 to date. Its Airbus rival, the A350, should enter service in 2013.

Each company seems to have a flagship able to keep it strong. But the A380 and the 787 may each be even more than usually dependent on an impeccable safety record. Meantime, a war of words continues intermittently about subsidies. Airbus has received direct funding. On the other hand, the benefits from basic research by NASA and the Pentagon have very considerably accrued to American firms. Britain's big firm in the field, British Aerospace Systems, is heavily oriented towards transatlantic bilateralism. It is a primary partner in the Joint Strike Fighter development, described as the largest military programme ever. It quite possibly represents the last generation of manned interceptors.

Last year European Defence and Aerospace Systems provisionally won in partnership with Northrop Grumman a contract to provide the USAF with next generation in-flight refuelling planes. Europe's experience with wide-bodied jets would surely copper bottom, too, a claim to lead the development of a long-range maritime bomber, should such a requirement be agreed. A European twin pillar in high technology cannot equate with the American in scale and diversity. Even so, the EU should remain in healthy enough contention in this challenging borderland between soft and hard power.

An Atlantic Nuclear Force?

In the second half of 2008, with the EU Presidency in his hands, Nicolas Sarkozy had an opportunity to promote his visions of closer ties with the USA and of Europe developing its own strategic identity. *A priori*, one could foresee these aims eventually coming together not least via a revival of the Atlantic Nuclear Force (ANF), a concept aired in the middle sixties. Essentially this would comprise the French and British strategic deterrents plus a sizeable fraction of the American and would proffer a nuclear shield to NATO Europe.

France and Britain each sustain strategic deterrence by means of several missile-bearing submarines. These missiles bear thermonuclear warheads of national design. Each country intends, when the time is ripe, to renew their deterrents along similar lines. Soon after De Gaulle's retirement in 1969, a lively interest was taken in Anglo-French operational integration in this field. Among the ranking politicians who aired this idea specifically were Georges Pompidou, Edward Heath and Franz-Josef Strauss.

Conceptually the Anglo-French and the American contributions to an ANF would be complementary. The small but indigenous European deterrents would be the more credible in the event of a heavy strategic attack on Europe being threatened, a contingency not inconceivable decades ahead if in the interim an appropriate world survival strategy fails to unfold. For the submarines in question could go rapidly into action to counter the devastation of their home region. This perspective was articulated by Georges Pompidou when, in 1964, he described the defence of Europe as "physically and geographically inseparable from that of France which is not the case with forces outside the European continent even if allied". Co-ordination of patrol schedules and contingency targeting plans should be entirely feasible. A merging of sovereign control would not be called for.

The much bigger and more diverse American nuclear umbrella would look the more convincing when responding flexibly to threats various at a lower level, threats *ipso facto* more likely. The crux surely is that a deterrence posture intended to cover Europe should rest on the complementality inherent in this contrast. It should also do something to rein in national nuclear sovereignties without abandoning that non-negotiable principle. Both these aims an ANF ought to be able to realize, provided procedures were in place both for doctrine development and for crisis management.

Anglo-French co-operation as here proposed might pave the way for further joint naval preparations for warlike or natural disaster contingencies, the core elements being the carrier forces each country will continue with. This could involve contingency planning and exercising for the ad hoc creation of NATO Europe task forces as crises develop. Presumably other members would particularly provide cruise missile platforms, a capability which would be much in demand in the event of major military action against a sizeable power. In this way, Europe might at last get proportional utility from its naval resources. Some 295,000 men and women serve in NATO Europe's navies and 510,000 in the United States Navy.[62] The contrast in ability to project power globally is considerably greater.

A further question revived by the political impasse is whether we should be, or perforce are, moving to a two-tier Europe, the inner formed around a Paris-Berlin collaboration axis of the kind Charles De Gaulle and Konrad Adenauer agreed on in the 1963 treaty. A sound basis for this seems not to exist as of now. For one thing, philosophic differences persisted through 2007 in regard to the management of the Euro currency, the Germans still stressing price stability and the French economic growth. Such differences receded come the banking emergency of 2008 but could resurface in some guise. In any case, distrust of the two-tier model persists across the EU.

On the other hand, Germany must be allowed elbow room commensurate with its massive and committed presence at the very heart of Europe. The best way forward could be through routinely close consultation within NATO with France and Britain plus the Low Countries. This might be facilitated by revamping the command structure of NATO Europe as follows: a Northern Command comprising Iceland, Norway, Denmark, Sweden and Finland; a Western consisting of the Low Countries, Germany, France and Britain; a Southern including Portugal, Spain, Italy, Greece and Turkey; and an Eastern made up of ex-members of the Warsaw Pact or of the USSR. The presumption might be that within each command representatives would discuss a wider range of topics than Defence as narrowly understood. The Low Countries might not readily act as a bloc if only because the Belgians are still committed to Euro-federation to an extent the Dutch no longer are.

Defined Geography

However, no internal restructurings are likely to receive wholehearted popular support until the enlargement question is finally resolved, one would have thought by referendum. Without this, the European Union is in danger of going the way of the Holy Roman Empire. But operating within quite firmly set geographical bounds, the Empire was able to limp along for ten centuries. The Union might not manage one.

The best basis on which to sustain a European pillar is a NATO and EU suitably in alignment geographically and not debilitated by hyperenlargement. That applies in all respects, including containing as best we may Muscovite pressures in the Caucasus. The breathing space created by Ireland's rejection of the EU Lisbon Treaty (an expedient remake of the 2005 Constitution) ought to be an opportunity to square things up.

Notes

1 N. J. G. Pounds, *An Historical and Political Geography of Europe* (London: George G. Harrap, 1947), p. 29.
2 *Keyhan International*, 12 October 1974.
3 Ali Gheissari and Vali Nasr, *Democracy in Iran* (Oxford: Oxford University Press, 2006), pp. 84–96.
4 "Silent No More", *The Economist* 388, 8588 (12 July 2008): 22.
5 Ali M. Ansari, *Iran under Ahmadinejad*, Adelphi Paper 393 (London: International Institute for Strategic Studies, December 2007), pp. 52–3.
6 John Kelsay, "Al-Shaybani and the Islamic Law of War", *Journal of Military Ethics* 2, 1 (2003): 63–75.
7 Seymour M. Hersh, *The New Yorker*, 7 to 14 July 2008.
8 Roy Jenkins, *Churchill* (London: Pan, 2001), p. 350.
9 Martin Gilbert, *Winston S. Churchill* (London: Heinemann, 1975), Chapters 31 to 33.
10 *Ibid.*, p. 563.
11 James Barker, "Policing Palestine", *History Today* 58, 6 (June 2008): 52–9.
12 Avi Shlaim, "Talking to the Enemy", *New Statesman*, 19 May 2008.
13 Elisha Efrat, *Geography and Politics in Israel since 1967* (London: Frank Cass, 1988), Chapter 3.
14 See the author's *The Future Global Challenge* (New York: Crane Russak for the Royal United Services Institute, 1977), pp. 291–3.
15 Judith T. Kenny, "Claiming the High Ground: Theories of Imperial Authority and the British Hill Stations in India", *Political Geography* 16, 8 (1997): 655–73.
16 Albert Einstein (Alan Harris, trans.), *The World as I see it* (London: Watts, 1940), pp. 96 and 105.
17 Walter Clay Lowdermilk, *Palestine, A Land of Promise* (London: Victor Gollancz, 1944), pp. 55–6.
18 Rhys Carpenter, *Discontinuity in Greek Civilisation* (Cambridge: Cambridge University Press, 1966), pp. 3 and 15–18.
19 Jaime Vicens Vives (Frances M. López-Morillas, trans.), *An Economic History of Spain* (Princeton: Princeton University Press, 1969), 108–10.
20 Arie S. Issar in Arie S. Issar and Neville Brown (eds.), *Water, Environment and Society in Times of Climate Change* (Dordrecht: Kluwer, 1998), Chapter 6.
21 Amos Frumkin *et al.*, in *Ibid.*, Chapter 5.
22 Tao Hanzhang (trans.), *The Art of War* (Ware: Wordsworth, 1993), p. 105.
23 H. H. Gerth and C. Wright Mills (ed.), *From Max Weber* (London: Routledge and Kegan Paul, 1970), Chapter XVII.
24 Joseph Needham, *Science and Civilisation in China*, 14 Vols. (Cambridge: Cambridge University Press, 1959), Vol. 3, pp. 459–60.
25 Richard Stephenson and Kevin Yau, "Oriental tales of Halley's comet", *New Scientist* 103, 1423 (24 September 1984): 30–2.
26 Gavin Menzies, *1421* (London: Bantam Press, 2002), pp. 323–30.
27 Nayan Chanda, "Sailing into Oblivion", *Far Easter Economic Review* 162, 36 (9 September 1999): 44–6.

28 D. Howard Smith, *Confucius* (London: Temple Smith, 1973), pp. 191–2.

29 David L. Hall and Roger T. Ames in Daniel A. Bell and Hahm Chaibong (eds.), *Confucianism for the Modern World* (Cambridge: Cambridge University Press, 2003), p. 129.

30 Vaclav Smil, *The Bad Earth* (London: Zed Press, 1984), p. 15.

31 Joan Robinson, *The Cultural Revolution in China* (Harmondsworth: Pelican Original, 1969), p. 42.

32 Karl A. Wittfogel, "A Stronger Oriental Despotism", *The China Quarterly* 1 (January 1960): 72–86.

33 John W. Garver, "Development of China's Overland Transportation Link with Central, South West and South Asia", *The China Quarterly* 185 (March 2006): 1–22.

34 He Li, "China's Growing Interest in Latin America and its Implications", *Journal of Strategic Studies* 30, 4–5 (August to October 2007): 833–62.

35 Padraig Carmody and Francis Owasu, "Competing Hegemons? Chinese versus American geo-economic strategies in Africa", *Political Geography* 26, 5 (June 2007): 504–24.

36 "Chinese Air Force in throes of a Cultural Revolution", *Aviation Week and Space Technology* 157, 19 (4 November 2002): 55–7.

37 Neville Brown (ed.), *American Missile Defence. Views from China and Europe* (Oxford Research Group, Current Decisions Report, 25, May 2000).

38 "An aberrant abacus", *Economist* 387, 8578 (3 May 2008): 96.

39 "Violence in Tibet also reflects class divisions", *Wall Street Journal* 28–30 March 2008, pp. 1 and 22–3.

40 George Walden, "The Resentful Dragon", *Daily Mail*, 2 August 2008.

41 Editorial, *Taiwan News*, 22 March 2008.

42 Herman Kahn, *The Emerging Japanese Superstate* (London: André Deutsch, 1971), pp. 128–30.

43 Jean Stoetzel, *Without the Chrysanthemum and the Sword* (Paris: UNESCO, 1955), pp. 16–17.

44 Ruth Benedict, *The Chrysanthemum and the Sword* (London: Secker and Warburg, 1947), Chapter 1.

45 Eugen Herrigel , "Japan's Art of Archery" in Shigeyoshi Matsumae (ed.), *Towards an Understanding of Budo's Thought* (Tokyo: Tokai University Press, 1987), pp. 112–36.

46 Richard Storry, *A History of Modern Japan* (Harmondsworth: Penguin, 1967), p. 139.

47 *Ashahi Shimbun*, 26 November 1995.

48 Changhua Wu, "China's green journey", *New Scientist* 199, 2668 (9 August 2008): 18.

49 Mark Leonard, "China's new intelligentsia", *Prospect* (March 2008): 26–32.

50 Thomas Metzger, "China's Challenge", *Philosophy* 26 (April/May 2000): 30–2.

51 Samuel P. Huntington, *The Clash of Civilizations and the Remaking of the World Order* (London: Touchstone, 1998), pp. 234–8.

52 Noam Chomsky, *Hegemony or Survival* (London: Hamish Hamilton, 2003).

53 H. G. Wells, *Anticipations* (London: Chapman and Hall, 1902), p. 241.

54 Michael Berendt, *The 2007 Annual Register: World Events*, Vol. 248 (Bethesda: ProQuest, 2007), pp. 403–4.

55 Will Hutton, "If Europe takes on too much we shall all lose", *The Observer*, 9 December 2001.

56 Karabekir Akkoyunlu, Adelphi Paper 392 *Military Reform and Democratisation* (London: International Institute for Strategic Studies, 2007), pp. 66–7 and 70.

57 Leader, "Turks Court Disaster", *The Times*, 30 June 2008.

58 "Bottom of the Heap", *The Economist* 387, 8585 (21 June 2008): 39–42.

59 Jonathan Dimbleby, "The New Fascist Empire", *Daily Mail*, 13 May 2008.

60 "With Allies Like These", *The Economist* 386, 8574 (5 April 2008): 73.

61 *INESAP Information Bulletin* 28 (April 2008): 25–34.

62 Figures compiled from *The Military Balance, 2008* (London: International Institute for Strategic Studies, 2008). Marine units are reckoned in.

16 | Survival Geography

Shortly after the Soviets invaded Afghanistan in 1979, the Royal United Services Institute published a prescient study. Its author, Geoffrey Warhurst, recalled how one biggish British expedition into that country was annihilated in 1840 and another badly mauled in 1880. Neither landscape nor social geography had altered greatly since. Nor had military mechanization yet countered the constraints these imposed. Moscow had let itself in for a decade of bloody "pacification" leading at best to stalemate tantamount to defeat.[1]

Warhurst warned how Moscow could feel goaded into a wider war as, towards 1990, "dissent at home and hostility abroad are orchestrated into a vast verbal assault" on the Soviet regime. He was thus recognizing that failure in Afghanistan would oblige the Kremlin to make a climacteric choice, close the Cold War down or hot it up. It fell to Mikhail Gorbachev to choose the former course, influenced also by the Strategic Defense Initiative. He thereby changed the geopolitical map of the world much more radically than he intended and more than anyone else ever has or likely will.

The New Geopolitics

What is harder to tell is what this change will mean in decades to come. One appreciates that certain states (notably India and Brazil) will become bigger players on the world scene. But what of relations between those five permanent members of the Security Council? Immediately one runs up against a profound nonsense in regard to the classical approach to geopolitics. It would have us treat the modern state as a unitary reality, "a natural given". Yet many, one can say all, states are artificial creations, insecure and unpredictable.[2]

Today this truism especially applies to Russia and China. The former feels sidelined by the latter and by how limited its warmish coastlines once again are compared with Soviet times, hemmed in though the Kremlin felt even then.[3] Meanwhile the latter is developing dramatically though at risk of overheating.

The fortunes of both France and Britain are considerably bound up with those of a European Union that aspires to become at least a kind of confederation but which is in real danger of frustrating its own aspirations through hyperenlargement. The

USA is widely seen as peaking out in terms of world power. It may be. But worth remembering is that a "high noon on the Potomac" notion waxed fashionable around the time the Vietnam war ended in 1975. What could be disastrous at this juncture is the USA seeking to offset relative decline, actual or anticipated, by going for the weaponization of Space.

Something both the USA and Europe have found difficult in 2008 has been relating the Russia with whom one has talked of functional collaboration on missile defence to the Russia that has forcefully escalated its contention with Georgia. However, the understanding that the then USSR was an Adversary Partner was also doing the rounds a good 30 years ago. Indeed, that inter-state relations almost always are studded with ambivalence has been a commonplace for centuries.

Given how global problems are mounting, we need Moscow's active partnership much more than at any time since World War Two. Securing it depends on the adversarial tendency in Moscow not making easy gains either through our overindulgence or through our overreaction.[4]

Gorbachev or Mugabe?

In terms of the old debate among geographers about whether circumstances present several possible courses of development or whether they predetermine the one way through, Gorbachev represents a triumph for possibilism. So, negatively, do various post-1945 dictators who have trashed clumsily and maliciously what had seemed good prospects for national progress. Take Robert Mugabe, President of Zimbabwe since it arose in hope in place of Rhodesia in 1980. Then its population was 7,500,000 of whom 270,000 were Whites. Today there are 12,000,000 of whom 70,000 are Whites. Perhaps five million Zimbabweans are economic and political fugitives abroad. Even the official inflation rate is currently over 200 million per cent.[5] A regional drought trend has been only a very secondary cause. All of which stands as history even if the power sharing agreement of September 2008 does prove to be for real.

Globalizing Freedom

In the wake of the youth revolt of the sixties, Zbigniew Brzezinski, later National Security Adviser to President Jimmy Carter, advocated blending the time-honoured "international" approach to world affairs with the "planetary" perspective radical youth extolled.[6] Today this imperative is still more urgent.

Episodically, H. G. Wells envisioned an "open conspiracy", a synergy between people possessed of planetary vision and interactive worldwide. By January 2008, David Milliband, Britain's Foreign Secretary, could speak of a "citizen's surge" towards democracy. He had in mind aspirations in that direction evident in the former Soviet bloc but also manifested elsewhere these two decades: Afghanistan, China, East Timor, Iran, Iraq, Kenya, Myanmar, Pakistan, Palestine, Sierra Leone, South Africa, Yugoslavia, Zimbabwe . . . and, latterly, Bhutan and Nepal. How consistently such aspiration has been requited is another matter.

Wells will have expected an "open conspiracy" to be led by radio hams and Esperanto pen pals. Nowadays cosmopolitan association is achieved via the Internet or intermingling through travel. For a while, foreign travel other than on holiday was very largely by those due to enter university or else graduands. Lately, however, more non-academics have travelled for work experience, language training and fun. The number of people with a working knowledge of English, the modern Esperanto, could reach three billion by 2050.

Focal as regards long-term immigration has been the largest of the Anglophone nations. Of 250 million people in the USA, 18 million are legally resident aliens while several million more are illegals. Immeasurable but evident, through the millennium dawn, was a planetary upsurge of enthusiasm (especially among the young) for the American lifestyle. An astute biographer of Osama bin Laden believed his subject had come, at long last, to identify with the Palestinians because this seemed his one hope of overriding a yearning within the Islamic masses "for all things American".[7] The unpopularity of the retiring White House incumbent has dented this effect less than once it might have.

Interactiveness should make it harder, for now at least, to get the young to wage war against one another. That consequence has to be mainly positive, provided it spreads with tolerable evenness the world over. But the downsides include the weakening of traditional communities and more scope for organized crime, political and general. Nor can talent drain be ignored. A particularly serious aspect currently is the emigration of medics from Black Africa.

Renaissance Legatees

Still, these negatives relate more fundamentally to the rootlessness of modern society gauged in relation to our evolutionary inheritance, human and pre-human. Taking the record back 15 million years into the Miocene, you uncover a saga of relentless struggle against daunting odds by tribal communities — usually several hundred strong, close knit and possessed of distinctive identities. We needs connect with the distant past aesthetically and via close societal bonding within which personal roles are sufficiently recognized, authentic and challenging to be satisfying. In which connection, the late J. K. Galbraith perceived divergence anew between "dreary painful or socially demeaning work" and the kinds which are "enjoyable, socially reputable and economically rewarding". He saw a "common gloss" being cast over this dichotomy via a "culture of contentment".[8]

Pertinent here are some forceful, albeit subjective, indictments of the Renaissance articulated within the literati last century. Eric Fromm, a celebrated Frankfurt School psychoanalyst and refugee from Nazism, addressed the vacuity he currently found so endemic. He traced it to your archetypal Renaissance virtuoso with his ill-judged sanguineness, impulsive enterprise and unsettling learning. Distracted by a *haute bourgeoisie* of imperious individualists, the masses may have felt less calm. Yet through self-imposed isolation, the virtuosos, too, languished.[9] Likewise, Simone Weil attributed today's "uprootedness" considerably to how the "Renaissance everywhere brought a break between people of culture and the mass of the population".[10] In 1992, the late Ted Hughes — then Britain's Poet Laureate — interpreted the narrative poems and tragedies Shakespeare wrote after 1590 as reflecting anxiety lest a "macho-reductionism" detached humankind from its emotional roots.[11]

These insights hardly lend themselves to historical evaluation. In so far as they might, History would endorse them less than fully. When transposing them to the contemporary West, one does well to accept that Galbraithian depiction of superficially common cultures. Elsewhere, notably in China, a sharp urban/rural divide in social values remains starkly evident. The Renaissance phenomenon can be seen as a polarization between what was then a small urban sector and a great silent rural majority.

Progress is never without contradictions. Societal togetherness and continuity cannot rest easily with freedom of opportunity. Yet the social democratic Left has always extolled both aims *à l'outrance*. A 1957 study from that perspective looked at Bethnal Green within London's East End. The authors found customary bonding remained strong, thanks to upward mobility being limited as yet and economic insecurity still chronic. Their proposals for squaring this circle via community housing policy by no means looked able to offset the widening opportunities for personal fulfilment lately attracting aspirant youth.[12] Then again, in his seminal 1963 essay on the ambiguities of economic growth, Mancur Olson spoke of a failure to recognize how often one of the more direct effects was more losers than winners.[13] His admonition could especially apply to change driven by modern IT. Yet the *sine qua non* of any polity seeking to promote world harmony must be a stable home base free from disintegrative tendencies.

As noted above, Winston Churchill and H. G. Wells were each consumed by awareness of how Science sharpened the choice History would make between triumph and tragedy. Wells posited a race between Education and Catastrophe, alternatives not all of us would juxtapose that simply. Churchill perennially endured a struggle within his own psyche between sunny optimism and his "black dog" of pessimism, a tension not inappropriate to the needs of 1940.[14] Looking to some such dialectic within the contemporary scene could be the key to answering the Western Question.

Catastrophe, Chaos and Long Waves

Something to ponder *en passant* is how far can the new mathematics inform such insights? Take the Catastrophe Theory enunciated by René Thom in 1972. It showed how a balance of forces acting three-dimensionally could flip. A standard instance is how critically a creature or a community at once angry and frightened can be poised between fight and flight. The saving grace is that a flip may be in what turns out to have been the right direction.[15] To the Ancient Greeks, *krisis* was a time when things went horribly wrong or else very right. Catastrophes were not always catastrophic.

In the sixties, Edward Lorenz of MIT pioneered "chaos theory", an account of how far initial conditions modulate dynamic progressions: "Now that Science is looking, 'chaos' seems to be everywhere. A rising column of cigarette smoke breaks into wild swirls. A flag snaps back and forth'.[16] Take two very similar particles closely adjacent. Slight differences in their situations could lead them down contrasting paths. Theory tells of their being drawn to different "attractors" in "state space". If a theoretical basis can be shown to exist after all for the popular and professional belief that global warming accentuates climatic erraticism, it will probably lie in the application of chaos theory to quite abnormal secular change in temperature or rainfall.

Lately "anti-chaos" has entered this debate. Apparently it draws inspiration from Erwin Schrödinger's celebration of Life's "astonishing gift of concentrating a *stream of order* on itself and thus escaping the decay into atomic chaos", alias skirting round the Second Law of Thermodynamics. Among the inanimate patterns similarly sustained are snowflakes.

All such theory highlights indeterminacy. In the world we know, initial conditions will be hard to delineate and abrupt changes hard to predict. Hardest of all will be forecasting conjunctions, the coinciding of criticalities in realms of enquiry that reductionist analysis would keep separate. Therefore such theory is likely to be applied to world affairs more at a philosophical level than through esoteric calculation. Take the sudden collapse of regimes, societies or — above all — cities not seldom to be observed in pre-industrial times. Any such event is nowadays deemed as likely to be the end result of a sustained adverse trend in rainfall, soil quality or whatever as it is to be the immediate consequence of an earthquake, barbarian foray or other abrupt occurrence.[17]

Among the perspectives complexity therefore seems to negate are neo-Marxian "world systems" theories about predetermined long waves of alternating progress or retraction, these measured in centuries. In any case, the models adduced fit ill the historical record even with some arbitrariness of definition. One felt constrained to criticize one such exemplar sharply in *History and Climate Change*.[18, 19] As regards the present outlook, so many of the trends under way look menacing it is hard not to believe some actually will be, especially through conjunctions. To say that is to emphasize randomness as opposed to cyclical or, indeed, situational determinism.

In Pursuance of Democracy

Since President Bush committed himself in 2001 to promoting democracy abroad, particularly in the Middle East, pundits have remarked how contrary such Enlightenment hubris is to the customary belief of American conservatives "in the power of culture and tradition and their distrust of social engineering".[20] A distinction influential in such quarters was the one the late Jeanne Kirkpatrick drew between the authoritarian and the totalitarian. The former arose out of traditional conservatism which putatively meant they were, in the final analysis, restrained whereas the latter were inherently committed to comprehensive upheaval at home and abroad.

Throughout one must bear in mind the stark choice the information explosion will present to every regime. Become progressively more open, accessible and accountable or ever more closed — i.e. in the ultimate fascistic, be the fascism Blue, Black, Red or hybrid. Correspondingly, relations between communities (ethnic, religious or whatever) living in proximity will become more accommodating or else more fractious, the latter being liable also to have illiberal consequences within.

At which point, one could conclude that the case for saying democracy untrammelled must and will prevail everywhere has been made without contest. However, it is not that simple. A recent study tested the modish contention that democracies *per se* were little disposed to make war on each other. It found how, in the period 1885 to 1992, this was twice as true for the more developed democracies, economically speaking, as for poor ones. Among the latter the connection was "at best weak".[21] In the era we are moving into all polities will be stressed by deprivation patterns, some novel or previously little considered.

Guided Democracy?

Half a century ago, economic growth had absolute pride of place as the yardstick of development. The nub therefore became how might post-colonial economies best achieve lift off without turning to Soviet Communism. Two alternatives were posited.

The one was to look to a distinctively national revolution promoted by a charismatic dictator, someone thought able to mobilize requisite funds and arrange *dirigiste* management. The other was to opt at independence for democratic institutions in association with a mixed economy involving public ownership of the "commanding heights" complemented by a large private sector operating, internally at least, quite a free market. Those espousing the former little exercised themselves about when or how a transition to the latter might be effected. China and India were sometimes juxtaposed.

Meanwhile one archetypal example of an intermediate form, guided democracy, was not infrequently cited. It was Turkey under Mustafa Kemal, alias Kemal Ataturk: the tough military commander who became first President of the Turkish republic formed after 1918. As President he was a modernizer. In 1925 he banned the fez as betokening societal backwardness. The next year he enacted a comprehensive legal code based on the Swiss. And so on.

Twelve years after his death in 1938, parliamentary elections were held but resulted in a defeat for forceful modernization. A decade later, the army intervened to insist on a progressive government acceding. Thus were things to proceed to the present time. Ironically though, the persisting missing link was a convincing economic lift off. Witness currently the struggle to launch a defence industry of sorts — e.g. a Turkish designed tank after 2035.[22] Ataturk is little referred to abroad these days. Nor does the Turkish army appear as sure of touch about political intervention as once it did.

Since the Soviet collapse, liberal democracy has certainly moved centre stage. Nevertheless, there is often insistence, from Right and Left, that a developing country's transition to it should be gradual and reversible, requisite guidance being proffered by head of state, party, priesthood or army. Yet this poses a real risk of consummation delayed too long because, whenever a target date approaches, things will turn too tense. History is replete with instances. Meanwhile guided democracies will very generally be too secretive on the security side and too concerned with shaping cultural identity, in earnest or as a diversion.

None the less, among the considerations liable to strengthen the hand of those reluctant to back full democracy forthwith is the "remarkable comeback" lately effected in the developing world and Russia by public investment and enterprise in "such commanding heights" as power generation, telecoms and aviation. Three quarters of the world's crude oil reserves are now state owned.[23] Also, banks are coming under state supervision and partial ownership more generally.

Retaining or regaining state control obviously buttresses the national sovereignty of a developing state. Otherwise it is open to conflicting interpretations. It may encourage corruption. Certainly it runs contrary to global trends since 1980. Nor may state ownership be vital to basic infrastructural development. Historically, huge programmes have been undertaken by private enterprise, usually with governmental guidance.

Nevertheless, the extra infrastructural programmes called for by mounting world stress might require more public involvement, not least in the running of key financial institutions. Arguably, too, this might often be easier in the context of guided democracy. From 2008 to 2017, China will probably spend 2.5 times as much per head on infrastructure as India, dynamic though the latter's economy now is in many ways.[24] China may well come to be seen as a major test case for guided democracy throughout. A lesser but very indicative one may be Thailand, sharply divided as of November 2008 between rural populism and a Bangkok monarchical technocracy.

Philosophic Deficit

In curiously short supply the last 50 years has been an input always pertinent in times of instability. One alludes to political and social philosophy of the kind so impacting the first half of the last century. Pedagogic studies of electoral behaviour are no substitute. Nor are anecdotal sketches by aspirant national leaders. Nor is cultural nostalgia.

This deficiency has been especially harmful to the developing world, profoundly agonized as it has been by such questions as federalism; local government; participation by large minorities; educational priorities; tradition versus modernity; women's rights; juridical modalities; national autonomy and identity; and regional collaboration. An example of the issues that arise is currently afforded by archipelagic Indonesia. To try and prevent democracy becoming too fractionated, no party which gains less than a fortieth of the national vote will qualify for parliamentary seats in the 2009 general election. In 2004 that would have disqualified 19 of the 27 parties standing.

In China, the rich corpus of thought emanating from Sun Yat-Sen (1866-1925) and colleagues was to be submerged by armed strife and by Maoism, more an insurgency doctrine than anything else. Sooner rather than later, too, India was obliged to move out from under the Mahatma Gandhi mantle. His approach had been one of introverted fundamentalism, too detached from modern choices. As Arthur Koestler caustically observed, "the spinning wheel found a place on the national flag but not in the peasants' cottages".[25]

One can take also Steve Biko of the African National Congress. Nothing can condone his dying from brain damage in policy custody in 1977. The fact remains he never made a convincing moral distinction between his "Black consciousness" movement and "White racism". And his scoffing at the "myth of integration" showed no awareness of how damaging retraction into yesteryear may be.[26] Then again, Edward Said, the distinguished Palestinian academic at Columbia University, wrote eloquently about pseudo-romantic stereotyping of the Oriental persona but not about the prerequisites for Palestinian social democracy.

Arguably, the most commendable endeavours to date have been by Julius Nyerere, Catholic convert and Edinburgh graduate, who was Tanzania's first President from 1964 to 1985. Not that as a thinker he knew untrammelled success. By 1971 he was acknowledging his flagship scheme for *ujamaa* cooperative villages was faltering.[27] Also the principle that general election seats should be contested only as between members of the ruling TANU party was a mockery. The same goes for his precept that African regional associations should be encouraged only on the basis that criticisms of fellow member states always be "in private and in a fraternal spirit".[28] Nevertheless, the gentle humanism informing his discussion of issues like equality and conflict resolution makes his thinking still germane. So does his passion for village life in a natural setting. One cannot easily think of a contemporary anywhere who carried those themes forward so well.

The International Financial Order

The overview essayed above of mathematical approaches to alternative futures indicated these tend to present humankind as poised at *the* great turning point between planetary togetherness and fascistic nightmare. In practice, of course, the outcome will be compounded from a succession of lesser crises. A moot point is how well the international payments system will cope. Alas that is a subject area about which it is all too easy to be awfully well-informed yet quite uncomprehending. This makes it hard for us laity to know whence to take advice on crisis management or structural reform.

The crunch issues were spotlighted by the too overlooked crisis of 1967 to 1971. Anxious to avoid the exchange rate fluctuations experienced inter-war, the Allies had agreed at Bretton Woods in 1944 on a fixed rates regime in which the main liquidity sources alongside gold would be holdings abroad of sterling and dollars.

By the late sixties, however, this formula was not working well. That decade liquidity growth was but half percentagewise the world trade expansion. This was partly because of constricted supplies of gold — still 40 per cent of the liquidity base. Meanwhile, the American deficit was becoming less a consequence of regular foreign aid and more of "Vietnam" plus a weakening trade balance. By then, too, sterling was overvalued, a situation not eased by the Suez Canal's closure from June 1967.

Britain's Prime Minister, Harold Wilson acted against the advice of senior colleagues by fighting to avoid sterling devaluation. Then in November 1967 he caved in, thereby precipitating four years of exchange rate chaos. In 1968 came the launch of Special Drawing Rights (SDRs), a means of payments enthusiasts saw as prospectively a truly international currency. However, by 1971 SDRs represented but three per cent of liquidity.

At the Smithsonian that December, the "Group of Ten" leading economies agreed a package providing for regular consultation and more flexible exchange rates, essentially the regime still obtaining. Within the parameters just delineated, it was working well enough as 2007 dawned. Yet trouble was looming in terms of how banks and insurance companies in the USA, Britain and elsewhere had been allowed to overstretch their commitments (most conspicuously as regards property mortgaging) by regulatory systems that were insufficiently refined and too laxly applied.

A big expansion of financial liquidity had for some years proved compatible with the maintenance almost throughout the economic system of the consistently low inflation rates demanded by received wisdom. It had done so because the "dot.com" revolution was effecting biggish gains in productivity very generally and also because of the export by Asia of large quantities of light consumer goods, of good quality but cheap. Where inflation became marked was in property, not least its residential sector. Extravagant housing booms were underwritten by mortgages that left the lenders very exposed once house prices were tumbling back to more reasonable levels. Then much as in 1929 when the crisis of confidence finally broke, it did so suddenly and massively, an all too classic catastrophe flip. The build back to full confidence may well take several years; and will almost certainly require the banking and insurance sectors to be subject to tighter regulation, nationally and — one trusts — multinationally. A pronounced difference with the 1967 to 1971 experience is that, this time round, there has been genuine fear that the whole banking edifice might collapse globally, effecting the end not just of capitalism but of civilization. The Governor of the Bank of England has spoken of the world being in its worst banking crisis since 1914. But then the Gold Standard proffered some stability.

One wonders whether, come calmer times, the Special Drawing Rights concept should be revisited. One could further surmise that any such departure might be backed by a precious metal. A feature of the present crisis has been a stronger demand for gold and silver, especially from those born the other side of 1940. Meanwhile, a supply side factor militating in favour of metallic underwriting is that production is currently dispersed quite well. With both gold and silver, not quite a quarter of the output of the top six among the national producers comes from the leader. The allusion is to South Africa and Peru respectively.

Oil Prices

The debate on all that will have to be left to specialists. But a double collateral should perhaps be explored a little. How far were higher oil prices also to blame for the crisis situation; and how far has their inflation been due to the Iraq war?

As regards the price sequence, the representative price for a barrel of oil in 2003 was $20 to $25 and (judging from a survey in the *Wall Street Journal* that January[29]) was expected to stay around that range through 2009. In fact it was *c.* $90 through the turn of year 2007–8 and during the early summer was to rise to $140. This was after the urban "surge" strategy had led to a significant improvement in security within Iraq though not before this verdict had been generally accepted. In October 2008, the price was $80, but falling.

On the question posed, Joseph Stiglitz and Linda Bilmes see no other extrinsic influences, barring Hurricane Katrina in 2005, that "could be given similar credit" for 2008 oil prices being so far ahead of 2003 expectations. They duly conclude that "a significant proportion" of the said differential "resulted from the war".[30] But is this not a field in which price forecasts several years ahead can be badly wrong (maybe two or three-fold?) even when it is throughout "business as usual" — meaning, the absence of international security complications or of natural disasters?

Besides, in the Middle East alone other international security questions were impacting. War with Iran was a possibility to reckon with. So, too, were several active conflicts. Besides, when someone like Barack Obama can say his aim would be to end within ten years all United States dependence on Middle East oil, the regional producers are bound to maximize profits while they can.

Currency Mediation

One must wait upon future needs, bearing in mind there is nothing currency and capital markets like less than extrinsic volatilities: tendencies this century will likely be studded with — all within a broadly Malthusian ambience. Strictly speaking, of course, Thomas Malthus was concerned just with intrinsic constraints on production increase. We neo-Malthusians have to factor in also extrinsic ones, not all of them calculable.

As one reviews the financial order, it is possible to identify certain territories (Britain, Switzerland, various offshore islands . . .) that have roles rooted in their historical geography. Iceland's overly fish-based economy has been hit catastrophically by the 2008 credit crunch much as it was slow to come out of the depression post-1929.[31] But overall the system cannot be well explained in terms of geographic

causation. You could fairly say the performance of international finance more deter-
mines the margins of interaction between humankind and its geographical ambience
than it is determined by them.

However, one must enter the caveat that the ability of nation states to act in
harmony is limited in this domain as in others by marked variations between them as
regards economic geography. This makes parity of contribution impossible to determine even when all are subscribing to the same economic philosophy.

Survival Costings

Monetary liquidity apart, the economic impact of the long global crisis confronting us
cannot be predicted with much assurance and will, in any case, be erratic. But it is still
needful to get a preliminary idea of the financial obligations which need be accepted
within a decade or two, particularly by those twenty nations (most of them "advanced"
economically) which produce over three-quarters of the world's Gross Domestic
Product and will have to assume the lead in developing crisis limitation strategies. At
present world GDP is *c.* 45 trillion dollars per annum. World population stands at 6.5
billion which is nearly four times the 1914 figure.

The added cost of commodity imports could medianally take an extra 10 per cent
of GDP for the more advanced nations several decades hence. One has to be talking,
too, about an extra several per cent per year to adapt their own economies, agriculture included. One can also well imagine official economic aid exceeding a trillion
dollars annually to cope with the measure of climate change already inevitable. At
present agricultural subsidies in-country are *c.* 350 billion dollars per annum worldwide.

If these guestimates prove to be somewhere near ball park, responding accordingly
will engender fierce debates about efficacy and equity, not least as reflected in taxation. Overall the situation ought to be manageable. However, the costs might
eventually have to be much higher if action is too long delayed. Increased security
expenditure would be part of the extra imposition, a large part if much armed strife
ensues. World defence budgets now total a trillion dollars per annum. Three times
that is what Stiglitz and Bilmes put the direct cost of the Iraq war over 10 years for the
USA alone. This reckoning excludes the oil price hike and other macro-economic
penalties. Even so, it still equals over a quarter of the 2005 American GDP. [32]
Comparable IISS calculations suggest cumulative outlays of perhaps 1.5 to 2.0 trillion by 2017. [33]

Peering as best one may beyond 2020, it does appear that a balanced Survival
Strategy will call for quantum jumps in capital expenditure on energy sources, urban
renewal and health care. One cannot be at all definitive about how these may be spread
as between public and private sectors or as between richer and poorer nations. But in
the nearer term at least, needful expansion of the public domain may be resisted by
electorates fearful of having to cover bad debt obligations at short notice and in
circumstances in which government borrowing is difficult. At present the United
States GDP is around 12 trillion dollars, just under a fifth of which is public expenditure. As of late 2008, the US Treasury is accepting a contingency commitment to
cover financial sector debts to the tune of half a trillion. Hopefully, an overview of neo-
Malthusian pressures globally will help us gauge the import of this apparent
imperative.

Malthusian Roller Coasting

So the twenty-first century is forecast to experience a succession of neo-Malthusian crises. Impacting already is how the progressive amassing of personal paraphernalia needing ever more room to keep and use is exacerbating urban congestion the world over. Cars are the most obvious example. Among other pressures ineluctably destined to assume more salience are Near Space congestion and looming shortages of various price-volatile non-ferrous metals.

Concern is also felt about the decline or collapse of ocean fisheries due as yet to over fishing more than climate. Two features in *Nature* highlight criticalities. One on large fish species found that "Since 1950 with the onset of industrialised fisheries we have rapidly reduced the resource base to less than 10 per cent . . . for entire communities . . . from the tropics to the Poles".[34] The second showed how meshes are constructed to catch preferentially the biggest members of a given species, those otherwise best adapted to keeping it healthy.[35] Alarming, too, is a study by the Netherlands Institute for Fisheries Research which found the number of European eels to be one per cent of the 1980 level.[36] Also disturbing is evidence that anti-submarine sonar may incidentally kill whales, porpoises and dolphins by distracting their sound location.

Suddenly food production generally is a global worry again. Increases in 2007 brought staple prices to between 60 and 100 per cent above 2004 levels. Early 2008 saw anxiety surge both sides of the Atlantic over how far this could be due to arable being used to grow crops for biofuel.[37] In March, Professor John Beddington, the British government's Chief Scientific Adviser, spoke of world food and energy needs rising by half by 2030. He apprehended a desperate food crisis breaking due to increasing drought but more to inducements or instructions to grow biofuel crops. He found cutting down rainforest to do this "profoundly stupid".

That energy is entering an era of generally short but decidedly variable supply has become the commonest of commonplaces. As a rule, this in itself may not be too constraining. At present 6 per cent of, for instance, the USA's GNP is spent on energy. Should that fraction treble, the added cost would represent but four years' recent economic growth. However, critical hydrocarbon shortages can onset abruptly. Besides, this sector still meets 80 per cent of global energy needs and is easily the chief source of anthropogenic global warming.

On which subject, Bjørn Lomborg continues to scorn what he sees as undue zealotry.[38] He now concedes that "we should take action on climate change" (p. 210) but censures those more insistent for "a kind of choreographed screaming" (p. 209). He sees the British government as "the most given to high pitched rhetoric on this score" (p. 210). So one is constrained to remind him, too, of an English cricketing maxim, "Bowl at the wicket, not the batsmen".

The Siberian Anticyclone

Regarding climate change, may one specifically flag concern as to whether the Siberian anticyclone has received the attention it should have. Throughout the winter season in recent centuries, it has been a conspicuous feature of the Eurasian climate scene. But as a "cold pool" phenomenon it could be vulnerable to being overridden in depth by the current warming trend. A vertical column of cold air between two pressure levels is denser therefore less tall than a corresponding column of warm, a consequence being that pressure falls more rapidly with height in the former case. A very preliminary scan of the 2003-6 readings indicates that the "cold pool" area has been warming each winter quite strongly.[39] If the Siberian winter High does wane, it is not unreasonable to expect that the requisite area of high pressure with descending air will be organized around the North Pole which is where it would be in the "planetary wind system" of an Earth with a uniform surface. This proposal is not necessarily inconsistent with modelling showing sub-polar Northern latitudes experiencing lower mean pressure as the Earth warms.[40]

During the early medieval era of relatively gentle warming (c. 750-1275 in Western Europe) something like the above does seem to have occurred. The scientific evidence from Northern Scandinavia is complex but supportive of this interpretation.[41] Moreover, the Norwegian Vikings were more reluctant to move north than to other points of the compass. They are not known to have visited Svalbard until 1194. Of the eleven settlements of township status in medieval Norway, Trondheim (at latitude 63°20') was the northernmost.[42]

The fortunes of the Siberian High can have consequences across the Northern Hemisphere in winter. Since the seasonal pressure maximum has, in fact, medianly resided over Mongolia (at 48°N, 100°E), these consequences are bound to be of especial moment for the Himalayas. If that great axis of Tertiary folding loses a sizeable fraction of its ice cover, there will be serious connotations for the flow regimes of the Indus, Ganges and Brahmaputra.

Energy Dilemmas

There was for too long a "global warming" stand off, India and China versus the USA. Each side used the other as an alibi for inaction. Mercifully, both are now edging away from that.[43]

But for all states, maintaining a consistent energy strategy will be near impossible in the decades ahead. What priority should national security receive as against short-term prosperity or global well-being? How does one cope with price instabilities, given the long lead times characteristic of capital-intensive energy installation? What account should be taken of new technologies? How much might one constrain the greenhouse effect?

The Nuclear Renaissance

Thirty years ago, James Lovelock and Lynn Marguilis emerged as prophets of a Green perspective. Their Gaia hypothesis proposed that Life on Earth constantly seeks to maintain conditions favourable to the prevailing ensemble of organisms. This adds significance to Lovelock's lately underlining the view he first expressed in 1979 that nuclear power was essential to squaring energy needs with curbing climate change.[44,45]

Lovelock adopts this pitch because the main alternatives Greens commend have weaknesses. Hydroelectricity is hamstrung by site shortage; and these days is felt to disturb the ecology overly — an issue which first assumed international prominence during the Aswan Dam dispute, 1955–6. Wind turbines and solar cells operate but part of the time; and the looked-for breakthroughs in energy storage have not materialized yet. One could add that, since global warming reduces the tropics to poles temperature gradients, it must weaken the general run of wind. Besides, wind turbine arrays look ugly except perhaps at sea. Tidal power, as far as it goes, may be a steadier bet. But for all the renewables, it would be helpful to have more information in the accessible public domain about overall "life time" costings, output distribution included.

The waste disposal argument against nuclear, Lovelock upturns. He calculates the CO_2 waste produced by reliance on hydrocarbons would (if duly recaptured) be two million times the volume in storage of equivalent fission waste held in vitrified rods.[46]

A cross which nuclear energy bears today is that, in the past, its problems have been played down too nonchalantly. Tolerance thresholds for the radioactivity residual from past tests in the atmosphere have had to be revised downwards.[47] There has been defaulting on reactor safety. These past 40 years, the commercial application of nuclear fusion has repeatedly been predicted to be 30 years away.

All the same, the International Thermonuclear Experimental Reactor (ITER) now under way does hold out more solid promise for perhaps 35 years hence. Additionally, new orders for fission reactors have lately been placed in the USA, France and Finland and are pending in perhaps ten other countries. Germane to this is that the latest fission reactor designs are more efficient and safer. Sliced costings now look favourable in relation to other energy price guestimates. Even so, the time span and large fraction of life-time cost devoted to reactor construction makes adjustment to variable demand especially hard to judge aright. Moreover, a secular uranium shortage is reflected in production having peaked in 1981; and in the price of uranium oxide rising six-fold in the five years from early 2001.

A Coal Revival?

The consumption of coal looks set to remain high after all. Well over five trillion tons of deposits have been identified worldwide, two-thirds in the former USSR. Countries still getting at least 80 per cent of their electricity from coal in 2003 included China, South Africa, Poland and Estonia.

So interest burgeons in how efficiently the carbon dioxide emissions from large-scale combustion can be "scrubbed" out prior to deposition underground. Witness the EU–China 2005 agreement to demonstrate feasibility by 2020. Hopes run high that 90 per cent of the CO_2 from power stations can be recovered at source, the added cost being perhaps a third. A promising research line is recapture with algae.[48]

Also over this span more will be learned about at last developing electrical batteries efficient and safe enough for long-distance road haulage. As yet, there are possibilities but hardly probabilities.[49]

Nuclear Proliferation

At present a quarter of the world's electricity comes from nuclear power. So what of the potential latent in this situation for nuclear warheads to proliferate? The grounds for being concerned rest on well tried arguments. Small deterrents can stand too exposed to pre-emption. And their weak strike capability might oblige them to focus forthwith on the adversary's capital city thus likely eliminating the very people one needs talk peace with. Command and control systems may not be sophisticated enough. Besides, there has long been a feeling that certain regimes in the developing world cannot be depended on to act at all rationally.[50] Denial of the Hitlerian Holocaust surely disqualifies the present Iranian President as a WMD custodian. But in this case, as in others, we must ensure that nuclear proscription does not induce alienated states to turn to biowarfare instead (see below).

Prideful Concern

An obstacle to progress throughout this domain is pride affronted. In 2004 the Director General of the International Atomic Energy Agency (IAEA) pointed out how in 1996 the International Court of Justice unanimously reaffirmed that established nuclear states were required to negotiate for "nuclear disarmament in all its aspects". He adjudged it "time to abandon the unworkable notion that it is morally reprehensible for some countries to pursue nuclear weapons but morally acceptable for others to rely on them".[51] Similarly in 1998 an adviser to India's Prime Minister had warned against the "great error" of assuming that "advocating the new mantras of globalization and the market makes national security subservient to global trade".[52]

Worth noting, too, is the interest Canada and Australia reportedly express in uranium enrichment facilities.[53] They stand first and second in the mining of uranium oxide, Kazakhstan being third. These states are recorded as having between them half the world's mineable reserves.[54]

Still, the Indian subcontinent and South-West Asia is where proliferation is today being played out earnestly. None of the existing military nuclear states therein (India, Pakistan and Israel) have ever felt able to sign the 1968 Non-Proliferation Treaty.

The Subcontinent

Pakistan resolved to seek a nuclear capability after India conducted a fission explosion in 1974, ostensibly with nothing bar "peaceful construction" in mind. Islamabad acted thus on the basis of *sub rosa* multilateral collaboration. A key player was A. Q. Khan. He began applying centrifuge technology he purloined when working in the Netherlands. His other linkages included selling nuclear technology to Libya for money and to North Korea for missile technology.

Meanwhile the Washington administration was giving its seal of approval to relevant Pakistani institutes, Khan Research Laboratories included. This was despite the Pressler Amendment passed by Congress in 1985 which said the US President should certify annually that Pakistan had not got "the Bomb". The background was that, until 1990, Pakistan was crucial to supporting the Taliban against the USSR in Afghanistan.

In 1995 New Delhi came close to testing a nuclear bomb but held back on Washington's insistence. Then in May 1998, a few weeks after the Bharatiya Janata Party (BJP) assumed office in New Delhi with a mandate to create a deterrent, India did conduct several tests. Pakistan swiftly followed suit, American overtures notwithstanding.

The next turning point was 9/11. Islamabad officially ceased hosting the Taliban. In exchange, Washington accepted that Pakistan had pardoned Khan's "unauthorised" proliferation activities. The notion of "unauthorised" was fanciful. The Pakistani Air Force delivered materials for him.[55]

Following the 1998 tests, the USA imposed sanctions on both countries. Yet whatever one says in principle about the instabilities of small deterrents, their mutuality in this respect may well have inhibited major war during their 2002 military stand off.

Between 2005 and 2007, the White House made a diametric policy switch regarding India and proliferation. Before 2006, the administration was prohibited from selling civil nuclear technology to India since it had not signed the NPT. Then Congress passed legislation allowing civil nuclear technology and fuel to be supplied though not if it conducted further tests. However, in July 2007, the President promised to help India build a nuclear repository and find other nuclear fuel sources in the event of a Congressional cut off.

Yet generally speaking, the West remains strongly placed to discourage or disrupt developing countries' endeavours to go military nuclear. One need not believe the 2003 invasion of Iraq was perfectly conceived to infer a causal connection with Iran's apparent suspension the same year of its WMD programme or Libya's formal renunciation that December of her nuclear weapons plans.

Strategic Biowarfare?

Unfortunately, such impact is bound to aggravate anxiety among emergent states lest the West stays irremediably far ahead across the military spectrum. The danger remains that eventually some regimes will decide they can only get on terms by going for WMD if not in nuclear form then in biological. As the then editor of the *Bulletin of Atomic Scientists* put it as this millennium dawned, "Evidence that nations might start or revitalise Weapons of Mass Destruction is meagre. By 2020, however, such might not be the case".[56] One cannot but ruefully endorse this with particular reference to bio-bombs, the consideration analysts all too regularly treat quite separately from mainstream strategic assessments. It is a hiving off we accept at our peril. To talk of eliminating the nuclear threat while ignoring the biological is downright irresponsible.

Granted it is always tempting in such matters to warn of a crunch 20 years away, not too far off to be visualized yet leaving time for precautionary measures. But in this regard it looks peculiarly valid. Relevant biotechnology is available now. But as one has seen in respect of unrestricted submarine blockade, area mass bombing and battlefield nuclear use, a decade or two can elapse between basic feasibility and systematic application.

A more urgently apprehensive interpretation was that rendered by the US Congress with the December 2008 report by its Commission on the Prevention of Weapons Destruction. This warned that, as things stand, it is "more likely than not" that there will be a WMD insurgency strike somewhere before 2014. Moreover, "Terrorists are more likely to be able to obtain and use a biological weapon than a nuclear weapon". In fact, "were one to map terrorism and Weapons of Mass Destruction today, all roads would intersect in Pakistan".

However, the other side of that coin surely is that a strategy enabling a democratic Pakistan to become a king-pin of stability in this regard would sharply reduce the overall threat for the next decade or two. The report also indicates that better security for relevant biological research sites (not least the 400 in the USA) could be contributory. By 2025 or so, however, the problem could be far more generalized.

Ultimately, a descent into what would pretty much be the penultimate warlike obscenity could be encouraged by a general retreat from reason and reasonableness.

It could in turn much encourage it. Specifically germane will be the advent of *the* bio age, an age in which our ethical values and our resolve in upholding them will be challenged by bewildering strides in life science. One has in mind human cloning, licit or backstreet; extended longevity; brain implants; genetic modification of plants and animals; the use of DNA/RNA to clone dinosaurs or whatever. All this will be against the background of the impoverishment and uglification of Wild Nature.

If the upshot is to undermine our appreciation of what a privilege it is to be human within the sweet oneness of Life, deep taboos against war by contamination may be compromised. The fact that biowar could proffer many gradations in terms of targeting, character and intensity could make it all too easy to shed customary inhibitions. So could the scope for clandestine delivery. So could the singular difficulty of arms control verification in this sphere.

Moreover, biowar could be a way back for insurgency and organized crime, modes of conflict currently being undermined by state-of-art processing of data collected by such means as voice, face or iris recognition; remote reading of number plates or passport chips; voice or Internet intercepts; and electronic tagging. Al Qaeda has had some interest in biological weapons along with chemical ones since 1995 at least. It seems its leading activist in these fields, Abu Khabab al-Musri, was killed in July 2008 in an American cross-border strike into the tribal zone of Pakistan.

Survival Basics

A precept progressive opinion has long subscribed to is that you cannot defeat insurgency without offering an appropriately generous settlement, politically and in terms of economic and social prospects. That was never absolutely true. But the counter-insurgency techniques to hand nowadays make it downright untrue. Nevertheless every government currently confronting insurgents will do well to treat it as valid in anticipation of the time approaching when the biowarfare option will swing the operational balance back towards committed insurgents.

Yet how 'appropriate generosity' will be gauged in relation to political necessities and aspirations can never be confidently predicted. Take the tripartite confederalism (Kurds, Sunnis, Shias) sometimes envisaged for Iraq, this to placate the Kurds and Sunnis. A Norwegian analysis has shown how this runs contrary to the history of the last five centuries, throughout which Baghdad has been very much the centrepiece of a unitary Mesopotamia.[57] This has been deemed imperative given the acute susceptibility of the twin rivers to silting. Indeed, it could be seen as authentic hydraulic despotism, except that Wittfogel himself overlooked the hydrological circumstances.

Another big upsurge ten or twenty years hence in the "long war" waged by insurgency and organized criminality need not emanate very particularly from one religious culture or, of course, any specific background. Inevitably, however, one addresses the situation in the *umma*, the community of Islam. For within it the Palestine question coincides "with a profound crisis of modernisation . . . that has produced a school of extremists whose influence extends beyond the Middle East into the Moslem diaspora in Western cities".[58] Setting in train steady progress on the Palestine situation has surely to start with ordinary Palestinians being accorded an ambience sufficiently relaxed to allow of reasonable discussion and positive reflection. We are regularly being advised that this is being denied by Hamas in the Gaza and by the Israeli army on the West Bank. What also needs to be appreciated is that

Fatah security on the West Bank can often disport itself in ways gratuitously arbitrary.[59] Taking the Middle East more generally, it probably has been fair to say it is a region about which our fears of genocidal conflict and hopes for liberalization need to be more coherently related.[60]

The London School of Economics has been the fount of a challenging neo-realism. Its gist is that our "instrumental" approach to insurgency war has come up against an "expressive" and "existentialist" opposition with far more historical provenance behind it. This latter blurs the distinction between war and peace, does not set store by cost–benefit analysis, anticipates the long duration of mainly devolved violence, and sees its concept of war as expressive of its whole life style. Only a limited convergence between these two approaches is effected through war.[61]

This mode of analysis does not blend smoothly with the pristine liberalism of "open conspiracy" and "planetary" perspectives. But these two ways of looking at a troubled world might be brought closer to good effect.

Multilateral Regionalism

At the end of World War Two, there was a surge of interest in having great river basins become settings for rounded regional development multinationally. The Danube and the Nile tended particularly to be mentioned. The inspiration throughout was the Tennessee Valley Authority (TVA), a New Deal recovery programme based on hydroelectricity. It was "inter-state", albeit simply within the American Union.

Writing in 1944 as TVA's Chairman, David Lilienthal (later, 1947–9, Chairman of the Atomic Energy Commission) commended the authority's achievements and prospects in relation to erosion control, malaria eradication and lacustrine amenities. He extolled its "grass roots" as "dreamers with shovels".[62] Pivotal to TVA was dam building. It was a riposte to Soviet claims that through their great dams new "Soviet Man" was arising.

These days even small dams are viewed ambivalently. Witness controversy over several being constructed on the Mekong. The Indus valley agreement on water sharing has continued despite wars between India and Pakistan but has not extended its remit. Of TVA one never hears. This is partly because of the altered perceptions of water control though also because, by the fifties, the impression many had was that "grass roots" democracy was too easily co-opted by the authority, formally or informally.[63]

Instead there have been repeated calls to revive that other big idea, the Marshall Plan. In 1987, US Congressmen proposed a mini-Marshall Plan for the Philippines. Come the early nineties, members of Congress advocated such schemes for Eastern Europe and the Global Environment. In 1996, Helmut Kohl suggested a Marshall Plan for the Middle East, as did Nelson Mandela for Southern Africa. Now Joseph Stiglitz and Linda Bilmes aver the Middle East could have had a Marhall Plan for the cost of the Iraq War.[64]

In 1992 Al Gore looked towards a Global Marshall Plan. However, one would have thought that attempting to go global thus would be to lose focus and cut across existing frameworks. Addressing the world crisis requires stronger regional structures, their strength partly deriving from the respective geographical spreads being suitably functional. For responding to climate change and ambient ecological disruption, river valleys can still be suitable settings: Mekong, Indus-Brahmaputra,

Tigris–Euphrates. . . . They can also be a good background for realizing other social objectives, provided the political geography is not too discordant.

Regional authorities could be lynch-pins, too, in the struggle to manage WMD. They proffer the best chance of ensuring emergent states never decide the only way to stay on terms militarily is bio-deterrence. The principle of assisting with their civil nuclear programmes individual countries who renounce national nuclear deterrence has already been articulated apropos North Korea and Iran. Yet it might be better if separation and enrichment plants and related facilities could be located in theatre while actively under the control of a representative and multipurpose institutional region- alism linked with the IAEA. This would not merely assuage pride. It could also allow of collective break out should the world scene ever turn so anarchic that there seemed no alternative barring bio-bombs. The more immediate risk of break outs by individual maverick states could still be sharply diminished.

Cohesion Deficits

All in all, adequate answers to the whole Western Question could hinge on more robust regionalisms being in place by 2025. Unfortunately our 2008 experience has hardly been of movement in that direction. ASEAN, usually seen as the firmest exemplar thus far, has been sorely taxed by Myanmar's repressive xenophobia. Meanwhile Southern Africa is a region desperately in need of constructive unity on Zimbabwe, AIDS, recurring droughts and much else. Thus far at least, misplaced pride has held back the governments concerned, especially the South African, most visibly in respect of Zimbabwe.

The Southern African Development Community (SADC) was formed in 1992, a declared aim being a common market between its 14 member states. Early in the twen- tieth century, it was involved in bringing peace to Angola and to the Congo, though how crucially is unclear. It failed miserably to censure massive irregularities during the Zimbabwean election of 2005. A majority on an SADC observer team actually opined the contest was free and fair. Had it done otherwise, things might have been better today for all concerned, Robert Mugabe included. All the same, SADC is a more defi- nite geographical expression than is the African Union. The departure of Mugabe could invigorate the community whether or not it had contributed to this consum- mation. The acid test for it would then be pro-active candour in respect of AIDS. Two-thirds of the world's 33 million sufferers (as of 2008) are in Sub-Saharan Africa.

Peculiarly worrisome in terms of ecological impoverishment is what we loosely term "the Arctic". Geographically, it is hard to delimit. You cannot do so satisfacto- rily using coastlines. Yet if you take the Arctic Circle (66°30′N) you exclude overmuch. Then if you go for the 60°N parallel instead, you include overmuch.

In March 2008, a workshop on "The Inhabited Arctic" was held at the British Academy. Of prime concern was the community decay consequent upon continual culture shocks and the emigration of aspiring youth. It may be that the UN could usefully publish at intervals interdisciplinary sets of papers reviewing the Arctic synop- tically. Other ecological themes similarly invite this approach. The condition globally of coral reefs is one.

Conciliatory Toughness

Just as there would be manifold ecology concerns without climate change, so would there be military ones without biowar. Nevertheless, the fact that the latter possibility looms calls for two-sided strategies, determinedly tough yet thoroughly conciliatory.

Starting on the latter tack, it is not hard to appreciate the Counter-Culture contention that "the Heavens were made for Wonder not for War". Do not weaponize Near Space in pursuance of, in particular, National Missile Defence. Less than might at first appear would be lost by such abstention. Missiles, ballistic or cruise, are not the most effective way to dispatch and disperse germs. Besides, very early on in the debates about how to engage salvos of incoming ballistic missiles, all sides conceded that 100 per cent effectiveness could never be adequately guaranteed. **(See 13 (2))**. Moreover, by renouncing Space weapons one would be stepping back from arms races far less constrained by Geography than would be the case on the Earth's surface, land or sea.

A commitment to "no thermonuclear first use" could be placatory, though how much so could be debated. What certainly merits attention is full disclosures all round of stockpiles of nuclear warheads and fissile materials. This subject the military nuclear states have consistently lacked candour about. Russia, China, the USA, Britain, Israel, India and Pakistan are regularly cited in this regard. May one also note that France's final decision to go military nuclear was covertly taken *c.* 1955, essentially by senior officials. Paris still seems far from true "transparency" about such matters.[65] As much had to be said, too, of Apartheid South Africa. A sudden invitation to the big nuclear complex near Pretoria was the prelude to the sole occasion in our 1986 fact-finding tour when one just felt set up. False assurances were gratuitously given about no military intent.

We are unlikely to crack the WMD problem unless the West sets a precedent of openness. Surely, too, a large multinational inspectorate will be needed for the biowarfare dimension, a body seeking access to all countries not just to profess equity but because insurgents or common criminals might opt for biowarfare almost anywhere. One can readily imagine such a monitoring force needing to be 20,000 strong, nearly an order of magnitude more than the total IAEA staff today. It is hard to see much else conferring credibility on the Biological Weapons Convention or any successor to it.

An objection which could be raised is that being so exercised about the bio threat confers on it a kind of satanic legitimacy, thus making it a subset of newly modish behaviour. Undeniably, too, some of the alienated will always see increased preparedness as a challenge inviting defiance. Yet allowing these considerations to weigh too heavily will reveal a failure to grasp how wide and far the biological revolution now breaking will extend. The way its ramifications will compound the current moral confusion could weaken inhibitions not just against biowar but, nastiest of all, bioweapons tailored to attack particular ethnic groups.[66]

Intervention Philosophy

Impossible though it is to predict just how the world scene will pan out the next several decades (not least as regards forceful interventions), one begins to see overlay between two cardinal aims. The one is preventing malign rulers from abusing their own people;

the other is constraining the spread of WMD, especially those elusive biologicals. The commonality lies in the promotion of open, liberal democracy. Unfortunately, this precept remains hard to apply consistently, given the irregular geography of the world we presently inhabit.

Peace-building military interventions in the post-Cold War era have to date consistently been in relation to polities that are smallish and readily accessible. How does one deal with ones larger or otherwise less manageable, bearing in mind that successful intervention is very liable to involve regime change? Can there be a doctrine for handling the aftermath?

An answer proffered above was to declare a State of Conflict, thereby generating manifold options. So what would follow? Would one be actively seeking regime change early on? And would this be admitted? How does one select tactics designed to unhinge a regime from its populace, not bind them together? And regarding the use of force, does one initially eschew territorial conquest in favour of precision strikes against, say, command and control nodes? How could Internal Arms Control best be brought to bear? How overt must one be about stratagems employed? Should not surveillance intrusion by unmanned aerial vehicles be ruled out except in the context of the declaration here envisaged? What part might cyberwar play? What principles can be laid down for nation rebuilding? What if the enemy is a democracy, in outward form or more substantially?

Action through the UN (and especially the Security Council) may underpin legitimacy and win understanding. Furthermore, sustaining the authority of the UN can bolster its subsidiary organs. The Intergovernmental Panel on Climate Change (IPCC) might well carry less weight if the UN waned for whatever reason. Any affiliated WMD inspectorate certainly would. Yet while top class intelligence would be imperative, obtaining and sharing it could prove vexatious in the UN ambience. The same goes for Internal Arms Control, not to mention nuclear dissuasion. In any case, managing so modulated a conflict could call for more political involvement in military operational decisions than has customarily been considered normal. That would mean tighter adherence to an overall master plan, less scope for devolved military initiative. Montgomery rather than Rommel.[67]

Fledgling Democracies

The Swedish development economist, Gunnar Myrdal, spoke of states which were "soft" in that "policies decided on are often not enforced if they are enacted at all; and, in that, the authorities, even when framing policies, are reluctant to place obligations on people".[68] He did not have the newer democracies particularly in mind. Nevertheless these can be peculiarly vulnerable to extremism from without or within when perhaps jungly borders are not tightly secured and civil authority imperfectly developed. A partial answer could lie in regional charters committing all signatory states to uphold liberty within themselves and help other members do likewise. Applied with resolve, these could deprive the violent of geographic sanctuaries.

War on Terror?

A *sine qua non* for answering the Western Question has to be an acknowledgement that we are embarked on a struggle much more convoluted than a straight down-the-line "war on terror". That truth is the more plain seen in relation to how skewed the semantics can be. It defies all precedent to describe as "terror" everything insurgents do but nothing any government does. It also cuts one off from a broad swathe of inter-mediate opinion very regularly present in dissentient communities — namely, those who back insurgents only at times when they themselves lose faith in compromise. Nor does one need instructions from government on word usage. There are strong grounds for describing as "terrorist" anybody turning to biowarfare. But that still does not mean the term should figure in relevant legislation.

None of which precludes one describing as "terrorist" many insurgent acts, par-ticularly in view of the trend away from discrimination made manifest in many insurgent strikes. One has oneself felt constrained to criticize radically a spread of guerrilla leaders — Yasser Arafat, Menachem Begin, Steve Biko, George Grivas, Ernesto Guevara, Mao Tse-Tung, Kamel Nasser, Josif Tito . . . One has also asked whether someone as gently magnanimous as Nelson Mandela could have remained paramount within the ANC had he not been cocooned on Robben Island. One has noted, too, how prone insurgent bosses in general are to (a) demand too much, ter-ritorially and otherwise, (b) dispatch idealistic youngsters on suicide missions they do not join themselves, and (c) repress their own people before and after achieving basic goals.

Still, the worst terror extant just lately has been state terror directed from Harare and Khartoum. Besides, the term "terrorist" as applied to underdogs and whosoever aligns with them can readily be turned about and ennobled as past derogatory terms have been: tafur, Sea Beggar, Roundhead, Whig, Tory, Quaker, *sans culottes*; Eskimo, Old Contemptibles, Vietcong, Blacks . . . In short, "terrorism" as currently employed is a word like "racism", a substitute for thought not a means of facilitating it.

Cults Without End?

Above all, we cannot assume the progressive weakening of *Al Qaeda*, militarily and politically, will ensure tranquillity for so very long. We could witness in the decades ahead a *melée* of cultish dissidence — anarchic, astrologic, fascistic, fundamentalist, millenarian, neo-Marxist, nihilist . . . Not least may this obtain in what are adjudged advanced nations. The cults will be filling, for the disaffected, a rootless void. A prime source of disaffection for many will be diminished fulfilment in the work place. Resentment on this score will be the stronger when and where the spread of qualified talent becomes much wider than the spread of really satisfying job oppor-tunities.

Then there are those downsides of mass society which unsettle all of us to a greater or less extent. By depriving Wild Nature of its overarching authority we deny ourselves a primordial setting. Weakening bonds within family and local community have a similar effect despite the advantages that can accrue in terms of individual opportu-nities. A degree of *anomie* from whatever causes is endurable. Comprehensive alienation is not.

To reckon with in this connection, too, will be the successive shocks brought on by energy crunches, commodity shortages, climatic disturbance, insurgency, regular warfare, currency discordances or whatever in combination. One cannot assume consistently constructive responses. Take the Wall Street crash of 1929. The ensuing slump was responded to positively from 1932 to 1933 with what would soon be known as Keynesian strategies in the USA of the New Deal and in Sweden. Contrariwise, much of central and Eastern Europe, East Asia and Latin America lurched into fascism. The irrational and the unreasonable triumphed.

A lot could depend on how the IT revolution pans out. Will it on balance be conducive to an Open Conspiracy of the young, engaging in creative thought about how to carry the world forward. Or will the outcome be more the fissuring of a generation as lonely individuals blog with their peers the world over, creating thereby quasi-tribal reversions with narrowly distinct identities. Sometimes such affiliations, sufficiently disconnected and insufficiently grounded in reason and reasonableness, could turn lethal.

One could further imagine, in a new less bookish culture, a renewed crisis of aim and object within universities: a crisis aggravated by the approach of absolute limits to scholarly knowledge in successive academic fields. Dissent on campus could well broaden into a questioning of the legitimacy of other leadership cadres, starting with those in political life. *A priori* it would be hard to deny some truth in the proposition that elected representatives are by definition unrepresentative. As the French used to say, two deputies, one of whom is a Communist have more in common than two Communists, one of whom is a deputy.

A good point of departure for analysis might be a revisit of the distinction the American theologian Reinhold Niebuhr drew between the Children of Light and the Children of Darkness. He saw the latter as "wise" because they were cognizant enough of evil within themselves to appreciate its universality. The Children of Light were "foolish" because they failed to comprehend evil, not least within themselves.[69]

Writing in the grimly revealing days of 1944–5, Niebuhr put the point starkly. Nevertheless, one should explore more closely the interplay between altruism and egoism, particularly as applied to politicians of whatever hue. Further insight might help to ensure that a data flow multiplying worldwide by several hundred per cent a year actively "empowers citizens and democratises societies."[70] Simply to assume it will, might guarantee it will not.

Astral Geography

Like other words of classical derivation, the term "Geography" must be allowed to evolve. Not least must it because, by 2030, say, we shall be exercised about Near Space no less than by biology or IT. Progress already in train will have philosophic as well as practical connotations.

Certain germane themes are aired in *Engaging the Cosmos*. These include developments in cosmology (Chapter 5 and 6); how the major faiths may interact with one another and with Science (Chapter 1); how far manned spaceflights should go (pp. 285–91); and the interception of asteroids or comets likely otherwise to hit us (pp. 302–4). About the religious obediences *per se*, one should at this juncture merely reiterate the truism that they will be unlikely to interact positively with Science and each other unless they respectively re-examine some dogmas. But the obverse of that is how the concept of Science as a repository of hard fact and fount of inescapable logic is being called into question on the frontiers of discovery. Astrophysics, for instance, is waxing more metaphysical with talk of alternative pasts as well as alternative futures; and with interest resurgent in there being more dimensions than the mundane three plus time. There may likewise be a greater readiness to accept scientific hypotheses that have explanatory utility even though they might not be directly testable.

Proposals that many other universes may somehow be extant can come into this tendency. For they hardly lend themselves to demonstration empirically, saving that astronomers will get to observe more definitively the event horizons around black holes, cosmic phenomena which may link to or spawn other cosmi.[71] [72] The multiverse proposition could well become received scientific wisdom this next decade. If so, its supposedly complete novelty could much accentuate the prevailing cultural flux. Nor can one ever discount the historical experience that times of great cultural flux tend not to be tranquil.

John Donne (1572-1631), one of England's "metaphysical poets", saw Copernicanism as a cosmology with "all cohesion gone" which, he wanly opined, undermined family and societal hierarchies leading to a situation in which "every man alone thinks he hath got to be a Phoenix". Taken at the full, however, the deeper impact of "multiverse" could be to reinforce feelings of planetary togetherness, common destiny and fate. In the sixties, after all, telemetry from the Apollo programme afforded a perception of Spaceship Earth most helpful to the ecology movement.

The Desertic Religions

A question always in the background is how far one can build on the commonalities between the three great religions (Judaism, Christianity and Islam) that emerged on the fringes of the Arabian desert.

In the geography of Holy Land history George Adam Smith wrote in 1894, he appeared oblivious of climate change. He was enthralled instead by the idea that the stark bareness of the desert encouraged single tribal gods, several of which were able, in due course, to translate into an all-encompassing monotheism. This notion he had gleaned from Ernest Renan (1823–92), the French historian of early Judaeo-Christianity.

The religious map of South Asia invites comparison. The Hindu pantheon could hardly have emerged in austere desert surrounds. Nor might have a god such as Vishna with his four arms and diverse incarnations. Nor, indeed, would a Buddha figure forever seeking harmony with a regenerative natural order. However, the presumptive corollary that monotheism enters not the rain forest holds less true. Within Islam, Sūfism has thrived in Iran but also South-East Asia. It was Smith's own opinion that desertic landscapes breed "seers, martyrs and fanatics".[73] Yet in this respect, too, causation has been contingent.

Natural Missilry

One cause and consequence of greater planetary togetherness ought to be a preparedness to defend our common home against asteroids and comets on course to impact. By 2009 NASA is due to have mapped 90 per cent of those Near Earth Objects which ultimately present a threat and are a kilometre or more across — i.e. liable to impact with a force of at least 100,000 megatons. Averagely, an NEO at least several hundred metres across will impact every 15,000 years.

Blast effects are absent in airless environments. But it should be possible to deflect such objects with the heat from close by nuclear detonations. The rub is, however, that many, probably most, asteroids are aggregations of rubble which unhelpfully means deflections in a spread of directions. Witness Japan's recent Hayabusa mission to Itokawa. Though small, this asteroid seems likely to be rubble assemblage.[74] It would, in any case, be extremely difficult to direct an interceptor missile towards a unitary asteroid, say, perhaps tens of millions of miles out when interceptor and target are closing at relative speeds of maybe five or ten miles a second.

Warning and maybe launch stations would likely be needed outside the Earth, at the geostationary altitude (40,000 km) and on the Moon and Mars, though probably not at the Lagrangian points. Not that one can easily believe even all this could ensure a near to 100 per cent success rate. Therefore contingency preparations should include provision for passive protection and disaster relief. Some of the effort could be national and regional but global management, too, would be requisite.

Global Contingency Cover

Such a commitment would closely relate to precautions against some other kinds of megadisaster. Supervolcanoes in California, Wyoming, Indonesia and New Zealand potentially threaten devastation of planetary life much as did that Toba explosion 72,000 BP. Closely akin geophysically is the earthquake hazard. Alarmingly, Tokyo lies just opposite what is by far the most pronounced submarine tectonic trough on this planet; and is 25 years overdue for the major earthquake it recurrently experiences. The longer the wait, the bigger the event?

In 2005, a Japanese government report recommended moving the capital from Tokyo. But nothing transpired. It would be fair comment that in these days of electronic data dissemination and storage a major Tokyo quake would cause less administrative chaos than it would have 25 years ago. But it would still be a most traumatic setback for Japan and therefore the world. The fact that this strong possibility receives little or no attention in texts on Japan's economic outlook is yet another example of our disinclination to relate specific threats to general prospects.

Every several 100,000 years or so, the Earth's magnetic field does a South–North flip, leaving the biosphere critically exposed awhile to hard cosmic radiation, not to mention the loss of magnetic field orientation by many forms of animal life. All this last happened 780,000 years ago. Lately the field has been weakening at 5 per cent a century. This may well not continue but if it should, a flip could be in prospect this millennium. Just how long the intermediate nullity would effectively last is unclear from the geological evidence. Decades or millennia? At all events, this would require response preparations much more esoteric and variegated than those for the impact and tectonic genre.

The UN-sponsored International Decade for Natural Disaster Reduction (1990-2000) is deemed to have encouraged "greater interchange" between Earth scientists and social scientists in specific disaster-prone regions. In Peru, for instance, volcanic events are being traced back through the Holocene, the last 10,000 years.[75] Yet this heightened awareness makes all the less excusable the absence of a co-ordinated warning network covering littoral states when the Sumatran tsunami struck in 2004. Sumatra, after all, has an awesome record of major tectonic events. Questions have also been asked about the inchoate and overly hyped inflow of international aid. This, too, reflected the absence of regional contingency planning.[76]

Within Indonesia, however, this awesome experience was actually good for inter-community relations. Witness the peace agreement the government in Djakarta signed with the Free Aceh Movement in May 2005. In Sri Lanka things went the other way.[77]

Limits to Growth?

The everyday benefit from thorough contingency planning could be a further strengthening of planetary togetherness plus the elevation within that sentiment of a canon of reason and responsibility. It could be part of a substitute in terms of fulfilment not only for warfare within our species but also for poverty, ill-health, ignorance and other customary ills.

Part but not all. This is because we must also look towards a definitive levelling off of the demands humankind makes on the natural environment. For one thing, we have surely to dismiss the idea that our species can go and colonize *en masse* Mars or other heavenly bodies, an idea episodically supported by a number of ranking astronomers. To Sir Martin Rees, the Astronomer Royal, that could be the best post-2100 prospect for the survival of our species in the face of genetically engineered epidemics disseminated by maybe "just one fanatic or weirdo with the mindset of those who can design computer viruses".[78] Freeman Dyson has seen it as a way to offset the "inexorable diminution of cultural diversity on Mother Earth".[79]

Such a migration would build up too slowly to ease our present discontents. For one thing, the logistic trade-offs would always be hugely burdensome. Nor could such frontiers be settled by free enterprise homesteaders or counter-culture utopians. The regimen would perforce be more Spartan than Sparta.

Terrestrial Salvation

So we must continue to seek salvation within the confines of Mother Earth. Keeping successive generations rational, reasonable and fulfilled will therefore not be easy with or without global programmes to manage natural calamities. Addressing the further outlook now could help motivate the resolution of the more immediate Western Question. However, longer-term Survival Studies will require a holistic approach stepping even further outside the realm of strategy as customarily conceived. This will in turn require the formulation of a planetary philosophy more interdisciplinary and generally robust than the likes of Wittfogel, Mackinder, or Mao Tse-tung ever achieved. Geography could be pivotal.

Notes

1 Geoffrey Warhurst, "Afghanistan – A Dissenting Appraisal", *RUSI Journal* 125, 1 (March 1980): 26–30.
2 Bernard Loo, "Geography and Strategic Studies", *The Journal of Strategic Studies* 26, 1 (March 2003): 156–74.
3 James Harris, "Encircled by Enemies: Stalin's Perception of the Capitalist World, 1918–41", *Journal of Strategic Studies* 30, 3 (June 2007): 513–48.
4 For further consideration of the West's reaction to the Georgia crisis see the author's "The Atlantic, Europe and Russia", *The Naval Review* 96, 4 (November 2008): 339–45.
5 *Zimbabwe News*, 9 October 2008.

6 Zbigniew Brzezinski, "The International and the Planetary" *Encounter* XXXIX, 2 (August 1972): 49–55.

7 Adam Robinson, *Bin Laden* (Edinburgh: Mainstream Publishing, 2001), p. 287.

8 John Kenneth Galbraith, *The Culture of Contentment* (London: Penguin, 1973), p. 33.

9 Eric Fromm, *Escape from Freedom* (London: Kegan Paul, 1942), pp. 40 *et seq.*

10 Simone Weil, *The Need for Roots* (New York: Harper and Row, 1971), p. 65.

11 Ted Hughes, *Shakespeare and the Goddess of Complete Being* (London: Faber and Faber, 1992), Appendix II.

12 Michael Young and Peter Willmott, *Family and Kinship in East London* (London: Routledge and Kegan Paul, 1957), Chapter XI.

13 Mancur Olson, "Rapid Growth as a Destabilising Force", *Journal of Economic History* XXIII, 4 (December 1963): 529–52.

14 Anthony Storr in B. H. Liddell Hart (ed.), *Churchill: Four Faces and the Man* (London: Allen Lane 1969), pp. 203–46.

15 E. C. Zeeman, "Catastrophe Theory", *Scientific American* 234, 4 (April 1976): 65–83.

16 T. N. Palmer, "A Non-Linear Dynamical Perspective on Climate Change", *Weather* 48, 10 (October 1993): 314–26.

17 Colin Renfrew in Chapter 21 of Colin Renfrew and Kenneth K. Cooke (eds.), *Transformations. Mathematical Approaches to Culture Change* (New York: Academic Press, 1979).

18 Andre Gunder Frank and Barry K. Gills (eds.), "World System Economic Cycles and Hegemonial Shift to Europe, 100 BC to 1500 AD", *Journal of European Economic History*, 22, 1 (Spring 1993): 155–83.

19 *HCC*, p. 87.

20 Piki Ish Shalom, "The Civilization of Clashes: Misapplying the Democratic Peace in the Middle East", *Political Science Quarterly* 122, 4 (Winter 2007–8): 533–54.

21 Azar Gat, *War and Human Civilisation* (Oxford: Oxford University Press, 2006), p. 589.

22 Burak Bekdil, "Why Turkish Efforts for Indigenous Development are too Ambitious", *RUSI Defence Systems* 11, 1 (June 2008): 106–7.

23 Ian Bremmer, "The Return of State Capitalism", *Survival* 50, 3 (June–July 2008): 55–63.

24 "Building BRICs of growth", *The Economist* 387, 8583 (7 June 2008): 92.

25 Arthur Koestler, *The Heel of Achilles* (London: Hutchinson, 1974), pp. 221–4.

26 Steven Biko, *I write what I like* (Harmondsworth: Penguin, 1963), pp. 36 and 55–61.

27 Julius K. Nyerere, *Freedom and Development* (Dar es Salaam: Oxford University Press, 1973), pp. 303–10.

28 Julius K. Nyerere, *Freedom and Socialism* (Dar es Salaam: Oxford University Press, 1968), p. 380.

29 Joseph E. Stiglitz and Linda J. Bilmes, *The Three Trillion Dollar War* (London: Allen Lane, 2008), Chapter 5, Footnote 6.

30 *Ibid.*, pp. 116–17.

31 Gunnar Karlsson, *Iceland's 1100 Years* (London: Hurst, 2005), Chapter 4.5.

32 Stiglitz and Bilmes, p. 31.

33 *The Military Balance, 2008* (London: International Institute for Strategic Studies, 2008, p. 20.

34 Ransom A. Myers and Boris Worm, "Rapid Worldwide Depletion of Predatory Fish Communities", *Nature* 623, 6937 (15 May 2003): 280–3.

35 David O'Connor, "Nets versus Nature", *Nature* 450 (8 November 2007): 179–80.

36 "Eels slide towards extinction", *New Scientist* 180, 2415 (4 October 2003): 14.

37 Editorial, "Bio-Foolishness", *Wall Street Journal*, 22nd April 2008.

38 Bjørn Lomborg, *Cool It* (London: Marshall Cavendish, 2007).

39 Data cleaned from 500 millibar "thickness patterns" in the Weather Log of the Northern Hemisphere published monthly as a supplement to *Weather*.

40 *Nature* 455, 7209 (4 September 2008): 5.

41 *HCC*, pp. 127–8, 144–51 and 155.

42 Olivind Lunde, "Archaeology and the Medieval Towns of Norway", *Medieval Archaeology* XXIX (1985): 120–35.

43 "Melting Asia: China, India and Climate Change", *The Economist* 387, 3583 (7 June 2008): 29–32.

44 James Lovelock, *Gaia: A New Look at Life on Earth* (Oxford: Oxford University Press, 1979), Chapter 2.

45 James Lovelock, *The Revenge of Gaia* (London: Allen Lane, 2006), Chapter 5.

46 *Ibid.*, pp. 91–2.

47 Catherine Caufield, *Multiple Exposures* (London: Secker and Warburg, 1989), pp. 246–7.

48 Nicholas Stern, *The Economics of Climate Change* (Cambridge: Cambridge University Press, 2007), pp. 592–7.

49 "In search of the perfect battery", *The Economist Technology Quarterly*, 08 March 2008, pp. 25–8.

50 Henry Kissinger, *Washington Post*, 8 September 1985.

51 Mohammed El Baradei, "Nuclear double standards", *New Scientist* 183, 2455 (10 July 2004): 17.

52 Jaswant Singh, "Against Nuclear Apartheid", *Foreign Affairs* 77, 5 (September–October 1998): 41–52.

53 David E. Sanger, *International Herald Tribune*, 28–29 July 2007.

54 David Cohen, "Earth Audit", *New Scientist* 194, 2605 (26 May 2008): 34–41.

55 Bruce Riedel, "South Asia's Nuclear Decade", *Survival* 50, 2 (April–May 2006): 107–26.

56 Mike Moore, "Unintended Consequences", *Bulletin of the Atomic Scientists* 56, 1 (January/February 2006): 58–66.

57 Reidar Visser, "Historical Myths of a Divided Iraq", *Survival* 50, 2 (April–May 2008): 95–106.

58 Michael Howard, "Are We at War?", *Survival* 50, 4 (August–September 2008): 247–56.

59 Raja Shehadeh, *Palestinian Walks* (London: Profile, 2008), pp. 207–15.

60 Matthew Sparke, "Geopolitical Fears, Geoeconomic Hopes and the Responsibilities of Geography", *Annals of the Association of American Geographers* 97, 2 (June 2007): 338–49.

61 Anthony Vinci, "Becoming the Enemy: Convergence in the American and Al Qaeda Ways of Warfare", *Journal of Strategic Studies* 31, 1 (February 2008): 69–88.

62 David E. Lilienthal, *TVA. Democracy on the March* (Harmondsworth, Penguin, 1944), Chapter 1, pp. 14–20.

63 Philip Selznick, *TVA and the Grass Roots* (New York: Harper Torchbooks, 1966, pp. 13–16.

64 Stiglitz and Bilmes, *The Three Trillion Dollar War*, p. xvi.

65 Mycle Schneider, "Nuclear Information in France, as transparent as mud", *INESAP Information Bulletin* 28 (April 2008): 85–6.

66 Wendy Barnaby, *The Plague Seekers* (New York: Continuum, 1999), pp. 136–40.

67 Stephen Bungay, *Alamein* (Aurum Press, 2002), pp. 32–3.

68 Gunnar Myrdal, *Asian Drama* (London: Allen Lane, 1968), 3 Vols., Vol. 1, p. 66.

69 Reinhold Niebuhr, *The Children of Light and the Children of Darkness* (London: Nisbet, 1945), p. 15.

70 Karl Donert, "Virtual Geography", *Geography* 85 (2000): 37–45.

71 Michio Kaku, *Parallel Worlds* (London: Penguin, 2005), pp. 253–5.

72 George Ellis, "Opposing the Multiverse", *Astronomy and Geophysics* 49, 2 (April 2008): 33–5.

73 George Adam Smith, *The Historical Geography of the Holy Land* (London: Hodder and Stoughton, 1894), pp. 29–30.

74 Erik Asphang, "Adventures in Near Earth Object Exploration", *Science* 312, 5778 (2 June 2006): 1328–9.

75. Martin R. Degg and David K. Chester, "Seismic and volcanic hazards in Peru: changing attitudes to disaster mitigation", *Geographical Journal* 171, 2 (June 2005): 125–45.

76 Benedict Korf, "Disaster, Generosity and the Other", *Geographical Journal* 172, 3 (September 2006): 245–7.

77 Jennifer Hyndman, "The Securitization of Fear", *Annals of the Association of American Geographers* 97, 2 (June 2007): 338–49.

78 Martin Rees, "Special Report: the Science of Eternity", *Prospect* 70 (January 2002): 50–4.

79 Freeman Dyson, *Disturbing the Universe* (New York: Harper and Row, 1979), Chapter 21.

Appendix A
Geodetics and Electromagnetics

Geometry was classically the science of Earth measurement. Today this particular interest is considered the preserve of a branch of Geometry, namely Geodesy. A calculation of primary importance is the slant range to the Earth's horizon for a body at a given altitude above Mean Sea Level (MSL). Applying at this tangential limit the theorem customarily (though probably incorrectly) ascribed to the ambiguous brilliance of Pythagoras (*c.* 572 to 497 BC), one finds the horizon for a ship's radar scanner 100 feet up will be 12.5 miles while for a satellite orbiting at 200 miles, it will be 1250.

The angular velocities at which bodies stay in orbit gradually decrease with height. Of especial interest for surveillance and communications is the altitude at which a satellite in equatorial orbit will be "geostationary" — i.e. always above the same point on the Earth's surface. It is 22,300 miles.

Next one should consider the path of a body moving "ballistically" which, in principle, means that after an instant initial impulse only gravity acts on it. In practice, pure ballistic flight cannot be achieved from an Earth launch for two reasons. The initial impulse comes from a rocket thrust which will be neither instantaneous nor steady. And there is air resistance especially nearer the surface. These two considerations combine to make "minimum energy" long trajectories steeper at the ends.

An important question legally and therefore operationally is what altitude should be designated the outer boundary of the atmosphere — i.e. the divide between national air space and international Near Space. An indicator now accepted very generally is the velocity at which an iron surface "ablates" — i.e. erodes . The greater the air density, the lower that velocity threshold is. Descending from 120 to 80 kilometres, it decreases from 58 to 6.5 km per second. This represents a much more rapid change, proportionally speaking, than occurs higher up or, indeed, lower down. The distinctiveness of this zone is highlighted by its being where meteors so visibly burn out.

During the early sixties, governments therefore came very generally to accept 100 km as being at least the *de facto* boundary between the national and international domains, the latter being designated a "universal common" (along with Antarctica and the High Seas) in the report to the UN in 1987 by the World Commission chaired by Gro Harlem Brundtland, formerly Sweden's Prime Minister.[1]

Rocketry and Range

A ballistic missile travelling 500 km horizontally on a minimum energy trajectory will attain an apogee (i.e. the high point of its trajectory) of 125 km. A Soviet SS-20 IRBM travelling 5000 km would have risen to 900 km. Moreover, the maximum horizontal range rises much more than impulse increments. A burn out velocity of 1600 metres per second (mps) will yield a range of 400 km; and thrice that speed, one of 2800 km. Direct exit from Earth to interplanetary Space requires 12000 mps. Half that speed could achieve direct exit from Mars.

Natural Electromagnetism

Electromagnetic radiation is the wave energy (heat and light) which emanates from all materials throughout the universe. It includes, of course, the "microwave radiation" residual from Creation's original "Big Bang". As it happens, the speed of travel of this energy (alias the "speed of light") is, to within one part in a hundred thousand, 300 million metres (or *c.* 186,200 miles) a second. Einstein proposed that although electromagnetic waves could be diverted by intense gravity, their velocity was the one great constant throughout Time and Space. He has yet to be proved wrong.

To comprehend the radiation process, one should start with the notion of a "black body", an idealized object that is a perfect absorber and emitter across all wavelengths —that is to say, not subject to any coloration effect. Its wavelength of peak emission will be an inverse linear function of how far its surface temperature is above "absolute cold". This end state, at which matter is drained of all energy, is at close to minus 273° Centigrade or Celsius. This temperature is duly rendered as "zero" on the specially contrived Kelvin scale. A degree Kelvin covers the same temperature range as a degree Centigrade/Celsius.

The mean temperature of the Earth's atmosphere at sea level is near to 300 Kelvin which connotes a "black body" peak emission at 9660 nanometres (billionths of a metre). A burning rocket might attain 3000 K. So might the surface of an intercontinental Re-Entry Vehicle descending to Earth. The surface of the sun is twice as hot; and its emission peak, 485 nm. Still, very few or no material aggregations, celestial or otherwise, are at all perfect black bodies. More or less all emit radiation within certain defined wavebands, a "coloration" which relates to the energy levels at which their respective surface molecules vibrate.

Beam Projection

Central to the "quantum mechanics" revolution then under way in physics was the "Uncertainty Principle" enunciated by Werner Heisenberg in 1927. Among its connotations is that photons, electrons and other minute entities have their individual movements governed not by rigid determinism but by probabilities. For many purposes, this does not effect "common sense" physics as "zillions" of entities will randomly vary in this respect around what thereby emerges as a well-defined median state. Something it does confirm, however, is the axiom that, regardless of project design, no beam of energy or free particles can be absolutely parallel. The divergence will be directly proportional to the wavelength though inversely so to the diameter of the projector. An Inverse Square Law is operative throughout. Double the distance a beam diverges from what may be tantamount to a point source and you reduce by three quarters its intensity.

Electromagnetic beams play a crucial part in modern communications. So do they in surveillance and tracking which may be done by sensing actively or passively. Active sensing involves transmitting pulses on a narrow frequency band which, being known, enables the probes to be identified as they return from echoing off a target. The fact of return reveals some target or other on a given bearing; and the time taken gives a measure of distance. It may also be possible to exploit the doppler effect (the apparent change in wavelength of a pulse coming off a moving target) to register its velocity relative to the projector. However, an active return cannot in itself confirm that a target locked onto is one being sought.

Passive sensing depends on the receipt and interpretation of electromagnetic emissions emanating from an operational target. The equipment required can weigh less and consume less energy and its use does not reveal one's own presence. Also it may be possible to establish target identity by recognition of its emission signature. However, the passive mode may not give bearing at all precisely. Nor does it register distance except perhaps approximately through triangulation between two or more sensing platforms. A human eye is a passive sensor.

The Electromagnetic Spectrum

The accepted delineation of the electromagnetic spectrum is set out in table below. The segment with wavelengths between 30 centimetres and one millimetre is known as the "microwaves".

Band	Frequency	Wavelength
Very low frequency (VLF)	0 – 30 kc/s	Above 10,000m
Low frequency	30 – 300 kc/s	10,000 – 1,000m
Medium frequency	300 – 3,000 kc/s	1,000 – 100m
High frequency (HF)	3 – 30 mc/s	100 – 10m
Very high frequency (VHF)	30 – 300 mc/s	10m – 1m
Ultra high frequency (UHF)	300 – 3,000 mc/s	100 – 10cm
Super high frequency (SFH)	3 – 30 kmc/s	10 – 1cm
Extremely high frequency (EHF)	30 – 300 kmc/s	10 – 1 mm
Infra-red	–	1,000,000 – 800nm
Visible light	–	800 – 400 nm
Ultra-violet	–	400 – 1nm
X-rays	–	1 – 0.001nm
Gamma rays	–	0.001 – 0.00001nm
Cosmic rays etc.	–	Below 0.00001 nm

kc/s	= kilocycles per sec	= 1,000 cps
mc/s	= megacycles per sec	= 1,000,000 cps
kmc/s	= kilomegacycles per sec	= 1,000,000,000 cps

A term now regularly employed to indicate frequency is the hertz, its meaning being a million. A gigahertz is a billion cycles per second; and a terahertz a trillion cps. At wavelengths below the millimetric range, talk of frequency is discontinued and one speaks just of wavelength measured in nanometers — i.e. billionths of a metre.

Longer wavelengths with more divergent beams sometimes facilitate preliminary search. Otherwise narrower beams are among the attractions of shorter wavelengths since they are more tightly directional and also, for a given energy flux, brighter. They therefore distinguish smaller targets and define all targets better. Similarly, the density of communication channels rises with the frequency. The whole VHF carries, even in principle, only 50,000 channels. Yet, save that signals may wander a little from whatever frequency is selected, the SHF should accommodate 50,000,000.

Generally speaking, however, the higher the frequency the harder it may be to apply the principle long basic to electromagnetic transmission — i.e. the modulation of electron oscillation within resonant electrical circuits so as to generate radiation that is "coherent", here meaning that the wavelengths within a given emission are in phase — i.e. all of a standard length and in systematic sequence. This is what allows of a radar's measuring range by pulse return and exploiting, too, the Doppler effect. Evidently, too, it is the essence of wireless signalling.[2]

Couple the advantages of shorter wave radio or radar with the greater difficulty involved in sustaining it, and one has a situation bound to engender fierce competition, especially in times of armed conflict. In World War Two, the Western Allies derived great advantage from the British invention, at Birmingham University in the autumn of 1940, of the "cavity magnetron": a device for the coherent generation, in confined spaces with low power inputs, of microwaves. The defeat by the Allied navies of Japan's *kamikaze* suicide strikes, 1944–5, considerably depended on the use of anti-aircraft shells, every one of which incorporated a miniature cavity magnetron cum radar. This technology could programme the shells for "proximity fusing" — i.e. exploding if passing within, say, 40 feet of their target.

Since 1960, much cachet has come to be attached to Light Amplification by Stimulated Emission of Radiation (LASER). A true laser will yield highly directional coherent light on a wavelength determined by the physical chemistry of the material being used. Until recently, however, there has been a wide gap as regards active transmission between radio waves and lasers. The former have not coped with frequencies much above 100 gigahertz while the latter have not extended much below 100 terahertz, the middle EHF and shorter-wave Infra-Red respectively.

Nevertheless, astronomers have been doing passive sensing within this gap for half a century.[3] Now tunable "quantum cascade lasers" are reaching down to several terahertz. Many terrestrial materials wholly or considerably opaque to visible light are transparent in the low terahertz range. Quite a number are dry organic — e.g. textiles.[4]

As regards increasing pulsed laser intensity through pulse compression, a number of countries are now investing in 100 to 1000 terawatt models (10^4 to 10^{15}W).[5] These intensities will not be fully translatable to Near Space and other austere environments. But their being registered elsewhere will be expressive of considerable progress under all conditions.

Atmospheric Absorption

By and large, absorption within the atmosphere intensifies as one progresses to higher frequencies. Thus through the Infra-Red absorption rates, though sharply variable, are often high. Then at about 1400 nanometres (nm), the remarkably good "window" begins which extends to 300 nm; and therefore fortuitously includes the peak transmission of our Sun. What we have evolved to know as visible light extends from 800 to 400 nm. After which, windows are very limited; and for wavebands shorter than 35 nm, the atmosphere is effectively opaque.

Vertical Atmospheric Zoning

The 100 km boundary between national and outer space is by no means the only recognized divide as one ascends through the atmosphere. More acutely geophysical is the "tropopause", the divide between the lower atmosphere (alias the troposphere) and the stratosphere. Molecular diffusion naturally occurs across this interface. But organized air currents rarely effect a transition. Therefore the stratosphere is almost completely devoid of water in any form.

Averagely, the tropopause is 11 km above Mean Sea Level in middle latitudes. Around the North Pole, eight is nearer the norm; and near the equator 16. Below it, clouds of particles of whatever kind are likely washed down by rain or snow within a few hours or, at most, a week or two. Conversely, the finer particles in any cloud above the tropopause might soon diffuse globally, then descend to Earth over several years.

Above the tropopause, the horizontal lamination continues. Between 20 and 30 km up occurs a relatively high concentration of ozone amounting to 500,000 tons or more, one part in 10,000 of the Earth's total atmosphere. Ozone is a variant of oxygen with three atoms per molecule instead of the usual two. Perennially, it has disintegrated and reformed in interaction with hard Ultra-Violet C radiation from the Sun, the consequent UV-C absorption warming the ambient atmosphere appreciably. Left to progress unimpeded to the Earth's surface, UV-C radiation would be extremely injurious to Life as we know it.

A subject of acute global concern therefore was the serendipitous discovery in 1985, by British Antarctic scientists, that this stratospheric ozone was being seriously depleted. Soon it was confirmed that ChloroFluoroCarbon (CFC) artificial sprays were the main cause. Mercifully, the Montreal Protocol of 1987 appears to have addressed this problem sufficiently.

At 50 km up, one enters the Mesosphere, the zone in which cooling with altitude resumes. At 80 km, this gives way to the ionosphere, a finely articulated zone some hundreds of kilometres deep which is electrically highly active because of the molecular and atomic dissociations caused by the incoming radiation in that exposed locale. Layers thus formed within it are reflective of wavelengths longer than the microwaves. This is the attribute that allowed of long-distance radio transmissions even before the advent of orbital satellites.

Notes

1   World Commission on Environment and Development, *Our Common Future* (Oxford: Oxford University Press, 1987), Chapter 10.
2   P. H. Boreherds, "Focussing a Coherent Beam of Light", *American Journal of Physics* 3916 (June 1971): 680–1.
3   Michael Rowan-Robinson, "Terahertz surveys", *Astronomy and Geophysics* 48, 4 (August 2007): 31-4.
4   Justin Mullins, "Forbidden Zone", *New Scientist* 175, 2360 (September 2002): 34–6.
5   André Gsponer, "Superlaser Development in Germany", *Inescap Information Bulletin* 19 (January 2002): 79.

Appendix B
Explosive Nuclear Release

One apprehension informing this study has been that biowarfare is the kind of lethal encounter that could soon concern us most. But at present nuclear war remains more at the forefront of our minds.

Nor is this remarkable. For millennia, firepower (sling shot, arrows, Greek fire, cannon balls . . .) was so precious a resource that the captains of war were preoccupied with applying it decisively. Then came the flash in the Hiroshima morning sky which transformed it into a superabundant resource we fear to use. In World War Two, some six million tons of high explosive were expended. The largest thermonuclear (i.e. hydrogen or fusion) detonation to date has been one the Soviets conducted above Novaya Zemlya in October 1961. It was rated equivalent to 60,000,000 tons of TNT high explosive. The Hiroshima fission bomb was 20,000 tons TNT equivalent, since known as a "nominal yield".

Very short wave radiation (x rays and gamma rays) from the flash of a nuclear detonation or from the radioactive fallout (i.e. debris) produced by it can have innumerable effects. It can, for instance, do genetic damage which passes on to human beings not even conceived at the time. All other life forms are likewise susceptible though the extent to which particular species are varies in a way that comes across as very random.

Nuclear Structure

An atom consists of electrons revolving round a nucleus. Every electron bears a negative electrical charge. Atoms may readily lose or gain electrons, becoming positively or negatively charged accordingly.

The particles which comprise the nucleus are neutrons and protons. These are of virtually equivalent mass, the difference being that a proton carries an electrical charge, a positive one matching an electron's negative one. A neutron carries no charge. All the particles within a nucleus are bound together by the "nuclear strong force", a fundamental attractive interaction which is very intense and very short range. A nucleus represents over 99.9 per cent of the mass of an atom in an electrically neutral state.

An element (e.g. iron, hydrogen, carbon) is a substance which cannot be modified by chemical means. Every atomic nucleus of a given element has the same number of protons, this being known as its Atomic Number. For that element, however, there will likely be several answers to the question of how many neutrons its nuclei have.

That number signifies what "isotope" an atom belongs to. Implicit in the notion that several isotopes can be characteristic of the same element is that their chemical behaviour is identical. The nuclei of an unstable isotope normally disintegrate with a characteristic frequency which is unaffected by chemistry, temperature or pressure.

The Atomic Weight of an element or an isotope is, in principle, indicative of the number of times its standard atom is heavier than a hydrogen one. By convention, however, the Atomic Weight of hydrogen is treated as 1.008 to allow of the value for oxygen being rounded to 16.00. Atomic Weights range up to *c.* 245.

Energy Release

Certain large nuclei are quite unstable. They can therefore be fissured by neutrons freed by previous disintegrations. A very rapid succession of disintegrations is perceived as a "chain reaction". Also emitted are alpha and beta particles. Comprising two protons and two neutrons, the former are equivalent to helium nuclei. Their being absorbed by large nuclei can change the latter's chemical identity. Beta particles are fast-moving electrons.

The process involves, too, energy release, this considerably in the form of short-wave electromagnetic radiation (gamma rays, X-rays, ultra-violet C . . .) inherently injurious to life. For fissioning involves not only a downward overall tendency in Atomic Weights. It is also associated with small decreases in aggregate mass, postulating no material transfers to or from the outside. This "mass defect" derives from matter bring converted to energy in accordance with Einstein's $E = mc^2$ formula, energy equals mass times the square of the "speed of light" (i.e. the velocity of electromagnetic radiation, *the* universal constant).

Fission could, in principle, bring about mass defect as far down the scale as Iron 56, the isotope with the largest binding force per nucleon (the generic term for protons and neutrons). Below this, the Einsteinian conversion can, again in principle, be achieved by fusion instead. In practice this has on Earth just meant by isotopes of hydrogen.

Applied Fission

The nuclear age dawned militarily with Project Manhattan, the American-British-Canadian programme (1942–5) to develop at Los Alamos, New Mexico, fission bombs. Then and ever since, two isotopes have been favoured both for fission bombs and for standard power reactors. They are Uranium 235 and Plutonium 239. The former was the core of the Hiroshima bomb and the latter of the Nagasaki. Availability apart, the requirement is for material fissile enough to allow explosive reaction though not so much so that it cannot be held waiting upon contingencies.

Uranium 235 is but 0.7 per cent of natural uranium, nearly all the rest being U238. Some reactors run on natural uranium. In general, however, the metal has the proportion of U235 increased significantly for reactor use; and raised to perhaps 90 per cent for military explosive purposes. During Manhattan, this "enrichment" was achieved via diffusion. A gasified sample of uranium was repeatedly recycled through meshes of nanometric porosity which arrested U238 more than they did the lighter U235 atoms. But the big facility at Oak Ridge, Tennessee, consumed electricity on a scale TVA would be needed to cope with. It was also vulnerable. These days gas centrifuges are preferred, the individual units being much smaller. A fully-fledged national programme might employ thousands, quite possibly dispersed to an extent.

A uranium nominal bomb has a critical mass around five kilograms. At that size, chain reactions can be rapid enough to generate energy explosively. But what principle of bomb design will allow of fissile material being kept stable until the time is ripe? For U235 it was alright to fire two half-critical charges into one another in a 'gun barrel'. But this was unsuitable for Plutonium 239 which produces more neutrons per fission. A Pu239 chain reaction triggered thus would start prematurely and so not be fully completed.

Therefore the engineering principle adopted for plutonium was "implosion". A slightly sub-critical Pu239 mass was enveloped in high explosive. Triggering this envelope could compress the plutonium core sufficiently for the instant needed to increase the "neutron capture" rate enough to induce an explosive chain reaction. The late Rudolph Peierls, the gentle physicist who led the British team at Los Alamos, described plutonium implosion as "the hardest of the problems solved". Nobody could be sure until the Alamogordo full-scale test that the solution would work.[1] Does this not suggest that even today a smallish would-be nuclear state could find making a plutonium bomb beyond them, without external assistance?

Something close to proof conclusive that it might do can be read into the outcome of the plutonium nuclear test North Korea carried out in 2006. Twenty minutes prior to the event, Pyongyang told Beijing the yield would be four kilotons. In fact, the seismic evidence is that it was 0.5 to 0.9 kilotons. This compares with the Pakistani test in 1998 which was two kilotons and the first French one in 1960 which was 60 kilotons.[2]

The waste from a reactor's consumption of uranium includes Plutonium 239. This isotope is dubbed "artificial" since, with its high neutron flux, it has but a 24,000-year half life. It has therefore been too unstable to survive in natural deposits across geological time. But although highly toxic, it can be isolated to sufficient purity by chemical means. A smallish uranium reactor with a capacity of, say, 200 megawatts, will generate annually enough Pu239 to make five or ten Nagasaki bombs.

Applied Fusion[3]

Fusion is the heat source of the stars, our sun included. The question has been how protons can combine, given that their positive electrical charges cause mutual repulsion — the Coulomb barrier. In stellar inner furnaces this "thermonuclear" process is facilitated by (a) thermal energies equivalent to 10 or 20 million degrees Kelvin, (b) the hydrogen involved being in the intensely concentrated and disaggregated state known as plasma, and (c) quantum-mechanical tunnelling through the Coulomb barrier. The heavier nuclei are created in supernovae, the huge explosions that are the penultimate phase of dying giant stars.

Terrestrially a fission nuclear charge serves to agitate the isotopic hydrogen into effecting explosive fusion. A hydrogen bomb might sometimes have a U238 outer casing since neutrons freed from the hydrogen could be moving fast enough to destabilize it, producing *inter alia* exceptionally high levels of radioactivity. This exceptionally macabre feature was not used in the Novaya Zemlya event mentioned above.

No country other than the "Big Five" already possessing thermonuclear warheads is anywhere near being able to develop deliverable ones independently. Some decades hence laser developments might allow of thermonuclear explosions *in situ* — "Doomsday devices".

A Nuclear Firebreak?

It was argued earlier — 13 (2) and 16 — that a signal contribution to arms limitation could be the West's committing itself militarily to thermonuclear "no first use" — i.e. not being the first to use hydrogen bombs. Here one should consider further the technical factors bearing on the usefulness or otherwise of this proposal.

An approximation to a cube root law obtains in the relationship between explosive yield and damage radii. Thus an eightfold increase in explosive power tends to double a defined radius. Evidently this narrows the impact gap between the most powerful fission bombs that might just about be fabricated and the least powerful fusion ones. None the less, it remains the case that thermonuclear warheads are developed upwards from the megaton range whereas fission ones have mainly been downwards through the kiloton range, sometimes using tampers to limit energy release. A quarter of a kiloton warhead was developed as early as 1960 for use in certain USAF air-to-air missiles, demolition charges and the Davy Crockett "atomic mortar" programme.

Comparisons might instructively be made between a ten kiloton fission warhead and a ten megaton thermonuclear one. A blast overpressure of several pounds per square inch is liable to cause generalized minor damage. The fission device might effect this at one mile or so and the fusion at 10 to 15. Much the same applies to the infliction in clear air of third degree burns. Radioactive fall out does not lend itself to radial comparisons but tends to be worse with fission in proportion to energy release. Fallout projected as high as the rainless stratosphere may then be distributed worldwide. All in all, the distinction being essayed between fission and thermonuclear is not that sharp but neither is it too esoteric.

Notes

1   Rudolph E. Peierls, *Atomic Histories* (New York: Springer-Verlag, 1997), pp. 111 and 261–2.
2   *Strategic Survey, 2007* (London: Routledge for the International Institute for Strategic Studies, 2007), p. 311.
3   For an overview of the peaceful application of thermonuclear fusion, see *Global Instability and Strategic Crisis*, Appendix B.

Index